Brain, Behaviour and Drugs

Brain, Behaviour and Drugs

Introduction to the Neurochemistry of Behaviour

DAVID M. WARBURTON

Reading University

JOHN WILEY & SONS

LONDON · NEW YORK · SYDNEY · TORONTO

Library of Congress Cataloging in Publication Data:

Warburton, David M
    Brain, behaviour, and drugs.

    1. Neurochemistry. 2. Neuro-psychopharmacology.
3. Drugs—Physiological effect. 1. Title.
[DNLM: 1. Behaviour—Drug effects. 2. Neurochemistry]
3. Psychopharmacology. WL104 W254b]
QP356.3.W37   612 .8 042   74–20789

ISBN 0 471 91991 8

Photosetting by Thomson Press (India) Limited, New Delhi
and printed in Great Britain by the Pitman Press Ltd., Bath, Avon.

To
Irene

Preface

This book is intended as a textbook for graduate and advanced undergraduate students in psychology, pharmacology and allied medical fields. It reflects a trend in psychopharmacology in which attempts are being made to understand behaviour in terms of the basic neurochemical systems in the brain. As a result of this approach the chapters are organized around the neurochemical systems believed to underlie specific types of behaviour rather than various classes of drug.

In order to give the uninitiated reader the necessary background, the first chapter gives an outline of the neuropharmacology of the drugs discussed in the later chapters. In addition, this chapter describes the current ideas on neural systems in the brain known to be controlling behaviour. The next five chapters are devoted to the basic neurochemical systems that have been identified so far and describes their importance to human behaviour. These chapters are on motivation, mood, attention, motor control and sleep. The last five chapters consider the ways in which these five systems can explain the phenomena of hallucinations and psychoses, drug dependence, memory, intelligence and anxiety. From these headings it can be seen that the book complements the standard textbooks in physiological and abnormal psychology.

Many individuals have contributed directly or indirectly to this book. I am indebted to many colleagues including A. Bentall, K. Brown, G. A. Heise, S. Lazareno, B. P. Moulden, D. S. Segal and K. Wesnes for their comments on various parts of the manuscript. Special thanks are due to my father, T. Q. Warburton, for his painstaking preparation of the references.

DAVID M. WARBURTON

Contents

1
Introduction

The last 50 years has seen an explosion in the pharmaceutical industry and the latest aspect of this has been the production of drugs which influence behaviour. The major step forward in the development of psychotherapeutic drugs was the discovery of the major tranquillizers in the early 1950's and this has been followed by the discovery of thousands of psychoactive drugs. Most of this work has consisted of testing chemicals which are structurally similar to those already in use and developing theories of drug action in terms of a drug's chemical structure and behaviour. More recently these research tactics have been extended to include research on the relations between biochemical processes in the nervous system and behaviour. This approach relies heavily on correlative research on the biochemistry of nervous function especially the chemical process involved in synaptic transmission. In order to assist the reader, these processes will be discussed in the first section of this chapter and it is intended that this section should be used for reference purposes while reading the other chapters.

A. NEUROCHEMICAL CONTROL SYSTEMS

This section places emphasis on the interconnection of neurons to form neural systems. In most parts of the nervous system the axons run in parallel forming fibre tracts. The cell bodies and dendrites of these neurons are concentrated in one place forming a nucleus and in the central nervous system some regions of the brain, such as the hypothalamus, may have ten or more nuclei which can be distinguished histologically. At a nucleus there is a convergence of neurons from other regions of the brain, and each group is carrying different information coded qualitatively in terms of the spike frequency. The nature of the information converging depends on the functioning of the synapses so that if it is a 'go' signal for that nucleus the neurons will have an excitatory transmitter at their synapses which will result in an excitatory potential, depolarization, on the cell bodies of the nucleus. Here these excitatory potentials will be added to any other excitatory potential and inhibitory hyperpolarizing potentials from other converging neurons will be subtracted. The sum of the excitation minus the inhibition determines the frequency of the spikes generated in the axons leading from cell bodies of the nucleus. This integrative function of specific nuclei suggests naturally that there are specific behavioural control centres in the brain where each centre corresponds to one nucleus.

2

Centre and network theories

The most important of these theories was proposed by Stellar (1954) for motivation, and it led to extensive research in physiological psychology using electrical recording, electrical stimulation and lesions as research tools. The basic assumption of the model (see Figure 1) was that 'the amount of motivated behaviour is a direct function of the amount of activity in certain excitatory centres in the hypothalamus' (Stellar, 1954, p. 6). The inputs to a motivational centre were from the internal and external environments, from cortical and thalamic centres and from a related inhibitory centre, which integrated various 'negative' inputs and depressed the activity of the excitatory centre. The sum of all the motivational information determined the output along the final common path to the effectors. Based on this model, a theory of behavioural centres was developed to explain all types of behaviour. According to this notion there were excitatory and inhibitory centres throughout the brain which controlled responding (Morgan, 1965) and the magnitude of any response was thought to depend on the balance of excitation and inhibition at the opposed centres. Some major centres were located in the hypothalamus and the appropriate positive and negative inputs converged on each one from cortical and thalamic centres.

In contrast to these concepts of strict localization, the neural network theory

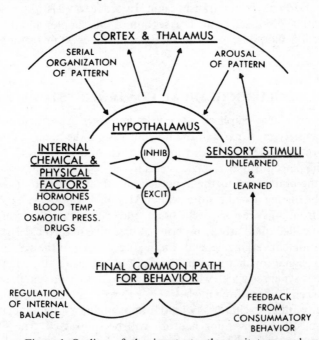

Figure 1. Outline of the inputs to the excitatory and inhibitory centres controlling motivated behaviour. From Stellar, E. (1954). *Psychol. Rev.,* **61,** 5–22. Reproduced by permission of the American Psychological Astociation

emphasizes that behaviour is the outcome of a pattern of activity in networks of neural units in series and parallel and does not depend on a single nucleus (Morgane, 1964). Instead of a linear deterministic chain, there are multiple interacting pathways combining to produce a behavioural pattern. These networks give the brain considerable resistance to disruption by malfunction of an individual pathway so that damage to a single pathway would not necessarily impair function radically. Functional impairment would only result from changes in the pattern of activity in whole network. This is a very important change of emphasis in our thinking about the central nervous system and it implies that the function of the constituent neural units in the network is constrained by their mutual dependence and interdependence. As a consequence, the whole dynamic complex maintains a coherence and integrity of action in spite of any unpredictable variability in the function of the units. This means that there can be individual differences in the neural units that do not change the input–output relations of the system.

This more global view of localization has its origins in the ideas of Flourens, Goltz and more recently, Lashley. Flourens (1824) believed that although specific regions had certain functions and properties, there was a unity in the CNS so that each part had an *action commune* (a joint function) in addition to its *action propre* (a specific function) so that loss of one part reduced the energy of the other parts but did not abolish the function. The functional network theory agrees there is an *action propre* to each anatomical nucleus in the sense that it enables integration of new information into the system, but that it also has an *action commune* in the output of the network.

In such a system it is probable that all inputs do not have equal influence on the final output but, rather, there are hierarchies of inputs to the network where those inputs closest to the output paths have more influence on the behaviour than others further away. It follows that damage to some pathways of the network will have more deleterious effects than others. In particular, lesions in the diencephalon and the hindbrain would have more effects on behaviour than lesions to the cortical parts of the network. These networks with their alternate pathways give the brain considerable resistance to disruption so that lesions may only produce a temporary loss of function, while reorganization of a network occurs. However, it would be erroneous to suppose that the contribution of an individual neuron to the network is negligible. On the contrary it is the average, but particular, characteristics of the individual neurons which results in the composite behaviour of a pathway. Similarly it is a fallacy to assume that, because a lesion to a pathway has little permanent effect, the pathway does not contribute significantly to the behaviour of the intact organism.

As a consequence of notions of functional control of behaviour by centres in the brain, theories of the biochemical basis of behaviour followed the same logic. It was argued that drugs modified behaviour by changing the levels of excitation and inhibition at specific sites in the brain. Thus, chlorpromazine was believed to act on the hypothalamic centres to reduce their level of activation while amphetamines produced increases in the level of activation (Duffy, 1962). This idea of drugs modifying levels of activity in neural centres carried over to the neurochemical studies of the brain where extensive work has been done to determine the levels of neurochemicals at specific sites and correlate these with behaviour. Thus the basic assumption underlying these studies is that the level of neural activity at a site will be a function of the levels of chemicals present. In particular, research has focused on presumed central nervous system transmitter substances, especially those established as peripheral neurotransmitters like norepinephrine (noradrenalin) and acetylcholine. Thus, drug-induced variations in the levels of transmitter substances were thought to change the output along some final common pathway for behaviour. Drugs could be either centrally excitatory or centrally depressant, depending on the changes they produced on the final output from the centre. The network theory makes a much stronger claim about the nature of the systems by arguing that there are a number of basic behavioural systems in the brain that have the same chemical characteristics, i.e. chemically coded, and the basic principles behind this argument will be discussed next.

Chemical coding of the network

The basic interconnections of the nervous system are believed to be genetically determined and the neural networks to be established during the development of the embryo. It is known that the advancing neuron tip is chemosensitive and the direction of growth depends on growth along chemical gradients (Sperry, 1963). This affinity alone would not explain the complete specificity of inter-connection present in the mature nervous system. Weiss (1968) showed that constitutional differences of individual muscles and receptors act on the neurons reaching them and create subspecies of peripheral neurons; modulation of neurons thus seems to occur. The latter impose their characteristics on the penultimate neurons and so on back to the central nervous system. To account for this, Roberts (1966) has proposed that contacting cells exchange material across their membranes so that both may synthesize constituents which initially are specific to the membranes of the other. As a result of this exchange, cells become interconnected and acquire biochemical similarity in proportion to the amount of materials exchanged. Thus the nervous system would seem to be organized by means of two kinds of interaction, affinity and modulation.

A consequence of chemical similarity of the pathways in a behavioural net-work is that it should be possible to trace these pathways by chemical techniques. Recent research has confirmed that some functionally identical pathways in the brain are chemically coded in the sense that they show similarities in terms of histochemical staining and response to chemical stimulation. Histochemical studies have been carried out by Shute and Lewis (1967) on the cholinergic systems and by Fuxe and Dahlstrom (Hillarp, Fuxe and Dahlstrom, 1966) on the monoaminergic pathways. The cholinergic pathways were traced primarily by examining the distribution of acetylcholinesterase containing neurons by means of acetylthiocholine staining. Acetylcholinesterase is the enzyme which inactivates acetylcholine at cholinergic synapses. Electron microscopy was used to show that most of the acetylcholinesterase was located at the fibre terminals. Microassay of these terminals demonstrated that areas staining strongly for acetylcholinesterase had high concentrations giving further evidence that these pathways were cholinergic (Lewis, Shute and Silver, 1964). The monoaminergic systems, norepinephrine, serotonin and dopamine, can be examined more directly because these transmitters fluoresce after treatment of the neurons with formaldehyde. Unfortunately, the latter pathways have not been mapped in such detail as the acetylcholinesterase because the neurons contain too low a concentration to be distinguished clearly unless the inactivating enzyme, monoamine oxidase, is inhibited. The detailed discussion of the course of each of these pathways will be postponed until the relevant chapters. However, it is not surprising that the basic systems have their origin in the more primitive portions of the brain. The cholinergic and monoaminergic pathways both run from the reticular formation through the diencephalon and both project to the septal and cingulate regions. In other parts of the hypothalamus there is no coincidence of the various systems. The co-occurrence at some sites suggests that the systems may have complementary function in behaviour, and this is supported by direct stimulation of the brain. Studies of the limbic and hypo-thalamic pathways have traced a cholinergic system involved in water balance, studied by Fisher and Coury (1962) and Coury (1967), and an adrenergic system controlling energy balance (Coury, 1967; Booth, 1967).

In addition, there appear to be at least three distinct systems originating in the reticular formation: an adrenergic system involved in behavioural arousal, see Chapter 3; a cholinergic system controlling sensory processing, see Chapter 4; and a serotonergic system important in sleep, see Chapter 4. Finally, there is some evidence for a dopaminergic pathway involved in motor control which interacts with the cholinergic reticular pathways.

These pathways appear to be fixed, after a critical period in development (Chapter 9) although some neurons can be modified permanently by experience once they have developed (see Chapter 10). However, we will examine how these systems respond to endogenous chemicals such as the hormones (Chapter 3), to repeated stimulation by chemicals with the occurrence of tolerance (Chapter 8), as well as the abnormalities of function that can occur as the result of certain drugs such as the hallucinogens (Chapter 7). As a groundwork for these later chapters the basic principles of synaptic function and the effects of drugs on them will be outlined in the remainder of this chapter.

B. CHEMICAL NATURE OF SYNAPTIC TRANSMISSION

It is nearly a hundred years since du Bois-Reymond (1877) suggested that transmission at the synapse was either electrical or chemical. The first strong evidence for a chemical transmitter substance was provided by Loewi and by Dale about fifty years ago, but only in the last twenty years has it been shown unequivocally that transmission at all synapses is a chemical phenomenon with a few exceptions. One of the strongest pieces of evidence is the synaptic delay, the 0·3 msec interval between the occurrence of the presynaptic potential and the postsynaptic potential. This interval would not be expected if transmission was electrical, but it is just right for the time required for the release and diffusion of a neurohumoral transmitter across the synaptic cleft about 200 Å wide, shown in Figure 2. Further evidence for chemical transmission is that although presynaptic potentials are excitatory, the postsynaptic potentials differ with some being excitatory and some being inhibitory. This would not be expected if electrical induction were involved but can be explained easily in terms of two different transmitters. At a quantitative level the amplitude of the postsynaptic potential is larger than the amplitude of the presynaptic potential, and bears no similarity to the magnitude of the potential predicted by an induction theory. Finally, if the postsynaptic membrane potential is altered electrically the forms of the postsynaptic potential are changed, which would not contradict a chemical transmitter theory but could not be explained by an induction theory (see Thompson, 1967).

This theory of neurohumoral transmission implies that a specific chemical

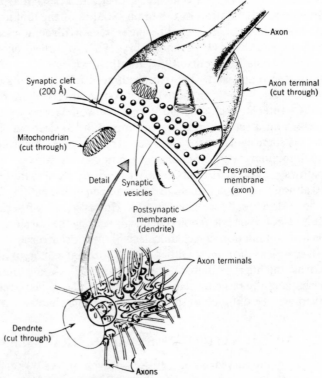

Figure 2. Structure of the synapse. From Stevens, *Neurophysiology*. Reproduced by permission of John Wiley and Sons, Inc.

is synthesized in the neuron, stored close to the synaptic cleft and released when the nerve impulse arrives at the axon terminal. The transmitter diffuses across the cleft and combines with a chemical receptor postsynaptically to produce a change in ionic permeability, giving the graded postsynaptic potential. After producing the potential the transmitter chemical must be removed rapidly from the cleft freeing the receptors for the next transmission. The next section will consider the whole process in some detail before considering how drugs can modify synaptic function. The transmitters used as examples will be acetylcholine, norepinephrine (noradrenalin), serotonin (5-hydroxytryptamine) and dopamine, since these are the major transmitters of behavioural importance. It should be pointed out that these four are not the only transmitters that are thought to be present in the nervous system. Other possible transmitter chemicals are histamine, gamma-aminobutyric acid (GABA) and glycine, but their involvement with behaviour is unclear at the present time.

Synthesis

The precise site of synthesis of transmitter substances is not known for certain but it is definitely within the neuron and possibly in the cell body. The cell body with the nucleus is the only region in the neuron where the manufacture of specific proteins can occur, including the production of the essential enzymes which catalyse transmitter synthesis. It is possible that the transmitter is associated with the tubular, endoplasmic reticulum growing out into the axon and transporting the chemicals to the terminals. At this terminal the tubules could pinch off to form the synaptic vesicles, the packets of transmitter, seen presynaptically under the electron microscope. Certainly there is strong evidence for axoplasmic flow in the neurons of the central nervous system carrying transmitters and enzymes (Ochs, 1972). Studies by Dahlstrom and her coworkers have demonstrated with histochemical fluorescence studies that when an axon nerve is blocked by crushing or ligation there is an accumulation of the transmitter norepinephrine on the cell body side of the blocked axon. It was possible from this to estimate the rate of transport of the central nervous system neurons as about 0·5 mm/hr. From other studies of transport in the peripheral nervous system in which the velocity can be as much as 10 mm/hr in the cat sympathetic nerve (Dahlström and Häggendal, 1966), and 17 mm/hr in the cat sciatic nerve (Ochs, 1972). In the latter review Ochs notes that this rate was found to be constant across several species such as rabbits, dogs and monkeys. The same rate of transport appears to occur in motor and sensory fibres and seems to be independent of the diameter of the nerve fibre or whether it is myelinated, or not.

The similarity of the rates at around 410 mm/day suggests that the same transport process is operating in all these fibres. A radioactive amino acid, leucine, injected into the nerve cell body of a motor neuron is incorporated into the protein being manufactured there, such as acetylcholinesterase, the inactivating enzyme for acetylcholine. This is the transmitter at the neuro-

8

muscular synapse (see Chapter 5). It has been found that acetylcholinesterase is being transported at a mean rate of 450 mm/day (Ranish and Ochs, unpublished but cited Ochs, 1972). In addition, there will also be synthesizing enzymes carried down like dopamine beta-hydroxylase, the synthesizing enzyme for norepinephrine, and choline acetylase for acetylcholine synthesis. As well as these, some of the materials carried down the nerve fibres by the fast transport system will be nonproteins like the transmitters, and it is known that some norepinephrine is transported down the axon to the terminal. However, Geffen and Rush (1968) concluded that axoplasmic flow replaces very little of the norepinephrine lost in normal synaptic activity and it follows from this conclusion that the bulk of transmitter is derived from synthesis occurring at the nerve terminal.

The mechanism of fast transport is under intensive study at the present time. The best hypothesis (Ochs, 1972) is that a 'transport filament' is synthesized in the cell soma and enters the axon to become the moving member of a sliding filament pair with the neurofilaments or microtubules as the stationary member. The molecules are bound to the transport filament and carried down the axon to the terminal by a process similar to muscular contraction where actin and myosin filaments slide past each other (see Huxley, 1969, for more details). Some of the predictions from the transport filament hypothesis have been supported, with one of the pieces of evidence being the importance of adenosine triphosphate, a source of metabolic energy which is also crucial for the sliding filaments in muscles.

The actual synthesis also depends on the presence of a source of metabolic energy, usually adenosine triphosphate (usually abbreviated as ATP), the specific enzymes necessary for synthesis, and the transmitter precursors taken up from the extracellular fluid. For example, the precursors required for

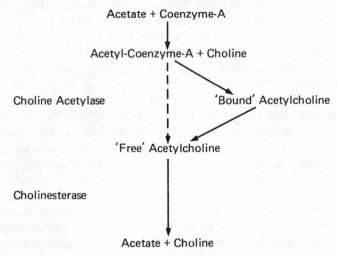

Figure 3. Diagram of the steps in the synthesis and inactivation of acetylcholine.

acetylcholine synthesis are an active acetate and coenzyme A which become acetylcoenzyme A and choline (See Figure 3). The specific enzyme is choline acetylase and there is evidence from *in vitro* studies of brain tissue that the activity of this enzyme is inhibited by acetylcholine (Potter, Glover and Saelens, 1968). End-product inhibition provides a very important mechanism for controlling the rate of transmitter production, so that synthesis is coordinated with the amount of activity at the synapse of the neuron. Thus it is not surprising that the rate of synthesis of acetylcholine can increase sevenfold

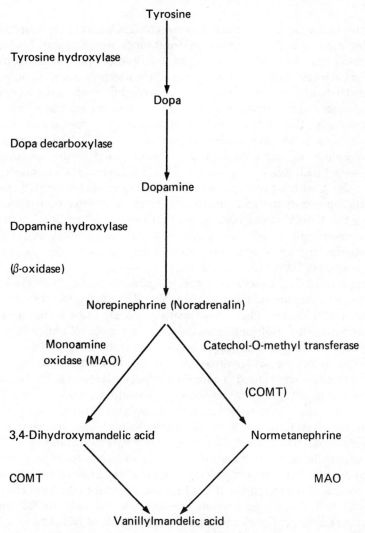

Figure 4. Diagram of the steps in the synthesis and inactivation of norepinephrine

during neural activity (Birks and MacIntosh, 1961). End-product inhibition also occurs in adrenergic neurons (Udenfriend, 1958) so that the transmitter norepinephrine inhibits tyrosine hydroxylase, the enzyme with the slowest metabolic rate in the synthesis of this transmitter (see Figure 4), and when adrenergic nerves are stimulated and norepinephrine used up the rate of norepinephrine synthesis is tripled. These two studies confirm that most transmitter synthesis occurs at the terminals with the manufacture of the synthesizing enzymes occurring at the nucleus in the soma.

Storage

The transmitter seems to be in at least three main forms in the neuron (Birks and MacIntosh, 1961): (1) stationary transmitter which cannot be depleted by repeated neural stimulation and is probably in the process of axoplasmic flow from the cell body; (2) depot transmitter which is stored bound and is presumed to be in the vesicles; and (3) surplus transmitter which is not contained in the vesicles and is unprotected from any inactivating enzyme which occurs intraneuronally. The evidence on the vesicle storage of transmitter has come from cell fractionation techniques (see De Robertis, 1967). This technique was used to isolate the nerve endings which were found in the mitochondrial fraction-separated using a centrifuge. This fraction was then subjected to osmotic shock with a hypotonic solution and the vescicles separated from the nerve ending membrane and subsynaptic component after centrifugation at 20,000 g for 20 min. It has been found that the vesicle fractions contain the highest concentrations of transmitter chemicals and synthesizing enzymes. These studies strongly support the idea that the vesicles are the quantal units for the storage of transmitter substance.

Electron microscopy studies of norepinephrine vesicles have shown that they have an outer limiting membrane with a dense core, and they are about 300–500 Å in diameter. The norepinephrine appears to be bound as a complex with adenosine triphosphate in the ratio of 4 moles of amine to 1 of the adenosine triphosphate. They also contain the synthesizing enzyme for norepinephrine, dopamine beta-hydroxylase, and so dopamine is probably taken up by the vesicles and oxidized to norepinephrine (see Iversen, 1970). In contrast, the synthesizing enzyme for acetylcholine, choline acetylase, is not located bound to the vesicle membrane but in the cytoplasm of nerve ending particles (synaptosomes) (Whittaker, 1970). The acetylcholine is then incorporated into the vesicles, although the precise method has not been demonstrated. It was emphasized in the last section that synthesis was under control of the transmitter itself, and this is extremely important in view of the limited storage capacity of the vesicles at the terminals. It has been estimated that the total reserves in adrenergic and cholinergic neurons are sufficient for only 10,000 impulses without synthesis, yet these same neurons are capable of releasing four to five times this quantity in an hour (Birks and MacIntosh, 1961; Sjärne, Hedquist and Bygdeman, 1969).

Release

Even when the membrane of the presynaptic axon is at its resting potential of -70 mV the synapse is not inactive. At frequent, but random, intervals small graded potentials are recorded postsynaptically (del Castillo and Katz, 1954). These 'miniature' potentials are believed to be due to the spontaneous discharge of the small packets (quanta) of transmitter from the vesicles into the synaptic cleft. It appears that the probability of release of a quantum is increased so that the number of quanta released is directly proportional to the size of the presynaptic potential and, of course, to the number of quanta available for release. In some neurons there may be about 1,000 quanta immediately available and about half a million altogether without synthesis (Elmqvist and Quastel, (1965) which is about enough for 10,000 impulses if synthesis were blocked.

It is found in some neurons that for every 15 mV increase in depolarization there is a tenfold increase in the frequency of a quantum release. Thus, for example, a 90 mV presynaptic depolarization would produce an increase of 10^6 in the frequency of released quanta. The exact mechanism by which the change in the potential causes release of quanta is unclear at this time, but the generally accepted story is that the vesicles migrate to the presynaptic membrane and discharge their soluble contents by a process of exocytosis (see Iversen, 1970). In exocytosis the vesicle fuses with the synaptic membrane which leads to the opening of both sets of membranes and the release of the transmitter and other soluble substances in the vesicles. One hypothesis for the molecular basis of exocytosis involves a contractile event similar to muscle. We have already mentioned that this involves actin and myosin, and similar proteins, neurin and stenin, have been discovered in mammalian brain (Berl, Puszkin and Nicklas, 1973). By the De Robertis cell fractionation technique it was found that most of the stenin was found in the fraction containing vesicles, while most of the neurin was obtained from the membrane fraction. These two proteins were found to combine in the presence of calcium ions to form neuro-stenin, which had similar properties to actomyosin in muscle. Berl and his associates have suggested that exocytosis involves the combination of neurin and stenin causing conformational changes in the membrane to release the transmitter. This sequence would be triggered by calcium which is released by depolarization in the same way as muscle contraction. After repolarization the two proteins would dissociate and the vesicle would separate to participate in re-uptake of free transmitter (see later sub-section) or be metabolized.

However, several eminent researchers are sceptical and argue that the vesicles may only be storing the transmitter and not directly involved in release (Kety, 1970; Robertson, 1970; Whittaker, 1970). In Robertson's model the vesicles fused in interconnected tubules in the resting state, and presynaptic depolarization results in them contracting into the vesicles, expelling their contents into the synaptic cleft, while in Kety's model the newly synthesized transmitter is concentrated in the presynaptic membrane ready for release. When release exceeds synthesis the vesicles would replenish the membrane stores ready for the next nerve impulse. This is supported by the finding of Whittaker (1970) that when radioactive acetylcholine is absorbed by cholinergic nerve terminals, the specific activity of membrane acetylcholine is higher than that of the vesicular acetylcholine. The acetylcholine released spontaneously has a greater radioactivity than that released during stimulation, and Whittaker has suggested that the fall in specific activity which occurs during stimulation is due to augmentation of the spontaneous release from the membranes by acetylcholine from the less radioactive vesicles.

Receptor combination

The transmitter substance diffuses across the cleft to the post synaptic membrane of the dendrites. Here it combines with transmitter receptor and simultaneously initiates the ionic flow of the graded postsynaptic potential. In most cases the form of the postsynaptic potential is determined by the time course of the transmitter concentration at the synapse. This reaches a peak very rapidly since release of all the transmitter by the presynaptic potential occurs immediately, and it then declines exponentially due to transmitter in-activation (see next sub-sections). There is no minimum number of transmit-ter–receptor interactions required to generate a postsynaptic potential so that the frequency of interaction will follow the same time course as the concentra-tion. It follows from this that if two presynaptic potentials occur in rapid suc-cession that the two transmitter concentrations will become added together and so the frequency of transmitter–receptor interaction, and thus the post-synaptic potential, will depend on the frequency of nerve impulses arriving at the terminal. Here we see that the quantitative information arriving at one side of the synapse is transferred accurately to the other side. Most of the studies that we will discuss are aimed at distorting this information by changing frequency of transmitter–receptor interactions with drugs, and examining the effects on behaviour.

The nature of the receptor which endows it with the property of chemical excitability has stimulated much research. Most experimenters believe that the receptor is some sort of protein, and de Robertis (1971) has recently discovered a proteo-lipid which has many of the appropriate properties. These studies began with experiments on the membranes of cholinergic nerve terminals. The M_1 fraction contains the nerve-ending membranes myelin and mitochondria. Treatment of this fraction with a nonionic detergent removed most of the limiting membrane leaving the junctional complex. Tests of the separated myelin, mitochondria and nerve-ending membrane showed that the radio-cholinesterase, radio-tubocurarine and radio-hexamethonium were taken up by the nerve-ending membranes while the separated junctional complex took up very little radio-cholinesterase, but just as much radio-tubocurarine and radio-hexamethonium. As tubocurarine and hexamethonium are cholinergic receptor blockers, while cholinesterase is not, this result suggests that the cholinergic receptors are localized in the functional complex and that cholinesterase has a narrower distribution, and is bound to a different macromolecule from the receptor. In spite of the claim of de Robertis (1971) and of others, these studies only demonstrate that there are some macromolecules that selectively bind cholinergic drugs.

Nevertheless, this sort of research is extremely important because it is the characteristics of the postsynaptic membrane which determine the nature of the ionic change produced by the transmitter. It is known that the same transmitter can be excitatory or inhibitory in action in different parts of the nervous system, and it is the receptors which determine whether depolarization or hyperpolarization occurs. In an attempt to classify the receptors, studies have been made using different sorts of blocking agents. It has been found that some neurons with norepinephrine as a transmitter, termed adrenergic neurons, can be blocked with either phenoxybenzamine or propanolol, and these have been called alpha-adrenergic and beta-adrenergic receptors respectively. However, this differentiation does not coincide with the excitatory–inhibitory classification so at the moment receptors are distinguished by the effective transmitter, its ionic action and the appropriate blocker (Iversen, 1970).

Enzymatic inactivation

Inactivating enzymes are located postsynaptically and intraneuronally in order to control the amount of free transmitter at the terminal. In the synaptic cleft it is essential for efficient transmission of information that the transmitter is removed rapidly after the postsynaptic membrane is polarized. The transmitter and receptors are in reversible action forming a transmitter–receptor complex. The amount of the latter will be a function of the 'free' transmitter and dissociation of the complex will occur as the free transmitter is eliminated by the enzyme. Acetylcholine is hydrolyzed by acetylcholinesterase to choline and acetic acid. (See Figure 3). In peripheral cholinergic neurons acetylcholinesterase is produced in the soma, and the most of it appears to be located postsynaptically around the terminals, but not in them. In adrenergic neurons there appear to be two enzymes involved; monoamine oxidase inactivates norepinephrine inside the neuron by removing amine groups by oxidation (oxidative deamination) while catechol-0-methyltransferase methylates norepinephrine in the synaptic cleft (see Figure 4). The importance of these enzymes can be estimated from the amounts of dihydroxymandelic acid, the deaminated metabolite, and normetanephrine, the 0-methylated metabolite. This has proved important in distinguishing the intraneuronal and extraneuronal action of drugs. However, it is important to point out that monoamine oxidase also deaminates two other transmitters, dopamine (see Figure 5) and serotonin (see Figure 6). This economy by the nervous system poses problems for distinguishing the biochemical systems involved in the behavioural changes produced by inhibitors of monoamine oxidase (see Chapter 3). Enzymatic inactivation of transmitter is probably the exception rather than the rule in the nervous system, and the most common method of inactivation is re-uptake (Iversen, 1970).

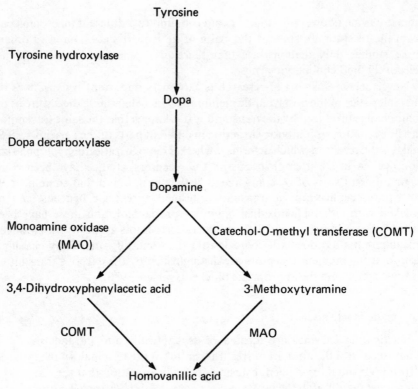

Figure 5. Diagram of the steps in the synthesis and inactivation of dopamine

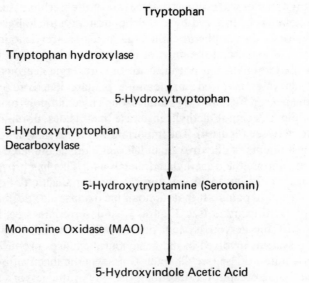

Figure 6. Diagram of the steps in the synthesis and inactivation of serotonin

Re-uptake

The other main mechanism for removal of the transmitter is the re-uptake of the transmitter presynaptically and the storage of it ready for release again In some transmitter systems (norepinephrine) it appears to be an extremely important mechanism, and can be demonstrated by injections of radioactive norepinephrine into the extracellular fluid, and measuring the accumulation of radioactivity in the presynaptic terminals (Glowinski, Axelrod and Iversen, 1966). In peripheral sympathetic neurons uptake can produce concentrations of radioactive norepinephrine which are thirty or forty times the concentration in the surrounding fluids (Iversen, 1967). It appears that 40–60 % of all the norepinephrine released is inactivated by uptake. This uptake system is an active process occurring against the concentration gradient. It is a saturable membrane transport mechanism requiring energy, is critically dependent upon temperature and sodium, and is situated in the neuronal membrane (Iversen, 1967). This process occurs at all adrenergic synapses tested, and accounts for the fact that very little free norepinephrine is collected when monoamine oxidase or catechol-0-methyltransferase are inhibited. Some re-uptake of acetylcholine probably occurs (Schuberth and Sundwall. 1967) but it is not very important because of the efficiency of acetylcholinesterase, e.g. all the acetylcholine in the brain could be hydrolyzed to choline in just over 10 sec (McIlwain, 1955). However, half of the choline produced by cholinesterase activity is taken up for re-use by cholinergic neurons.

C. DRUGS AND SYNAPTIC TRANSMISSION

It was stressed in the introductory section that the book will be organized around the relations between transmitter systems in the brain and behaviour. As a result this section will consider the manipulation of synaptic transmission by means of chemicals. The emphasis will be on the action of drugs which modify synapses which have acetylcholine, norepinephrine and serotonin as their transmitter substances, since these systems have been examined in some detail from the behavioural point of view. Since there are chemicals which modify every step in synaptic transmission outlined in the last section, this section will follow the organization of the last one.

Synthesis

In the last section it was mentioned that synthesis depended on the presence of precursors, synthesizing enzymes and ATP. In adrenergic neurons dopa is converted to dopamine and this is converted to norepinephrine. This pathway can be disrupted by means of alpha-methyldopa, which is converted to alpha-methyl-norepinephrine (Kopin, 1968). This compound is stored in place of norepinephrine in the vesicles and is released by nerve stimulation. As this false transmitter had reduced depolarizing properties it blocks synaptic transmission. Alpha-methyldopa is, strictly speaking, a non-specific competitive inhibitor of decarboxylase enzymes and besides interfering with dopa decarboxylase synthesis of dopamine and norepinephrine it will interfere with the synthesis of serotonin by blocking the conversion of 5-hydroxytryptophan to serotonin by 5-hydroxytryptophan decarboxylase. A more specific inhibitor of norepinephrine synthesis which interferes with the rate-limiting step in synthesis, the conversion of tyrosine to DOPA, is alpha-methyl-*para*-tyrosine, which acts like tyrosine and so competes with real tyrosine for the enzyme tyrosine hydroxylase. As a result, there is a marked fall in norepinephrine levels in the adrenergic neurons, including those in the brain (Spector, Sjoerdsma and Udenfriend, 1965) but there much less effect on serotonin neurons. This selectivity of action is extremely important in comparing the behavioural importance of the norepinephrine and serotonin systems.

Drugs which interfere with the synthesis of acetylcholine have been used very little in behavioural analysis. As we shall see later, the hemicholiniums interfere with synthesis by blocking re-uptake and reducing the amount of choline available, but they are more properly considered later. The only other method of interfering with acetylcholine synthesis is by inhibiting the synthesizing enzyme, choline acetylase. One group of compounds that have a high degree of inhibitory activity are the styrylpyridine analogues (Cavallito, White, Yun and Foldes, 1970). It appears that the acetylcoenzyme A reacts with the choline acetylase, but the transfer of the acetyl group from the enzyme complex to choline is blocked. As yet very little work of relevance to behaviour has been done with these drugs, except for some evidence that these compounds block neuromuscular transmission (see discussion after Cavallito et al., 1970).

As we have seen, the synthesis of serotonin (5-hydroxytryptamine) can be blocked by a decarboxylase inhibitor which prevents the synthesis of the transmitter from its precursor, 5-hydroxytryptophan (see Figure 6). However, the rate-limiting step seems to be the production of this immediate precursor from tryptophan (Jéquier, Lovenberg and Sjoerdsma, 1967). A relatively specific inhibitor for serotonin synthesis is *para*-chlorophenylalanine (Koe and Weissman, 1966) which appears to block tryptophan hydroxylase, the synthesizing enzyme for 5-hydroxytryptophan (Jéquier et al., 1967). However, it may also prevent 5-hydroxytryptophan reaching the active site on the synthesizing enzyme 5-hydroxytryptophan decarboxylase, because it inhibits synthesis even after loading with 5-hydroxytryptophan (Koe and Weissman, 1966). Loading with the precursor 5-hydroxytryptophan is one of the methods of increasing serotonin (see Chapter 6), since the decarboxylase is a very active enzyme, and so the precursor is immediately transformed into serotonin. Since it passes the blood-brain barrier easily, it can be injected intraperitoneally and used to raise the levels of serotonin. Increase of serotonin outside the central nervous system does not seem to affect behaviour, which is just as well because bananas contain between 3 and 5 mg of serotonin each! (Sjoerdsma, 1965).

Dopamine synthesis can also be modified by administering its precursor *l*-dopa as one would expect from Figure 5. It has been demonstrated that *l*-dopa produces a marked increase in the levels of brain dopamine, although there is also some increase in the levels of norepinephrine (Carlsson, 1959). This technique has proved important in the treatment of Parkinson's Disease, believed to be due to a fall in dopamine in a neural network involved in motor control (see Chapter 5).

Storage

After synthesis, transmitters are stored bound to protein protecting them from any intraneuronal inactivating enzymes. In cholinergic neurons the cholinolytics such as atropine and scopolamine compete for the storage receptors and as a result increase the amount of free actylcholine (Giarman and Pepeu, 1964). A well known drug, reserpine (see Chapter 3), blocks storage of norepinephrine, which results in a depletion of the transmitter because it is degraded by the intraneuronal enzyme, monoamine oxidase. Functionally blockade of storage receptors by reserpine is equivalent to competition for tyrosine hydroxylase by alpha-methyl-*para*-tyrosine. The major difference between the two actions is that reserpine also depletes serotonin (Brodie and Shore, 1957) dopamine (Zbinden, 1960) and histamine (Adam and Hye, 1964; 1966). A similar non-specific but shorter acting drug is tetrabenazine (Zbinden, 1960; Pletscher, Brossi and Gey, 1962).

Release

Before a transmitter can become functional it must be released into the synaptic cleft. Another well-known drug, amphetamine, enhances the release of norepinephrine from the neuron (Glowinski, Axelrod and Iversen, 1966). This action can be deduced from the marked increase in 0-methylated metabolite, normetanephrine, resulting from the breakdown by catechol-0-methyltrans-ferase in the synaptic cleft. In the cholinergic system the major agent affecting acetylcholine release is botulinum toxin, which seems to block the mechanisms for release in the presynaptic membrane (Zacks, Metzger, Smith and Blumberg, 1962). As a result of this blockage there is progressive reduction in activity at all cholinergic neurons, most obviously at the neuromuscular junction (Burgen, Dickens and Zatman, 1949).

Postsynaptic receptor interaction

In neuropharmacology the term 'receptor' is usually used to refer to the postsynaptic receptors. We have already mentioned that there are several types of receptor for each transmitter. In the cholinergic system one group of receptor blockers are the muscarinic cholinolytics (anticholinergic) drugs such as atropine and scopolamine whose similarity to acetylcholine enables these drugs to compete for the muscarinic postsynaptic receptor sites, and so reduce the number of receptor–transmitter interactions which pass on information from one side of the synapse to the other. Muscarinic synapses have been identified in the brain, for example basal ganglia (McLennan and York, 1966) and cortex (Krnjevic and Phillis, 1963b), and the effects of atropine are discussed in Chapters 2, 4 and 5. In addition to muscarinic blockers there are a separate group of nicotinic blockers, e.g. curare, curare-like drugs, and hexamethonium. Curare and tubocurarine block receptors without depolarizing (del Castillo and Katz, 1957), and their effect on the nicotinic neuromuscular synapses is discussed in Chapter 5.

A second class of drugs acting on the receptors are the mimetics which imitate the action of the transmitter. The most important cholinomimetic is carbamyl-choline chloride, carbachol, which acts like acetylcholine on the postsynaptic cholinergic receptors because of its chemical structure but does not combine with receptors of the inactivating enzyme for acetylcholine, acetylcholinesterase, and so has a longer effect than acetylcholine. As well as depolarization of the postsynaptic membrane and generation of graded potentials, carbachol is known to release acetylcholine from the terminal, perhaps by displacing it from the intraneuronal stores (McKinstry and Koelle, 1967). Carbachol activates both nicotinic and muscarinic receptors and is thus an 'all purpose' agent for investigating cholinergic neural systems controlling behaviour. As it does not pass the blood–brain barrier it must be injected directly into the brain to have any effect on the neurochemical systems. Its use is discussed in Chapters 2, 4 and 5. Specific agents for investigating nicotinic and muscarinic cholinergic receptors do exist, and two of these are nicotine and acetyl-α-methylcholine, but it is not known how important this distinction is for the central nervous system where there are known to be muscarinic, nicotinic and mixed receptors on the Renshaw cell, see Chapter 5.

As we pointed out in the previous section, there are also two sorts of adrener-gic synapse found in mammalian nervous systems, and adrenergic blocking drugs have been discovered which are highly specific by acting on either alpha-adrenergic receptors or beta-adrenergic receptors. The common alpha-adrenergic blockers are chlorpromazine, dibenzyline, dibenamine and phento-lamine, and the most common beta-adrenergic blocker, propanolol. These compounds were used to distinguish the two adrenergic systems involved in food intake which are discussed in Chapter 2. Chlorpromazine, which is one of the alpha-adrenergic blockers, deserves special mention because it is the first major tranquillizer discovered and has been used extensively in the therapy of agitated states (see Chapter 3). Although it is considered throughout as a catecholamine blocker (see Chapters 3 and 5), it is also effective as an anti-cholinesterase and as an antihistamine agent. Interpretation of its action as an adrenergic block should strictly speaking be supported by data on the effect of other alpha blockers, but this has rarely been done in behavioural studies.

Not many adrenomimetics have been developed because of the low activity of the extraneuronal inactivating enzyme for norepinephrine. Thus norepine-phrine can be injected close to synapses to act directly. Norepinephrine occurs in two forms, isomers, and one of these rotates polarized light to the left (levorotatory) and the other turns the light to the right (dextrorotatory). More properly it should be the levorotatory isomer of norepinephrine, called *l*-norepinephrine, that is used since the dextrorotatory isomer is biologically inactive, but most of the psychological experiments have used a mixture of the two because laboratory synthesis is not as specific as neuronal synthesis. The same principle holds true for dopamine where *l*-dopamine is the biologically active form. Adrenomimetics which act on both noradrenergic and dopamine receptors are *l*-epinephrine and *l*-isoproterenol, which are natural and synthetic catecholamines, respectively. Amphetamine is also thought to be an adrenomimetic, as well as acting as an adrenergic releasing drug, and as a monoamine oxidase inhibitor, as we shall see in the next section on drugs which block enzymatic inactivation.

Enzymatic inactivation

It is an important biochemical control that monoamine oxidase deaminates 'free' norepinephrine within the neuron. There are quite a number of compounds which inhibit the action of monoamine oxidase and they achieve this by preventing the formation of enzyme–substrate complex of the monoamine oxidase transmitter that normally occurs prior to deamination. Most of the commonly used monoamine oxidase inhibitors like iproniazid, tranylcypromine, pargyline and amphetamine combine irreversibly with the enzyme. The result of this inhibition is a marked rise in the brain levels of norepinephrine, serotonin and dopamine (Everett, Wiegand and Rinaldi, 1963). However, there is no observable change in behaviour of a normal animal and there is no prolongation of the effects of the nerve stimulation (Kanijo, Koelle and Wagner, 1955). This suggests that the increased norepinephrine, serotonin and dopamine are localized intraneuronally. Fluorescence staining using the Falck and Hillärp technique shows that there are increases in fluorescence intensity at the nerve terminals, axons and cell bodies showing increased concentration of transmitter confirming this suggestion (Norberg, 1965). This action of a monoamine oxidase inhibitor can also be inferred from the decrease in 2-3-hydroxymandelic acid, the deaminated metabolite of norepinephrine produced by extraneuronal inactivation (see Figure 4).

Intraneuronal inactivation is not an important mechanism in the cholinergic system, and there is no evidence that it occurs at all. The most important control of the free acetylcholine extraneuronally is enzymatic inactivation by acetylcholinesterase. As a result, any compounds which interact with acetylcholinesterase will have a profound effect on cholinergic function. It seems that the acetylcholinesterase molecule has two important parts, the anionic site and the esteratic site, and it is the esteratic site which is intimately involved in the hydrolysis of acetylcholine. It is believed that anticholinesterases combine with the esteratic site and reduce the number of molecules of acetylcholine that can be hydrolyzed. Physostigmine is a reversible inhibitor of acetylcholinesterase, and it competes with acetylcholine for the esteratic sites (see

Stein and Lewis, 1969) and slows down the inactivation. As a consequence the brain cholinesterase activity, the ability to hydrolyze acetylcholine, and falls to a minimum after about 15 min. and takes about an hour to recover after a behaviourally active dose, e.g. 0.05 mg/kg in the rat (Irwin and Hein, 1962). The central effects of physostigmine can be compared with neostigmine, a quarternary compound that passes through the blood–brain barrier poorly in normal animals, and so has little effect on brain cholinesterase (Irwin and Hein, 1962) while reversibly inhibiting peripheral cholinesterase (see Chapters 4 and 5). Another group of anticholinesterase compounds are the organo-phosphorus inhibitors (Holmstedt, 1963) which become attached to the enzyme and phosphorylate the esteratic site of the acetylcholinesterase molecules rendering it inactive so that a new site must be synthesized. These compounds have been investigated extensively because of their toxic properties for man and insects, and as therapeutic agents. The first synthesized compound was tetraethyl pyrophosphate in the mid-19th century, but the major discoveries began at the end of the 1930's, and continued throughout World War II. Of the more than 2,000 compounds synthesized, most mention will be made of di-isopropylfluorophosphate and sarin, which have been studied behaviourally (see Chapters 4, 5 and 7).

The consequence of inhibiting acetylcholinesterase is an increase in the acetylcholine in all body regions that the drugs reach. Drugs, like physostigmine and diisopropylfluorophosphate, which pass through the blood–brain barrier, produce a massive accumulation of the transmitter in the brain (Stewart, 1952). It has been shown that brain acetylcholine did not increase until there was over 40 % inhibition of the enzyme (Aprison, 1962; Holmstedt, 1967) and it has been argued that between 40 and 60 % inhibition of acetylcholinesterase the enzyme loses control of its substrate. There is considerable evidence cited that there is a similar critical level for neural change. In addition, there is clear evidence that increasing acetylcholine does not have a monotonically increasing change in neural function. It seems that small increases in acetylcholine facilitate cholinergic transmission, but an excessive accumulation of acetylcholine will hold the synapse in a depolarized state and prevent it from being activated (Crossland, 1960). Metz (1958) found that for the respiratory reflex the critical level for the change from facilitation to impairment occurred around 60 % inhibition of acetylcholinesterase. There seems to be a corresponding change in behaviour at 40 % and 60 % inhibition as we shall see in Chapters 4 and 7.

As we have seen already, extraneuronal inactivation in the adrenergic and serotonin systems is by catechol-0-methyltransferase, which 0-methylates the transmitters. A drug which does impair this process is tropolone, but it has little neural effect because of the efficiency of the re-uptake inactivation process, and so the major method of enchancing monoamine function is by blocking re-uptake.

Re-uptake

The re-uptake mechanism was inferred from experiments which demonstrated that radioactive norepinephrine was taken up and stored presynaptically (Glowinski and Axelrod, 1966). From this observation it follows that drugs which prevent the accumulation of radioactivity probably impair the re-uptake process in the presynaptic membrane resulting in increased norepinephrine being available for interaction with the postsynaptic membrane. Drugs which seem to block this active uptake mechanism are imipramine, desipramine, and cocaine. Normally, if norepinephrine is injected close to adrenergic neurons the transmitter is taken up and fluorescence staining has shown that it had accumulated in the nerve terminals and axons (Hillarp and Malmfors, 1964). However, a dose of cocaine blocked the accumulation of the norepinephrine and it looked as if this was a competitive blockade because the block was partially overcome by increasing the concentration of nonepinephrine (Hillarp Malmfors, 1964). It is thought that re-uptake is not such a specific incorporation process as vesicular storage (Giachetti and Shore, 1966) because of the wide range of drugs which block it. For example, serotonin re-uptake is blocked by norepinephrine (Palaič, Page and Khairallak, 1967), but it is not known if this occurs in normal circumstances. Serotonin re-uptake is also blocked to some extent by imipramine and desipramine.

The presence of a number of neurochemically coded behavioural networks in the central nervous system has important implications for research in psychopharmacology. In particular, it has crucial consequences for the specificity of drug action. Any chemical agent injected into the body must be absorbed into the bloodstream and be carried to an active site for it to be behaviourally effective. For most of the drugs that will be discussed, this means transportation to some place in the central nervous system. The outcome of the drug's action at this site will be a change in the information carried by a set of neurons either by modifying axonal transmission or by changing chemical transmission at the synapse. The mechanisms involved in axonal transmission seem to be the same for all neurons, while we saw in Section B that a number of different transmitter chemicals have been identified in the central nervous system. It follows that drugs acting on one transmitter system will act on fewer neural pathways than those acting on axonal transmission. As a result all the drugs discussed in the rest of the book are considered in terms of their action on synapses and their consequent modification of the pattern of activity in a neurochemical network. Logically the pattern of activity can either be enhanced or impaired resulting in a greater or lesser intensity of behavioural responses. Increases in some behaviour will be the result of increases in the frequency of transmitter–receptor interactions. Transmitter mimetics are considered as if they were transmitters. It is important to remember that all drugs do not produce a monotonically increasing effect with increasing dose. Inhibitors of inactivating enzymes have different effects at low and high doses because high doses result in such large amounts of transmitter that the receptors are continuously bound to transmitter, and the number of transmitter–receptor interactions are decreased (depolarization block). Thus the function of a behavioural network does not depend on the levels of transmitter *per se*, but on the dynamic interaction of transmitter and receptor throughout the network. This conception of function contrasts with the static 'levels of transmitter' notion of the centre theory, and this is reflected in the changed emphasis where behaviour is compared with the amount of functional transmitter measured in terms of transmitter turnover rates, the rates of synthesis and breakdown. It can be seen that a drug could increase release so that more transmitter–receptor interactions occurred, but so long as synthesis kept up with release the change in behaviour would not be correlated with any change in transmitter levels.

Of course, there are other reasons why in most cases there is no clear relationship between levels of a chemical and behaviour. In some neural systems there appears to be a safety margin in levels of transmitter and its enzymes to protect it against malfunction. In the brain cholinergic system there seems to be a 40 % safety margin in the amount of inactivating enzyme, acetylcholinesterase, so that neural function is not disrupted until this critical level is reached and acetylcholine levels rise dramatically (Aprison, 1962; Holmstedt, 1967). Behavioural effects parallel the changes in neural function and so there are critical doses of anticholinesterase which inhibit the enzyme activity to 60 % of normal and produce changes in responding (see Chapter 4). As we saw in Section B of this chapter, there is only enough transmitter in sympathetic neurons for about 10,000 neural impulses without further synthesis (Birks and MacIntosh, 1961), but that activity induces synthesis. In the central nervous system brain amines can be depleted to about 10 % of normal before behavioural changes are observed (Haggendahl and Lindquist, 1964). Part of this safety margin may be related to neural redundancy in the sense of function not being dependent on any one synapse. Disruption in a nucleus will only be detectable when a proportion of synapses are malfunctioning. Similarly, conduction within a network will not depend on a single fibre but on the integrity of a proportion of fibres in the tracts. The analysis of the behavioural function of neurochemical networks is complicated by the fact that a biochemically specific drug injected into the body will be generally distributed, so that it will be reaching and acting on several neural networks which have the same transmitter. This problem is multiplied because no drug is completely specific biochemically at all doses. The behavioural changes that are observed will depend on the regional distribution of the drug, the dose reaching each site, the different sensitivities of the networks, and the specificity of the behavioural test. It is well known that a profile of the mode of action of any drug can only be compiled by multiple testing (Russell, 1964). In any behavioural test both a main effect and a number of unwanted effects, or 'side effects', will be observed. A teasing out of the basic neuro-chemical systems will depend to a great extent on the skill of the behavioural scientist in devising suitable tests. Part of the problem on drugs inducing changes in several networks can be eliminated by direct injection into specific loci in the brain using a cannula, and stimulating single networks. If a transmitter mimetic is injected, then false information can be introduced into a nucleus and the effects on behaviour evaluated by means of the test battery. This technique enables different sorts of questions raised by the network theory to be asked. Instead of wondering 'What is the effect of the level of activity in a brain centre on behaviour?' we can now ask 'Which neural system in the central nervous system is dominantly active during a specific sort of behaviour?' Some answers to this question are given in the remaining chapters.

2
Control of Homeostatic Motivation

In the last chapter we discussed the properties of a neural network controlling behaviour and the advantages this would have over specific centres. This chapter will discuss two types of behaviour, water and food intake, important in the regulation of internal states of the organism. Pioneering work on the mechanisms underlying these two types of behaviour was carried out by Claude Bernard, Walter Cannon and Curt Richter and it is appropriate to consider the thinking which was behind their research. The notion of multiple regions involved in the control of internal states is in the tradition of Claude Bernard who continually stressed the point that control of the internal milieu was not limited to one specific part of the body, but was the result of a number of anatomical units with different but related functions (Bernard, 1865). Bernard (1859) regarded the internal milieu as a physiological unit consisting of all the bodily fluids, and which was a closed system with indirect contact with the external environment. His famous statement that 'the stability of the internal environment is the condition of a free and independent life' (Bernard, 1859) means that internal adjustments of the chemical balance in the internal environment ensure that the organism is less dependent on the constant fluctuations of the physical conditions in the external environment. These ideas were elaborated by Cannon (1939) in his discussion of the biological processes involved in maintaining the relative stability, or *homeostasis*, of the internal environment. It is kept within a normal range by means of feedback mechanisms which trigger the biological processes whenever the imbalance exceeds the normal limits. In many cases the biological processes are physiological and do not involve behaviour, but Cannon (1939) recognized that there were circumstances in which physiological homeostasis was inadequate to maintain the internal stability and then appropriate behaviour was triggered. Richter (1942–43) went on to investigate a wide range of behaviours whose motivation was homeostatic in origin. For example, he showed that rats that could not control their sodium metabolism due to an adrenalectomy preferred drinking water with higher sodium chloride concentration than normal rats. These adrenalectomized rats die when deprived of the extra sodium chloride or if their gustatory nerves are cut so that they cannot discriminate between the two solutions. Thus the change in mineral intake was an important part of the biological response of the organism to compensate for the loss of the adrenals, the normal physiological regulation. As we shall see in the rest of the chapter, variations in the organism's intake of chemicals are homeostatically motivated and are coordinated with the physiological mechanism involved in homeostasis. Motivated behaviour consists of both appetitive and consummatory components. By appetitive behaviour we mean the pattern of responses that precede the attainment of a goal and consummatory behaviour is the final act in the sequence, i.e. the goal behaviour. Thus an animal deprived of food will perform food seeking (appetitive behaviour) until water (goal) is found, and then drinking (consummatory behaviour) occurs.

A. CONTROL OF WATER BALANCE

Water is distributed in the body within the cells (intracellular), surrounding the cells (extracellular) and in the blood stream (humoral). The relation between intracellular fluid depends on the amount of sodium ions in the extracellular fluid. The intake of salt (sodium chloride) increases the sodium ion concentration of the extracellular fluid and produces *relative dehydration*. The body attempts to control this increase in three ways: by water intake, by conservation of water by the kidneys due to the release of antidiuretic hormone from the pituitary, and also by increased excretion of sodium ions from the kidneys. When bleeding occurs there is fluid loss but no change in the sodium ions, and this is called *absolute dehydration* (Fitzsimmons, 1961). It results in water intake and enhanced kidney absorption due to antidiuretic hormone release. Corresponding to these two states are *hydration* and *cellular hydration*. In hydration there is an increase in both extracellular and intracellular fluid and it is controlled by inhibition of antidiuretic hormone. Cellular hydration results from depletion of the extracellular sodium ions and is characterized by increases in the intracellular fluid and reduced blood volume. The diminished blood supply to the kidney triggers the release of renin, which leads to the formation of angiotensin. The angiotensin acts to maintain blood pressure and stimulates the release of aldosterone from the adrenals, which potentiates the reabsorption of sodium ions from urine into the blood. Cellular hydration and relative dehydration are usually explained in terms of the activation of osmoreceptors which activate some regulatory mechanism that restores the tonicity of the body cells. It is interesting that sodium-depleted animals continue to drink water and secrete antidiuretic hormone. This implies that there is also a mechanism for controlling the fluid volume which is operating to some extent independently of an osmoregulatory mechanism. The volumetric mechanism is probably the controlling factor in absolute dehydration where there is no change in tonicity. The present evidence suggests that the volume effects are mediated via the kidneys. Reduced blood volume activates angiotensin production and angiotensin results in drinking even in hydrated animals (Fitzsimmons, 1971). This additional mechanism is clearly important in situations, such as bleeding, when sodium ion concentration is not an adequate index of fluid balance.

The search for the location of osmoreceptors and angiotensin receptors involved direct injection of solutions into the brain. In 1947 results of injecting

26

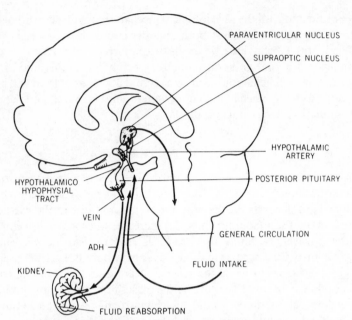

Figure 7. Central and peripheral control of water balance.
The supraoptic and paraventricular nuclei contain osmoreceptors which receive information about the amount of
fluid in the arteries. From House, E. L., and Pansky, B.
A Functional Approach to Neuroanatomy. Reproduced by
permission of the McGraw-Hill Book Co.

hypertonic sodium chloride into the carotid artery on antidiuretic hormone
secretion was reported by Verney (1947). It was found that the increased
osmotic pressure in the internal carotid artery distributing blood to the hypothalamus resulted in increased release of the hormone from the pituitary, which
enhanced water reabsorption from the kidneys (see Figure 7). Verney postulated
that there were osmoreceptors in the hypothalamus sensitive to the toxicity of
blood in the internal carotid artery. After a series of studies with direct injection
of hypertonic salt solutions Andersson (1952, 1953) hypothesized that there
were apparently two distinct sets of osmoreceptors in the goat; one set in the
paraventricular region close to the third ventricle which controlled water intake,
and another set close to the median eminence which controlled antidiuretic
hormone secretion. Electrical stimulation of the paraventricular region also
initiates drinking in the satiated goat (Andersson and McCann, 1955) and rat
(Miller, 1960). However, Andersson (1966) has recently summarized much of
his own work and has concluded that there is only one set of sodium ion receptors close to the third ventricle. Stimulation of these receptors with hypertonic
saline solutions but not other hypertonic solutions elicit drinking and antidiuretic hormone release. These may be the receptors sensitive to changes in
the sodium ion concentration and trigger osmoregulation. In addition,
Andersson (1966) found that angiotensin introduced into the third ventricle

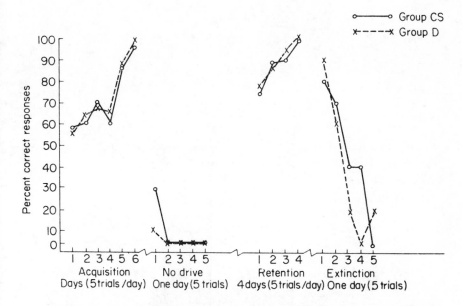

Figure 8. The various curves show the comparability of acquisition, relearning and extinction of a T-maze response under conditions of $23\frac{1}{2}$ hours water deprivation and of direct cholinergic stimulation of the lateral hypothalamus. From Khavari, K. A., and Russell, R. W. (1966). *J. comp. physiol. Psychol.*, **36**, 339–345. Reproduced by permission of American Psychological Association

stimulates drinking and antidiuretic hormone release. However, angiotensin combined with hypertonic saline resulted in marked enhancement of both drinking and hormone release. Doses of angiotensin injected via a cannula directly into the subfornical organ elicited drinking with a latency of less than 30 sec. suggesting that there was no spread of the chemical to stimulate a neighbouring site (Simpson and Routtenberg, 1973).

Control of water balance

The control of the water balance is the result of integrating a mass of information from the internal and external environment of the organism, because as well as information on the tonicity of the body fluids, information on the fluid content, distension of the stomach, taste, smell and dryness of the mouth are used (Grossman, 1967). Similarly the organism's reaction to a negative water balance is comprehensive, consisting of increased reabsorption by the kidney tubules as well as the behavioural responses of water seeking (appetitive response) and drinking (consummatory response). It follows from this that if there is a neural network involved in the control of water balance it must integrate the behavioural and physiological responses. In the next section evidence will be presented which strongly supports the theory of a cholinergically coded network in the brain controlling water balance.

28

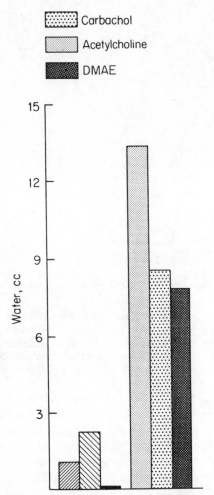

Figure 9. Effect of acetylcholine and carbachol stimulation of the hypothalamus on water intake during a 1-hr test. The histograms to the left show the lack of effect of norepinephrine, epinephrine, and dopamine, respectively. From Grossman, S. P. (1964). *Int. J. Neuropharmacol.*, **3**, 45–48. Reproduced by permission of Pergamon Press

Motivational characteristics

The first evidence of cholinergic involvement came from a study by Grossman (1960) in which he found that drinking in satiated rats was induced by the introduction of crystalline carbamylcholine chloride, carbachol, a cholinomimetic, and by acetylcholine mixed with an anticholinesterase to prevent enzymatic inactivation. Typically the rats drank about 12·0 ml of water in a 20 min. test period (see Figure 9) showing that there was a similarity between normal deprivation-induced drinking and carbachol-induced drinking. However, it was also crucial to demonstrate that carbachol in the hypothalamus would induce appetitive behaviour as well as consummatory responding. Grossman (1960) also showed that the chemically stimulated animals would learn to lever press to get water. In a more thorough study Khavari and Russell (1966) compared the effects of central cholinergic stimulation and water deprivation. They showed an amazing parallel between the rates of acquisition, relearning and extinction of a T-maze response (see Figure 8). In addition they demonstrated that subjects who had learned with water deprivation showed identical behaviour when they were motivated by cholinergic injections in the lateral hypothalamus. This interchangeability of the learned response from one mode of activation to the other shows conclusively that there is an identity between the two states, as far as behaviour is concerned. A study by Chun-Wuei Chien and Miller (Miller, 1965) has extended the similarity between natural and carbachol-induced deprivation by showing that the lateral hypothalamic nucleus is involved in the homeostatic system regulating water balance. They found that carbachol injections changed the excretion of urine, decreasing the volume and increasing the concentration. The latter finding showed that the decrease in volume was caused by reabsorption produced by the action of the antidiuretic hormone on the kidney tubules.

Pharmacological properties

A series of studies have been concerned with demonstrating whether those carbachol effects were due to the cholinergic properties of the lateral hypothalamus or whether the effects were unrelated to the neurohumoral properties of the drug. This latter idea is unlikely in view of the effects of acetylcholine itself and even less likely because Stein and Seifter (1962) found that injections of muscarine into the lateral hypothalamus induced drinking whereas nicotine did not. These central effects were blocked by pretreatment with peripheral blocking of the muscarinic blocking agent, atropine sulphate. In support of this study, central injection of atropine sulphate, a cholinergic blocking agent, reduced the water intake of water-deprived rats (Grossman, 1962) which also gave additional evidence for the identity of deprivation and carbachol-induced drinking. This issue was finally settled by the work of Chun-Wuei Chien and Miller (Miller, 1965) which showed that injections of physostigmine into the brain of a very slightly deprived rat increased the amount of water drunk in 30-min. test period from 0·3 ml to 10·5 ml. Clearly there is a brain cholinergic system with muscarinic properties involved in the homeostatic control of water balance. The anatomical locus of this system is discussed in the next section.

Neural network

The experiments by Fisher and Coury (Fisher and Coury, 1962; Coury, 1967)

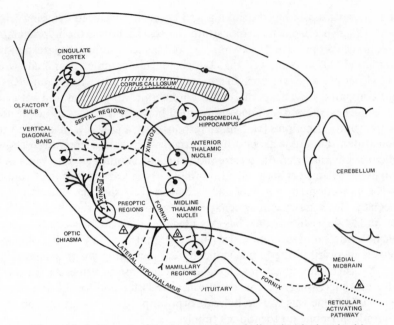

Figure 10. A diagram of the brain pathway believed to be involved in the control of water balance and traced by cholinergic stimulation. Three other brain regions also sum to trigger the drinking system although they are not cholinosensitive. These are the temperature-sensitive neurons in the hypothalamus, the osmoreceptors of the supraoptic and paraventricular nuclei and the reticular activating pathway. From Fisher, A. L. (1964). *Sci. Amer.*, **216**(6). Reproduced by permission of W. H. Freeman and Co.

have shown that the perifornical region is only one of many brain regions implicated in the control of water intake. These areas fall within a medially oriented, generalized 'Nauta' circuit including the septal area, the dorsomedial hippocampus, the fornix, mamillary regions of the hypothalamus, mamillo-thalamic tract, anterior nuclei of the thalamus, cingulate gyrus and preoptic regions (see Figure 10). These sites are all nuclei consisting of clusters of synapses and they represent points at which information is being integrated into the neural network although the nature of the information is not known at the present time. However, it is possible that stimulation effects in the limbic circuit, involving the septal area and dorsal hippocampus, may represent the activation of non-specific modulatory systems rather than effects on drinking specifically. Discussion of this idea is presented in a later section on the cholinergic arousal systems and Coury (1967) has some other speculations about their function.

In contradiction to this circuit theory of thirst, an alternative explanation has been discussed (Fisher and Levitt, 1967; Routtenberg, 1967). It was suggested that carbachol's effects may be on the receptors lining the ventricular wall. The multiple effective sites may be due to diffusion into the ventricles, and thence by ventricular flow to this single centre. This would explain a number of puzzling facts. (1) Carbachol applied to the axons of the fornix elicits drinking, although carbachol is presumed to act at synapses (Routtenberg, 1967). (2) Some of the most consistent sites for eliciting drinking are in the limbic system (septal area and dorsal hippocampus) where electrical stimulation is ineffective in inducing drinking (Fisher and Levitt, 1967). In fact, septal lesions have the same effects as carbachol stimulation (Routtenberg, 1967), and it seems strange that stimulation and ablation have identical effects unless the lesions are irritative (see Reynolds, 1965). (3) The areas in which carbachol produces drinking are medial and dorsal to ablated regions which cause adipsia in the hypothalamus (Routtenberg, 1967). (4) There is a minimum latency for drinking of about 3 min. or more at any site in the presumed circuit, which is more easily explained by diffusion than delay of action at an effective site (Fisher and Levitt, 1967). (5) Blocking of carbachol stimulation at a site by contralaterally applied atropine (Levitt and Fisher, 1966) is difficult to interpret except by postulating diffusion (Fisher and Levitt, 1967, Routtenberg, 1967).

Despite this strong indirect evidence, there is little positive evidence for ventricular spread to a single site. In several studies direct injection of carbachol in the lateral ventricles produced no drinking (Myers and Cicero, 1968; Khavari, Heebink and Traupman, 1968). In these studies the smallest dose injected was 0.25 μg by Khavari et al. (1968), and it has been argued by Routtenberg (1968a and b) that large doses may have produced a depolarization block and that smaller doses are required to produce drinking. It is worth noting, however, that this dose is about half of the optimal carbachol dose for inducing drinking by stimulation of the hypothalamus (Miller, 1965; Russell, 1966) and the limbic system (Levitt, White and Sander, 1970).

The only work which has reported drinking with intraventricular doses of carbachol is a brief report by Fisher and Levitt (1967) where a 3 μg injection of 1 part carbachol to 4 parts eserine elicited the intake of 8–17 ml after a latency of 10 min. The effect suggests that the drug is diffusing from the ventricle to some effective site, and unpublished studies in our laboratory suggest that it takes about 10 min. for a drug to flow from the lateral ventricle to the third ventricle.

In a careful study of the drinking elicited by carbachol introduced into sites in the limbic system (Levitt, et al. 1970), it was found that there was no significant difference in the volume drunk or latency as a function of distance to the ventricle (range $0.0–2.0$ mm). In addition, there were a number of sites distributed throughout the limbic system which had low threshold which argues against diffusion from this placement to a single effective site in the hypothalamus (Levitt et al., 1970). Some of these active sites are not close to the ventricles and obviously provide strong evidence against the ventricular involvement hypothesis (Fisher and Levitt, 1967). For example, one effective region is in the cingulate cortex. The thickest portion of the corpus callosum separates this site from the ventricles, Feldberg (1963) has reported that diffusion through white matter is extremely slow. The balance of evidence is strongly in favour of a neural network controlling water balance rather than a single centre, and that the transmitter in this network is acetylcholine and the receptors are muscarinic.

Specific modulatory systems

As well as the basic network controlling water balance, there appear to be at least one region in the brain, the amygdala, which enhances the amount of drinking and eating provided there is some activity in the network. Grossman (1964a) observed that carbachol injected bilaterally into the posteroventral amygdala of satiated animals produced no changes in water intake. However, if the animals were deprived, marked enhancement of drinking was observed in the injected animals compared with the unstimulated, deprived, control animals. Thus the drinking network must already be activated if amygdala modulation is to occur and sufficient activity can be produced by as little as 3-hr deprivation (Russell, Singer, Flanagan, Stone and Russell, 1968). This study also showed that there was an upper limit of activity produced by 23-hr deprivation above which modulation could produce no further increase. The modulatory effects were also related to the dose of carbachol injected into the amygdala. These data also argue against the Routtenberg diffusion hypothesis since injection of drugs into this site, close to the lateral ventricles, did not elicit drinking, unless there was some deprivation.

Species variations in drinking system

These basic data were obtained from studies of the rat, but differences in response to drugs have been observed between species and within species. Most variations in response to a specific dose have been attributed to differences in the amount of drug reaching the active site. Brodie (1962) suggests that the intensity of response to many drugs is inherently similar in various species, but that there are marked variations in the metabolism and biotransformation processes and their rates. However, the possibility must not be overlooked that there are inherent differences in the circuits, both in location and in their chemical natures. Some experimental evidence supports this idea.

We have already described the cholinosensitive network in the rat brain which follows the anatomically defined Papez-Nauta circuit (Fisher and Coury, 1962; Coury, 1967). In contrast, in the cat, drinking has not been observed when comparable sites have been stimulated cholinergically (Hernandez-Peon, 1963). In fact, sleep has been observed at some of these sites (see Chapter 6), rage at others, while the remainder were not responsive. Several explanations can be given for these data: (i) the drinking system is coded differently in the cat (Miller, 1965); (ii) two cholinergic systems are running close together and there are differences in dominance between the systems (Miller, 1965), so that carbachol induces sleep in cats because this system is dominant over drinking; (iii) the drinking network has a slightly different route in the two brains.

Another study which could be used to support the hypothesis of chemical coding differences from species to species has come from studies of motivation circuits in the rabbit. In one experiment the Papez-Nauta circuit was stimulated cholinergically and food and water intake measured (Sommer, Nevin and LeVine, 1967). It was found that 4·3 μg decreased water intake but induced eating, but this dose is ten times the dose used by Miller (1965) in rat studies, and in fact Sommer et al. show that between 0·6 and 1·07 μg do increase water intake in rabbits by about 12 ml. Thus comparable doses have the same effects in rabbits and rats when injected into sites in the water intake network. It can be argued that doses of this magnitude would produce a receptor blockade of a cholinergic drinking circuit, depressing drinking, and a blockade of the cholinergic inhibition of eating, producing increased food intake. It cannot be argued as Sommer et al. (1967) did, that the doses must be made comparable in terms of micrograms per gram of body weight.

In summary, the evidence for species differences in chemical coding is not strong in the sense that the existing data can be explained in terms of networks known in the rat. Nevertheless, it is a logical possibility to be considered when drug effects differ from species to species. Much more likely are intraspecies variations in drug response that result from differences in circuit sensitivity, including greater or lesser turnover rates of a transmitter, enzyme or both, and variations in the sensitivity of receptors.

B. CONTROL OF FOOD INTAKE

It is obvious that the control of food intake and thus the regulation of the bodily requirements of carbohydrates, fats, amino acids, vitamins and minerals is going to be vastly more complicated because of the large number of chemicals involved. It is equally obvious that wild animals manage to regulate their diet by self-selection, in normal circumstances. Numerous 'cafeteria' studies in the laboratory have shown that most rats will select a stable diet of pure carbohydrate, fat, amino acids, even though there are individual differences in the amounts ingested (Young, 1941). Studies of this sort suggest that there may be separate receptors in the body sensitive to carbohydrates, fats and amino acids. The strongest evidence is for central receptors sensitive to circulating glucose, and these studies will be considered first.

Glucostatic theory

One of the earliest chemical hypothesis about the control of food intake was that of Carlson (1916) who proposed that the critical factor in eating was the level of blood glucose which controlled gastric mobility. However, evidence has accumulated showing that neither gastric contractions nor the level of blood glucose was important in regulating eating (Mayer and Thomas, 1967), otherwise diabetics would not have such a strong appetite. Instead, Mayer (1953) suggested that the rate of use of glucose by glucoreceptors in the ventromedial hypothalamus was the critical determinant of eating. Thus, if the use of glucose falls, either due to a fall in the level of blood sugar or due to failure of glucose metabolism, then food seeking and intake will be more likely to occur. Mayer (1955) measured the arteriovenous differences in glucose concentration reaching the brain (Δ-glucose) and found they were correlated directly with the level of satiety. Thus low glucose use coincided with hunger feelings and gastric contractions, while a high level was evident when the animal was satiated.

As evidence in support of the ventromedial region as the locus of the glucoreceptors, it was shown that this region has a special affinity for glucose circulating in the blood because it takes up radioactively labelled glucose more than other hypothalamic regions, and the amount of uptake is inversely proportional to the level of satiety (Forssberg and Larsson, 1954). In addition, Mayer found that injections of gold thioglucose, but not any other gold thiosalts, damaged these cells. Animals whose cells were damaged showed increased food intake and became obese (Marshall, Barnett and Mayer, 1955). Unfortunately, injections of isotonic glucose directly into the ventromedial hypothalamus did not terminate eating in deprived rats (Epstein, 1960; Wagner and de Groot, 1963), even though the concentration was very high. However, electrical recording studies showed that as the blood level of glucose rises the firing rate in ventromedial neurons increases, and it would be predicted that this will inhibit firing in the food intake network. Studies supporting this idea include one by Anand, Chhina and Singh (1962) that showed that increases in blood sugar and arteriovenous differences increased the activity of VMH neurons and decreased the firing rate of inhibitory neurons to the lateral hypothalamus, the structure thought to be important in initiating eating. It has been argued (Edelman, Schwartz, Kronbite and Livingston, 1965) that the firing of the glucoreceptors depends on the metabolism of a specific quantity of glucose in a glial cell. The amount of firing along the fibre tract to the lateral hypothalamus will be proportional to the number of glucoreceptors firing. If the glucoreceptors have differing sensitivites depending on either the intrinsic rate of glucose metabolism in neuron–glial complex, or on the supply of glucose to each complex, then as the Δ-glucose increases more receptors will fire and the inhibition on the lateral hypothalamus will increase, making eating less probable. Conversely, as glucose use (Δ-glucose) decreases then few glucoreceptors will be activated and inhibition of the regions controlling food intake will diminish.

One question that has not been considered yet is whether feeding is terminated by the absorption of blood sugar from the stomach or whether there is a short-term control of some other sort. At one time it was thought that oral and gastric factors, such as mechanical stimulation, terminated the consummatory behaviour before absorption had occurred (Janowitz and Hollander, 1953). Recent evidence is less conclusive because blood glucose levels start to rise when animals begin to eat or even when sweet substances are put in their mouth (Nicolaïdas, 1969). It is not clear whether this is due to conditioning, stimulation of chemoreceptors in the oropharyngeal or gastric regions, or some other mechanism. Certainly there appear to be peripheral glucoreceptors (Jacobs, 1962; Mayer and Thomas, 1967). Sharma and Nasset (1962) stimulated the wall of the intestine with glucose solution and with specific amino acids and found that the activity of the same neurons in the mesenteric nerve were increased by glucose and some by amino acids, demonstrating the presence of peripheral interoreceptors for chemicals. As far as other mechanisms are concerned, there is some evidence for non-specific diffusion of radioactive glucose from the oropharynx to the brain bypassing all known anatomical pathways (Kare, Schechter, Grossman and Roth, 1969).

Other chemostats

It seems that glucose use provides a short-term mechanism for responding to the day-to-day variations in energy use. However, it has become evident that there is probably long-term regulation based on fat sensitivity, lipostatic control. Mayer and Thomas (1967) suggested that the mobilization of stored fat will change glucose use so that food intake is reduced until weight loss occurs. However, this hypothesis is inadequate because obesity results in increased lipolysis which decreases glucose use, and so weight gain should be a self-accelerating process and not a self-limiting process (Kennedy, 1966). There is no evidence about the nature of the mechanism involved in the lipostatic control. It is not mediated via the ventromedial hypothalamus because animals with lesions in this area do regulate their body weight, although it is at a higher level than control subjects (Teitelbaum, 1955). Perhaps it is mediated via taste sensitivity because obese animals (Teitelbaum, 1955) and men (Hashim and van Itallie, 1965) are hyperresponsive to the palatability of their food. In fact, a review by Schachter (1968) has shown that obese people are relatively insensitive to variations in the internal, physiological cues correlated with food deprivation, but highly sensitive to food-related cues from the environment.

Controversy has continued up to the present time about whether protein intake is controlled. Evidence against the position has come from demonstrations that rats do not select diets with adequate amounts of protein in certain circumstances, for example, when highly palatable foods are present (Scott and Quint, 1946). However, a recent study by Rozin (1968) has shown that there is a minimum protein requirement for rats, and that each individual selects a diet to exceed this minimum. In this study rats were allowed access to liquid sources of carbohydrates, fats and protein, and when their intake had stabilized the protein solution was diluted. The rats markedly increased their intake to compensate for the dilution, even when the diluted solution was made less palatable with quinine.

If there are mechanisms for controlling protein intake, then how do they operate? It is unlikely to be via palatability as protein sources are relatively unpalatable, and so it must be regulated via central or peripheral chemoreceptors. The study of Sharma and Nasset (1962) mentioned earlier demonstrated

36

that amino acids excited receptors located in the wall of the intestine, but there was no evidence that injections of amino acids into the bloodstream increased firing of the VMH neurons in the same way as glucose (Anand, Chhina and Singh, 1962). In conclusion, evidence for central receptors sensitive to amino acids circulating in the blood stream is lacking.

Control of food intake

In addition to eliciting drinking from the lateral hypothalamus, Grossman (1960) observed that feeding could be produced in satiated rats by introducing crystalline norepinephrine into the same region. These animals would consume

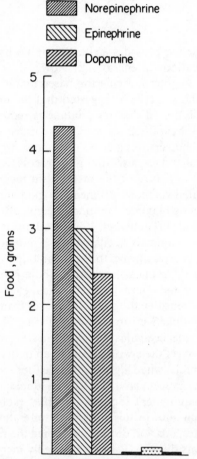

Figure 11. Effect of norepinephrine, epinephrine and dopamine stimulation of the hypothalamus on food intake during 1-hr test. The histograms to the right show the lack of effect of cholinergic stimulation. From Grossman, S. P. (1964). *Int. J. Neuropharmacol.*, **3**, 45–58. Reproduced by permission of Pergamon Press

4–5 gm of laboratory diet in a 50-min. test session, which represents about a quarter of their food intake (see Figure 11). There was no compensation for this increased intake of food, and there was increase in weight the following day. Repeated daily stimulation resulted in a more rapid weight gain than control animals over a 2-month period. However, the animals did not become obese like animals lesioned in the ventromedial hypothalamus, perhaps because of a long-term regulatory mechanism which operates after excessive intake has resulted in some deposits of fat (Grossman, 1969). Thus there seems to be some adreno-sensitive nucleus in the lateral hypothalamic region which overrides the normal short-term satiation mechanisms of the animal. Electrical stimulation of the same region induces food and water intake, and sometimes non-specific gnawing (Miller, 1960) while unilateral lesions induce temporary adipsia and aphagia (Gold, 1967) and bilateral lesions produce long-term aphagia and adipsia (Teitelbaum and Epstein, 1962). Thus the lateral hypothalamus appears to be important for the control of eating as well as drinking.

Motivational characteristics

Although we have demonstrated that norepinephrine introduced into the lateral hypothalamic region produces consummatory behaviour, it is important to demonstrate that the stimulation is activating a motivational system which is similar to deprivation induced food intake and not just non-specific gnawing. Grossman (1960) demonstrated that stimulated animals would lever press for food pellets. In other respects there do appear to be some qualitative differences between deprivation-induced eating and norepinephrine-elicited eating. Booth and Quartermain (1965) observed that norepinephrine-stimulated rats ingested more mash when the mash was flavoured with saccharine than when the rats were deprived of food for 3 hr and injected with saline. When the mash was adulterated with quinine the relationship was reversed so that the rats ate more when they were deprived and injected with saline than when they were injected with norepinephrine. The writers comment that repeated testing with the same mashes seems to abolish the effect but unfortunately no data is provided. In addition, studies of Miller and his associates (Miller, 1965) have shown that the norepinephrine injection mobilized glucose regardless of food intake, suggesting that the stimulation is activating the homeostatic mechanism which normally mobilizes the body's energy resources in response to food deprivation. Unfortunately, for this simple picture, carbachol has an even greater effect, and at the present time no explanation of this result has been given.

Pharmacology of food intake

The first dose response studies with crystalline implants of norepinephrine-induced food intake were performed by Miller, Gottesman, Kay and Emery (1964) using liquid food ('Metrecal'). It was found that normally rats preferred this food to their normal laboratory pellets, but that carbachol-stimulated rats preferred water. However, doses as low as 24×10^{-10} M (i.e. 0·8 mg) of norepinephrine reduced consumption of the liquid diet. Miller et al. (1964) discuss the doses in comparison with reports of endogenous levels and argue that they were of the same order of magnitude as those present in the hypothalamus. A similar dose response curve for l-norepinephrine was obtained by Booth (1968a) using dry laboratory food, while d-norepinephrine, the amine not present in the body, had no significant effect. Another catecholamine, l-epinephrine, proved to be just as effective as l-norepinephrine, while isoproterenol, a beta-adrenergic receptor stimulant, was ineffective, which suggests that the effect was mediated by alpha-adrenergic receptors (Booth 1968b; Slangen and Miller, 1969).

In his first study Grossman (1962) demonstrated that the alpha-adrenergic blocking agent, ethoxybutamoxane injected intraperitoneally, blocked eating produced by centrally injected norepinephrine, while centrally injected dibenzyline, another alpha-adrenergic blocker, reduced eating after deprivation. Since this study more powerful drugs have been developed, and it was found that the alpha-adrenergic blockers, phenoxybenzamine and phentolamine, prevented eating after hypothalamic injections of l-norepinephrine, while the beta-adrenergic blocker, propranolol, had no effect (Booth, 1968a). Booth concludes that the evidence is consistent with the hypothesis that alpha-adrenergic modulation of postsynaptic activity in the hypothalamus is involved in the control of food intake in the rat.

Anatomical location of network

The results of careful analysis of the adrenosensitive sites from which eating may be elicited show that they form a coherent group extending through a portion of the substantia innominata at the level of the anterior hypothalamus (Booth, 1967). From here, pathways spread dorsally forwards to the lateral septum (Grossman, 1964b; Booth, 1967; Coury, 1967) and backwards through the thalamus to just behind the habenula close to the ventricles (Booth, 1967; Coury, 1967). Booth (1967) commented that there were just as many ineffective as effective sites close to the ventricles which argues against the ventricular spread hypothesis of Routtenberg (1967). These active sites are located along the septohabenular tract of the stria medullaris, which originates in the septal region and projects to the habenular complex (Gurdjian, 1926). The efferent fibres of the habenular complex may influence the motor systems involved in eating (Booth, 1967). In addition, there was one series of effective sites subfornically to the rostral ventromedial hypothalamus, and another also running ventrally to the olfactory tubercle at the borders of the preoptic area. From this careful analysis it is clear that the adrenosensitive loci do not coincide with the structures in the lateral hypothalamus that were thought to be involved in food intake on the basis of lesions and electrical stimulation studies, but are more medial to it. Booth (1967) suggests that these structures may mediate the recovery from lateral hypothalamic lesions (Epstein, 1971). As we saw previously, these lesioned rats show the same pattern of sensitivity to flavour as norepinephrine-stimulated rats. Evidence in support of this hypothesis has come from a study in which injections of norepinephrine into the lateral ventricles of rats recovering from lateral hypothalamic lesions abolished the aphagia and in some cases produced hyperphagia. In the same study (Berger, Wise and Stein, 1971) the alpha-adrenergic blocker, phentolamine hydrochloride, suppressed eating in both normal rats and rats who had recovered from lateral hypothalamic lesions, whereas propranolol, a beta-adrenergic blocker, induced overeating in some animals recovering from the lesion. This finding suggests that there may be two pharmacologically distinct adrenergic systems involved in food intake, an alpha-adrenergic system mediating eating, and a beta-adrenergic system mediating satiety.

In a study before this one, Leibowitz (1970a and b) also found evidence for two opposing adrenergic systems; the alpha-adrenergic 'hunger' synapses were in the ventromedial region, as Booth (1967) had suggested, while the beta-adrenergic 'satiety' synapses were in the lateral hypothalamic region (Leibowitz, 1970b). This seems paradoxical in terms of the lesion data discussed earlier, which showed that the ventromedial area was important in terminating food intake, while the lateral hypothalamus was crucial for the initiation of eating. This apparent paradox is resolved when it is realized that the ventromedial area is important in receiving information related to the termination of food intake from various interoreceptors, and as a result it inhibits the lateral hypothalamus. The transmitter at the synapses of these inhibitory neurons on the lateral hypothalamus is norepinephrine, and the postsynaptic receptors are beta-adrenergic. On the other hand; information relevant for initiating eating activates adrenergic inhibitory synapses on the cells in the ventromedial hypothalamus, whose receptors are alpha-adrenergic, and eating occurs as a consequence of the inhibition of inhibitory neurons from the ventromedial to the lateral hypothalamus to the LH, i.e. by disinhibition. Unfortunately, the lateral hypothalamus has no adrenergic cell bodies, so Leibowitz (1970b) has suggested that the cell bodies for these alpha-adrenergic neurons are in the lower brain stem and this region receives 'hunger-related' inputs from the peripheral nervous system, and from the lateral hypothalamus. Thus there are two types of adrenergic pathway involved in opposite aspects of food intake, and so a non-specific adrenergic stimulant will act on both structures. This implies that an effective anorexigenic drug should have predominantly beta-adrenergic effects, and this will be examined next.

Anorexigenic drugs

The high proportion of obese individuals is one of the indices of developed countries. At the present time it is fashionable to be slim, and so considerable research has been devoted to discovering appetite depressants. The group of drugs that have been found to be most successful are structurally related to amphetamine, which was the first drug tested (Ersner, 1940). About 70–75 %

of the subjects tested lost weight with these compounds, although about a third of these also lost weight when given a placebo (Fazekas, Ehrmantraut and Kleh, 1958). In studies on animals there is a marked depression of food intake by amphetamine injected intraperitoneally (Siegal and Sterling, 1959).

The mechanism underlying this effect has been the subject of some controversy. Some of this discussion has centered around the following paradox. Norepinephrine injected into certain central sites elicits food intake, but peripheral injections of amphetamine, which increases the amounts of functional norepinephrine in the brain, reduces eating. An injection of amphetamine into an adrenosensitive site in the anterior lateral hypothalamus did not elicit eating in satiated rats (Booth, 1968a). However, injections of amphetamine into the same region depressed food intake (Booth, 1968b). The dose response curves for the effect were similar to those obtained after peripheral injections, except that the hypothalamic dose of amphetamine was one tenth that of the intraperitoneal dose. The explanation proposed by Booth for this finding was that amphetamine increases the amount of norepinephrine present in the synapse by increasing the amount released by normal activity in the feeding neurons, and also by inhibiting re-uptake. As a consequence, an excessive quantity of transmitter accumulates producing a depolarization block in the food intake network. However, the findings of Leibowitz (1970b) suggest that the amphetamine may be acting to increase the functional amounts of norepinephrine at the beta-adrenergic synapses in the LH mediating satiation. She tested this possibility by injecting either amphetamine, and/or amphetamine and the beta-adrenergic blocker, isoproterenol. The amphetamine alone suppressed eating as Booth (1968b) had found, but the beta-adrenergic blocker prevented this. Tests on the adrenergic synapses in the ventromedial hypothalamus showed that amphetamine had a weak hunger stimulating effect. Thus systemic amphetamine will act on both the alpha-adrenergic and beta-adrenergic systems, but the dominant effect will be anorexia.

C. MODULATION OF NETWORK FUNCTION

In this section studies will be discussed where chemical stimulation modifies the amount of eating or drinking that would occur when the animal is given access to food or water. There are three sorts of effect to be considered: (i) the interactions between sites in the same functional systems; (ii) interaction between sites in different systems; (iii) the action of specific modulatory systems.

The interactions between sites in the same functional system have been mentioned in passing, and they will first be summarized here. Russell and Khavari, discussed in Russell (1966), found that unilateral stimulation of the lateral hypothalamus results in significantly less drinking than bilateral injections, and that the mean latency of water intake was shorter in the bilateral condition, with no overlap for the individual scores in the two conditions. This difference supports the hypothesis that response output depends on the amount of excitatory information in the circuit. About the same time Levitt and Fisher (1966) found that atropine sulphate introduced into a number of loci in the water balance network reduced water intake elicited by carbachol injected into other sites in the same system, including contralateral loci. From this study it was clear that the blockade was bidirectional because it could be produced by injections anterior and posterior to the carbachol stimulated site. This demonstrated that this system at least was not merely a descending pathway funnelling information to some caudal part of the brain, but that 'the neural basis of the thirst drive consists of complex alternative and reciprocal pathways of neurons susceptible to activation by cholinergic stimulation' (Levitt and Fisher, 1966, p. 560).

At the present time this sort of interaction has not been demonstrated in the network controlling food intake, but mutual inhibition between the two systems controlling water and energy balance has been found. Thus, adrenergic stimulation of the lateral hypothalamus inhibited deprivation-induced drinking, while carbachol injections into the same regions inhibited eating resulting from food deprivation. This mutual inhibition is also found at the septum (Grossman, 1964b), and the amygdala (Grossman, 1964a; Singer and Montgomery, 1968). Thus the neurotransmitter released at one neural locus would seem to activate the next unit in the network, and simultaneously inhibit adjacent nuclei mediating any competing behaviour. These findings raise interesting pharmacological possibilities where the receptor blocker of one system should result in the removal of inhibition from the system mediating competing behaviour. Grossman (1962) obtained no direct evidence for this suggestion with stimulation in the lateral hypothalamus, but Liebowitz (1971) convincingly confirmed and extended this prediction. She found that the alpha-adrenergic blocker, phentolamine, enhanced water intake in deprived rats, but that the beta-adrenergic blocker, propranolol, suppressed water consumption in water-deprived rats. The beta-adrenergic stimulant, isoproterenol enhanced water intake, showing that only the alpha-adrenergic system oppose the cholinergic drinking system while the beta-adrenergic system has similar effects.

Of particular interest is the function of the ventral amygdala in food and water intake. In satiated subjects no effects of drugs on either food or water intake were observed (Fisher and Coury, 1962; Grossman, 1964a). However, in slightly water-deprived subjects carbachol increased water intake (Grossman, 1964a) so that a rat deprived for 3 hr now drank as much as if he had been deprived of water for 23 hr (Russell et al., 1968). Both norepinephrine and atropine reduced water intake in deprived subjects (Grossman, 1964a), but dibenzyline significantly increased intake. The drugs had the opposite sorts of effects on food intake, and atropine produced a slight increase in food intake (Grossman, 1964a). In a further examination of this modulation Singer and Montgomery (1968) replicated the norepinephrine potentiation of deprivation-induced eating, the norepinephrine blockade of deprivation-induced drinking, the carbachol potentiation of deprivation-induced drinking, but not the carbachol reduction of deprivation-induced eating. The authors explain the latter finding by suggesting that the carbachol dose was too small to be effective.

3
Biochemical Basis of Mood

Most of us have experienced elation and depression at one time or another, but most of our existence is spent in a narrow range of mood and behavioural arousal. However, there are some people who exhibit behavioural patterns of either excessive activity, mania, or overwhelming lethargy, depression, and more rarely those who oscillate between the two extremes (manic-depression). Of the two simple cases, mania is the least common. The typical manic's excessive activity is manifested in all types of behaviour, walking, talking, dancing and singing. In some cases this excessive energy can be expressed in aggression and sexual activity. His thought processes are confused, with flights of ideas, tangential associations, and no inhibitions in expressing these thoughts. The behaviour and thoughts are a reflection of his basic elated mood, his buoyant optimism, inflated self-esteem and lack of insight.

Depression is not such a simple desorder. It is a varied collection of symptoms that may result from a clear physical disorder, exogenous depression, an obvious psychological cause, reactive depression, to an illness without any obvious physical or psychological precipitation (i.e. neurotic depression, agitated depression and retarded depression). In spite of these different diagnostic categories, the various subclasses all exhibit common features which are the opposite of those seen in mania. As a group they are unlikely to initiate any activity, often to the point of being stuperous, and the only exception to this rule is the attempt to commit suicide. The thinking processes are usually normal, but the depressive realizes that he has lost his previous capacity to experience life as he did before and so feels that he has lost his personality. He feels generally despondent and may have feelings of guilt for past doings, real and imagined.

The revolution in psychiatric therapy came with the development of the first tranquillizers, promethazine and diethazine in the 1940's. These phenothiazine compounds were originally synthesized as antihistaminic agents for the treatment of allergic symptoms, but one of their problems was the side effects of sedation. However, in 1950 a French surgeon, Laborit, recognized the importance of these effects and used them as a premedication for surgery. A year later another phenothiazine, chlorpromazine, was developed, which had a large toxicity to efficacy ratio. By 1954 the drug had been used by psychiatrists in France, Britain and the United States, and was revolutionizing the life in the chronic wards. In many wards curtains were hung for the first time, physical restraint was eliminated for all but the most exceptional cases. In the same year

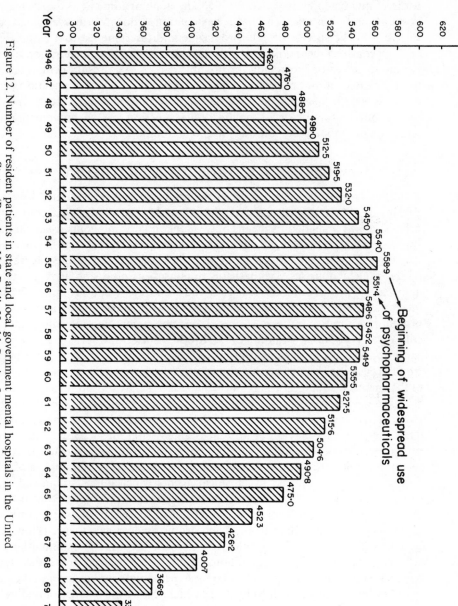

No. of Patients (in thousands)

Figure 12. Number of resident patients in state and local government mental hospitals in the United States. (Based on U.S. Public Health Service figures.)

psychiatrists in the United States learned of another tranquillizing drug, reserpine, which had been derived from a shrub, *Rauwolfia serpentina*, and used for treating psychotics in India. By 1956, the trend of admissions to psychiatric hospitals had been reversed, and it has continued downwards ever since (see Figure 12). The success of chlorpromazine and reserpine led to a completely new approach to mental illness, with the emphasis on the amelioration of the symptoms instead of searching for possible causes. As a result, drug research has been directed towards discovering more effective psychotherapeutic agents by studying the neurochemical effects of reserpine and chlorpromazine. It turned out that both drugs modified the function of the norepinephrine, serotonin and dopamine pathways in the midbrain.

A. ERGOTROPIC–TROPHOTROPIC THEORY

In the early 1950's the function of the midbrain had been studied by Hess (1954b) by means of electrical stimulation, and he concluded that emotional behaviour was regulated by means of two functionally antagonistic subcortical systems which he termed the ergotropic and trophotropic systems. The ergotropic system was defined as a set of pathways concerned with the integration of the skeletal muscles, sympathetic responses, arousal and psychic states involved in positive action, whereas the opposing trophotropic pathways integrates the central prosympathetic system with somatomotor activities to promote restorative processes. Based on this dichotomy, Brodie and his colleagues (Brodie and Shore, 1959; Brodie, Sulser and Costa, 1961) have suggested that the ergotropic transmitter was norepinephrine and the trophotropic transmitter was serotonin, thus putting forward one of the first theories of the chemical coding of behaviour. Changes in behaviour produced by some chemical agents were considered as changes in the relative excitability of the two systems as a result of the agents influencing transmission at the synapses. Thus, heightened behavioural arousal resulted from an increased activity in norepinephrine neurons relative to the activity of serotoninergic neurons, while sedation resulted from the opposite change. As a result, predominance of one system could be produced by chemical agents which either stimulated that system or blocked the antagonistic one. Behavioural arousal occurred as a result of ergotropic predominance produced by adrenergic stimulation or with serotonergic blocking. In contrast, trophotropic predominance with its lowered behavioural responsiveness resulted from either serotonergic stimulation or adrenergic blocking (Brodie and Shore, 1959). It follows from these ideas that patients whose illness is characterized by hyperactivity and exaggerated emotional responses may be helped by drugs which restore the balance of midbrain activity towards normal by suppressing the ergotropic system, adrenergic blockers, or by stimulating the trophotropic system, serotonergic stimulation (Brodie et al., 1961), also patients suffering from diminished affect and motor retardation might be treated with drugs suppressing the trophotropic system, serotonin blockers, or activating the ergotropic system, adrenergic stimulation.

This hypothesis stimulated intensive animal research which attempted to correlate the behavioural changes induced by drugs with the levels of various amines. Unfortunately, few drugs are completely specific in acting on only one transmitter system, so that relationships have to be established by eliminating known alternatives. One method is to compare the time course of behavioural change with the duration of the biochemical changes, and a second is to use drug combinations to confine the drug effects to one system. The application of these techniques to the ergotropic–trophotropic theory will be outlined by considering the behavioural consequences of drugs (a) increasing activity in central adrenergic neurons, (b) decreasing activity in central adrenergic neurons, as well as similar changes in central serotonin neurons.

Behavioural effects of increased adrenergic activity

As an example of how research proceeds in this area, let us consider studies of the effect of amphetamines on biochemistry and behaviour. This drug was first used for the treatment of depression in the mid-1930's, and extensive studies of its biochemical mode of action have been carried out. It has become clear that this group of drugs releases free norepinephrine from the endogenous stores in the neurons, and block the inactivation of norepinephrine by cellular re-uptake (Glowinski and Axelrod, 1966). As a result of action of the extra-neuronal enzyme, catechol-0-methyl transferase, inactivating free norepinephrine, large doses of amphetamines (e.g. 10 mg/kg of amphetamine sulphate) deplete the cerebral concentration of norepinephrine gradually while increasing serotonin and changing dopamine very little (Smith, 1965). It must be remembered that these estimates of concentration represent static measures taken at specific times, but that changes in behaviour are believed to be the result of the dynamic processes of norepinephrine release that produce the depleted concentrations (Scheckel and Boff, 1964). Thus it is meaningless to attempt a correlation between the biochemistry and behaviour by testing the animals behaviourally, and then doing the biochemistry as some workers have done (see Russell and Warburton, 1973).

The effects of amphetamine on behaviour will be considered in relation to operant responding. In general the effect of amphetamine on operant behaviour is to enhance the tendency to respond rather than eliciting new behaviour. This distinction made by Stein (1964b) is important because it emphasizes that changes in norepinephrine function modulate ongoing behaviour, in contrast to the experiments in the last chapter where norepinephrine elicited food intake. There are numerous examples of this enhancement ranging from studies of general activity (Smith, 1965; Heise and Boff, 1971) to complex schedules of reinforcement (Clark and Steele, 1966).

This generalization must be qualified by referring to some important studies which show that the baseline rate of responding seems to play a part in determining the direction and magnitude of the drug effect. In a series of studies on avoidance responding, Verhave (1958) found that untrained animals who had very small spontaneous level pressing rates, e.g. 2–6 presses per hour, were unaffected by 1·0 mg/kg of methamphetamine, and only slight increases were found with 2·5 mg/kg. However, when animals had received $2\frac{1}{2}$ hr training the effect of the same doses of amphetamine was to produce a marked increase in lever pressing rate. Thus amphetamine did not increase responding unless there was a minimum tendency to respond. On the other hand, other studies suggest that when the baseline response rates are high, then increased rates are no longer observed, and in some cases the response rate decreases (Dews, 1958). Dews proposed that amphetamine only increased responding when the baseline rates are low, that medium rates are not increased, and high rates may be decreased. This extension of the previous generalization holds over a wide range of operant situations (Kelleher and Morse, 1968). In a number of cases high response rates increased rather than decreased, although the increases were proportionally less than the augmentation observed with low response rates (Stein, 1964b). Kelleher and Morse (1968) examined the proportional changes in responding after a fixed dose of amphetamine, and found there was a decreasing linear relation between response rate as a percentage of the control rate as a function of the control response rate. Thus they found that the proportional increase in response rate was an inverse function of the control rate up to a critical control rate and for greater rates the proportional decrease in response rate was directly related to the control rate. Some ideas about the neurochemical control of this system are embodied in the Theory of Reward Thresholds.

Theory of reward thresholds

This theory was put forward by Stein (1964a, b) and by Olds (1962). It pointed out that in an operant situation many responses occur prior to the culminating sequence of approach, press and release of the lever that delivers reinforcement. It would be predicted that the stimuli associated with all of these responses would have a tendency to activate a hypothalamic reward system; those that had been remote from reinforcement would have less tendency than those closely associated with reward (Stein, 1964a). A decrease in the reward threshold by drugs would increase the tendency for all stimuli to activate the reward system, but remote stimuli would be increased proportionally more. This would explain why the frequency of occurrence of previously low probability responses was found to be augmented proportionally more than high probability responses during the increased release of norepinephrine by amphetamine (Stein, 1964b; Olds and Olds, 1964). Olds and Stein postulated that the neural pathway involved is the reward system of the median forebrain bundle, and, although Olds (1962) does not believe that this system is the same as the ergotropic system of Hess, he proposes that functioning of the reward system depends on norepinephrine. In order to obtain support for this notion, a number of drugs have been tested for their effects on the threshold for self-stimulation (Stein, 1962). In early training each lever press delivered trains of current of fixed intensity to the posterior hypothalamus; after stable rates were obtained, threshold training was begun. Each press then produced a train of current, and also reduced the intensity by a small step. A second lever was available which reset the current at its original intensity. In this way the animal reset the current when it was at a non-rewarding level, i.e. subthreshold. The threshold were found to be stable and could be obtained reliably during testing over many months. Using this preparation, Stein (1962) found that amphetamine lowered the threshold supporting the proposal that norepinephrine was involved in the function of the reward system of the median forebrain bundle.

Another drug which increases norepinephrine function at synapses is imipramine, an important member of the tricyclic antidepressants. These compounds do not interfere with the synthesis storage, release or enzymatic inactivation of the brain norepinephrine, but block the presynaptic re-uptake of this amine (see Section A of Chapter 1). This slowing up of the uptake 'pump' would result in an increased amount of free norepinephrine available for interaction with the postsynaptic membrane receptors, and also potentiation of any treatment which resulted in norepinephrine being released into the synaptic cleft. Thus Stein (1962) demonstrated a marked enhancement of the effects of methamphetamine and imipramine on self-stimulation, although imipramine had no effect on its own, probably because of inactivation of norepinephrine by catechol-0-methyltransferase.

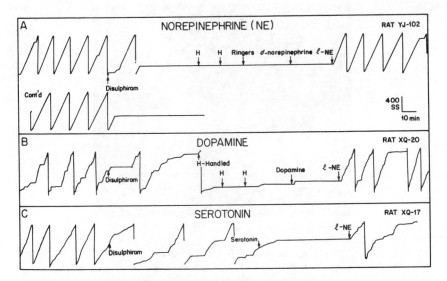

Figure 13. Suppression of self-stimulation (SS) by disulphiram (200 mg/kg), intraperitoneal and reversal of behavioural suppression by intraventricular injection of *l*-norepinephrine [5 μg in (A) and (B), 20 μg in (C)]. Equivalent doses of *d*-norepinephrine, dopamine, or serotonin do not restore self-stimulation. The curves are drawn by cumulating self-stimulations over time; the pen resets automatically after 100 self-stimulations. From Wise, C. D., and Stein, L. (1969). *Science*, **163**, 299–301. Reproduced by permission of the American Association for the Advancement of Science

48

In a more recent study, Wise and Stein (1969) administered disulphiram, which inhibits the synthesis of norepinephrine, to rats with electrodes in the median forebrain bundle. Depletion of norepinephrine suppressed self-stimulation responding, but the impairement was abolished by intraventricular injection of *l*-norepinephrine, the isomer occurring naturally in the body, but not *d*-norepinephrine (See Figure 13). Neither injections of dopamine nor injections of serotonin restored the self-stimulation responding. Thus the reinstatement of the suppressed behaviour depended on the replenishment of the functional depots with the naturally occurring isomer of the transmitter. This implies that *l*-norepinephrine is the transmitter released during 'normal' transmission in the median forebrain bundle.

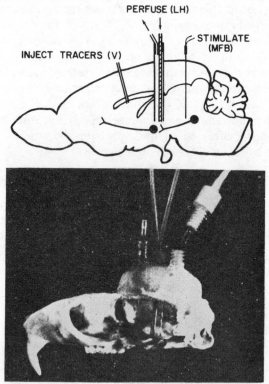

Figure 14. Diagram of perfusion experiment showing relative locations of stimulating electrode in the medial forebrain bundle (MFB), perfusion cannula in lateral hypothalamus (LH), and needle for injection of radioisotopes in lateral ventricle (V) on an outline of rat brain. The Figure shows the arrangement of the various devices on a rat's skull (Stein and Wise). From Stein, L., and Wise C. D. (1969). *J. comp. physiol. Psychol.*, **67**, 189–198. Reproduced by permission of the American Psychological Association

As a direct test of this hypothesis, Stein and Wise (1969) implanted electrodes in the median forebrain bundle, and a push–pull cannula in the lateral hypothalamus (see Figure 14). Injections of radioactive norepinephrine were made intraventricularly, and after 3 hr, when the norepinephrine had been taken up, perfusion was begun in the hypothalamus. Background levels of radioactivity were recorded in the perfusate of unstimulated animals, but applica-

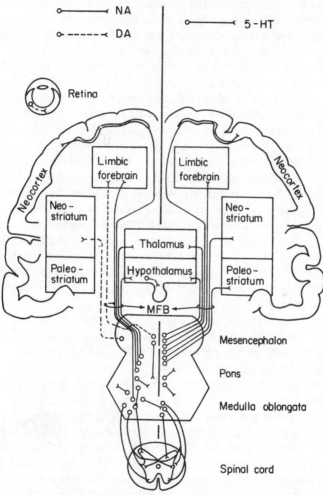

Figure 15. Schematic representation of the bodies and their axonal projections of dopaminergic (DA), noradrenergic (NA), and serotonergic (5-HT, 5-hydroxytryptamine) neurons in the brain and spinal cord, as determined by fluorescent histochemistry. Dopamine- and norepinephrine-containing neurons are shown on the left, 5-hydroxytryptamine-containing neurons on the right. From Andén, N-E., et al. (1966). *Acta physiol. Scand.*, **67**, 313–326. Reproduced by permission of Scandinavian Society for Physiology

tion of rewarding stimulation produced large increases in the release of radioactivity. Electrodes which did not produce high self-stimulation rates were ineffective in releasing radioactivity. A chemical analysis disclosed that most of the radioactivity was due to 0-methylated metabolites of norepinephrine, like noremetanephrine, produced by enzymatic inactivation of the norepinephrine by catechol-0-methyltransferase. This study gives strong support for the idea that norepinephrine is the transmitter involved in the function of the hypothalamic reward system, and so in the behaviour controlled by it.

Independent support for this work has been provided by a fluorescence staining technique developed at the Karolinska Institute in Sweden (Hillarp, Fuxe and Dahlstrom, 1966), whose precision has enabled extensive and accurate mapping of the neural pathways containing these amines. In the reticular formation, two groups of norepinephrine neurons have been found: one set descends in the spinal cord while the second ascends in the medial forebrain bundle and terminates in the hypothalamus, neocortex and parts of the limbic forebrain, including the septal area and hippocampal formation (see Figure 15) Parallel to these neurons are others with serotonin-bearing axons whose cell bodies are in the mesencephalon and end in the neocortex, hypothalamus and limbic forebrain. The major differences in distribution are the higher concentrations of norepinephrine in the medial and rostral parts of the hypothalamus and the higher concentrations of serotonin in the caudate nucleus and putamen. By means of electron microscopy, the intraneuronal distribution of the amines has been found to be highly characteristic. The soma and axons have low concentrations, whereas the synaptic terminals have 50 to 100 times more of the amine in the form of granules which are believed to be stores of the inactive, or bound, amine.

From this it seems clear that there is an adrenergic system in the brain which is important in behavioural activation, as Brodie had proposed, but this system is not the same as the ergotropic system of Hess, although they may be related. Instead the neural pathway controlling behavioural arousal in the rat seems to be the adrenergic fibres ascending from the reticular formation along the median forebrain bundle.

Although this evidence is compelling, one piece of evidence was still missing. It would be expected that injections of norepinephrine into the brain would produce behavioural excitement. However, Mandell and Spooner (1968) summarized the results of eleven studies of chicks, mice, rats and kittens in which norepinephrine, given so that it passed the blood–brain barrier, produced a monotonic dose-response pattern of behavioural depression. In order to examine this paradox, Dr. David Segal, in the laboratory at University of California, Irvine, modified an intracerebral infusion apparatus previously used for introducing small quantities of cholinergic drugs into the hippocampal formation (Warburton and Russell, 1969). He then used the apparatus for slowly increasing the levels of brain norepinephrine by infusing it slowly into the lateral ventricles. Segal measured both locomotor activity in an activity cage and continuous avoidance responding in a lever pressing box. He found

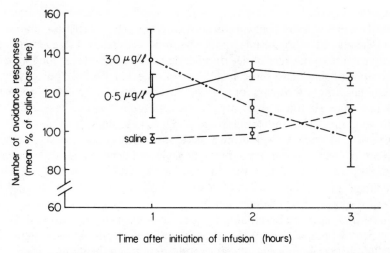

Time after initiation of infusion (hours)

Figure 16. Effect of continuous infusion of two doses of DL-norepinephrine and saline on avoidance responding (Sidman). Results are expressed as the mean percentage of change in avoidance responding when compared with saline baseline (each animal was used as his own control). The dose of 0·5 μg/μl resulted in a consistent increase in avoidances throughout the 3-hr test session. At 3·0 μg/μl, there was an initial increase in avoidance responding followed by a decreasing tendency to respond. The brackets indicate the standard error of the mean for each point. From Segal, D. S., and Mandell, A. J. (1969). *Proc. nat. Acad. Sci.,* **66,** 289–293. Reproduced by permission of National Academy of Science, Washington, D.C.

that initially small increases in norepinephrine produced excitation, but as the levels increased their activity decreased and they appeared sedated (Segal and Mandell, 1969). Thus the time-response curves (see Figure 16) were an inverted U shape and since the amount of norepinephrine in the brain was increasing linearly with time the dose-response curve was also an inverted U shape. The most likely explanation for the effects can be given in terms of an impairment of neural function when excessive transmitter produces a depolarization block of the postsynaptic receptors (see Section C, Chapter 1).

Behavioural effects of decreased adrenergic activity

Adrenergic activity in the central nervous system has been reduced by using drugs which block synthesis, prevent storage, and drugs which prevent norepinephrine reaching the postsynaptic receptors. From the studies just discussed, we would expect that these procedures would decrease lever pressing in operant situations. This prediction has been confirmed using the three types of drug mentioned. In this section attention will be focused on reserpine and the reserpine-like drug, tetrabenazine. Both drugs produce reduced levels of norepinephrine, dopamine and serotonin in the brain, but only when the levels were below about 10 % of normal in rats were any marked functional disturbances observed. It is believed that it is possible to deplete the large inert store of amines without behavioural consequences, while reduction of the small labile functional store results in behavioural depression (Haggendahl and Lindquist, 1964). In order to decide whether the effects on behaviour were due to norepinephrine or to serotonin, Carlsson injected DOPA and 5-hydroxy-tryptophan, the precursors of the two transmitters into mice pretreated with reserpine. He found that only DOPA restored normal activity in the mice (Carlsson, Lindquist and Magnussen, 1957). These findings were confirmed by Everett and Toman (1959) using monkeys, and they concluded that serotonin does not play an important role in the behavioural deficit resulting from reserpine. Instead, they argued that it was the central catecholamines, either norepinephrine or dopamine, which were crucial for the normal activity and general responsiveness of the organism.

These findings were based on gross observation of the animals and did not allow precise comparisons of the effects of different drugs. At the beginning of the 1960's the techniques of operant conditioning were being used extensively by psychopharmacologists to examine drug effects. One of the operant methods used in the analysis of reserpine and tetrabenazine effects was continuous, non-discriminated avoidance (Heise and Boff, 1962). In this schedule rats were trained to postpone a shock for 40 sec. by lever pressing. A well trained rat pressed the lever regularly at intervals of less than 40 sec. and so avoided shock. If the animal did not respond within 40 sec. he received a shock every 20 sec. until the lever was pressed again. Reserpine and tetrabenazine had a marked depressant effect on the lever pressing rate and the magnitude and duration of the depression were proportional to the dose injected (see Figure 17). A second finding in these studies was that iproniazid, which increased the intra-neuronal levels by inhibiting monoamine oxidase, had no effect on responding, showing it was the amount of transmitter reaching the postsynaptic receptors that was the crucial factor in the behavioural effects. This was dramatically confirmed (see Figure 17) by injecting iproniazid before tetrabenazine, which produced a dramatic increase in response rate compared with the decrease produced by tetrabenazine alone (Heise and Boff, 1960). After about 1 hr of rapid responding the rate returned to normal, and then the typical decreased response rate was observed. The explanation for this finding can be found in the biochemical changes in amounts of transmitter at the synapses. Iproniazid inhibited the inactivating enzyme intraneuronally so that when tetrabenazine prevented storage of norepinephrine there was a large amount of free transmitter available for release. There was a clear correlation between the released, i.e. functional norepinephrine and the behaviour, while the levels of serotonin remain consistently higher than normal (Scheckel and Boff, 1966).

In a series of studies following up this work (Scheckel and Boff, 1964, 1966) used another set of drug combinations to examine the behaviour-amine correlation. In the first series they used three inhibitors of norepinephrine synthesis, alpha-methyl-*para*-tyrosine which inhibits the synthesis of DOPA from tyrosine, RO 4–4602 which inhibits the conversion of DOPA to dopamine and di-

53

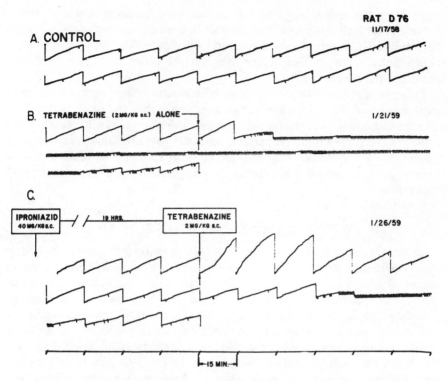

Figure 17. Cumulative recorder tracings illustrating the effect of iproniazid pretreatment on avoidance responding after tetrabenazine.

Abscissa: Time: 15 min. intervals are indicated by the time at the bottom and by the resetting of the pen (long vertical lines). Ordinate: Cumulative lever presses since pen reset to base line. Rate of lever pressing is proportional to slope of the cumulative response line. Each vertical pip on the response line indicates a 0·5 sec shock. A. Control. Avoidance rate was approximately 48 responses per 15-min. interval. B. Effects of standard test dose of 2·0 mg/kg tetrabenazine administered at arrow. C. Performance in the tetrabenazine blocking test by a rat that received 40 mg/kg s.c. iproniazid 19 hr before the tetrabenazine injection. From Heise, G. A., and Boff, E. (1960). *J. pharmacol. exper. Therap.*, **129**, 155–161. Reproduced by permission of the authors and Williams and Wilkins Co

sulphiram which prevents the synthesis of norepinephrine from dopamine. Disulphiram was just as effective in preventing recovery after tetrabenazine as RO 4–4602 showing that dopamine was not the crucial transmitter system in depression. In addition alpha-methyl-*para*-tyrosine which lowered norepinephrine levels and has little effect on serotonin (Spector, Sjoerdsma and Udenfriend, 1965), also prevented recovery. This same research strategy was used by Scheckel and Boff (1964) to examine the effects of imipramine, another antidepressant drug, on animals injected with tetrabenazine. Rats were pretreated with a dose of imipramine, which itself had no effect on behaviour when injected alone. The two drugs combined produced increased lever pressing about 2½ hr after the tetrabenazine injection at a time when the norepinephrine levels were decreasing, but the other amines were increased (Zbinden, 1960). Scheckel and Boff concluded that the effects were due to the release of norepinephrine. The interpretation was tested by depleting the norepinephrine with alpha-methyl-meta-tyrosine before injecting the other two drugs, and this prevented the stimulation.

These studies show that reserpine's behavioural effects seem to correlate with changes in the norepinephrine system, rather than the serotonin system, and are consistent with Brodie's hypothesis by showing that decreases in the activities of central adrenergic neurons resulted in behavioural depression. The studies using iproniazid and imipramine with reserpine confirm the sub-hypothesis that increases in the adrenergic function produce behavioural excitement.

B. CATECHOLAMINE HYPOTHESIS OF MOOD

In terms of parsimony most researchers have ignored the serotonin system and emphasized the importance of norepinephrine systems in the control of mood (Bunney and Davis, 1965; Schildkraut, 1967; Schildkraut and Kety, 1967). For example, Schildkraut and Kety (1967) have postulated that depressed states in man are due to a diminished amount of functional norepinephrine, an increase results in euphoria, while an excess produces a manic state. This hypothesis drew together the evidence derived from the behavioural and biochemical studies in animals discussed in the previous part of this chapter with the psychopharmacological action of these drugs in man. The logical steps involved in this sort of strategy can be outlined as follows, using the format of Schildkraut (1969).

(1) Drug X has a behavioural effect in animals.
(2) Drug X produces specific biochemical effects.
(3) Drug X produces similar behavioural effects in man, allowing for any species differences in behaviour.
(4) Thus the biochemical effects produced in animals provide data relevant to the behavioural effect of drug X in man.
(5) If the behavioural effect of drug X in man has the same characteristics as pathological behaviour, then the biochemical changes produced by drug X in animals may also provide data relevant for the understanding of abnormal behaviour.

Thus for reserpine the behavioural and biochemical effects have already been described, but it is important to consider how far the reserpinized animal can be used as a model for clinical depression. Obviously it would be convenient at this point to cite a thorough well-controlled study of the psychological effects of reserpine on normal subjects, but unfortunately these are non-existent. Supporting data for the effect of reserpine on man come from studies of hypertensives, where a proportion, about 10–15 % (Klerman, 1966), of these patients show 'depression' when treated with reserpine.

In general these studies show that there is apathy, and/or motor retardation, similar to the findings in animals establishing step 3 and leading on to step 4. However, although apathy and motor retardation characterize the clinical depressive, the hypertensive treated with reserpine does not show the severe pessimism and feelings of guilt of pathological depression, and there has been some debate about whether the conditions are identical (Ayd, 1958), although most clinicians feel that they are indistinguishable (Efron and Kety, 1966; Schildkraut and Kety, 1967).

The effects of amphetamines have been much more intensely studies because of the belief that this drug improves performance, especially under conditions of fatigue or monotony. Controversy exists because of the confounding of the objective and subjective assessment of improvement. In a series of studies, Smith and Beecher (1959, 1960a, b) measured the effect of amphetamine sulphate on athletic performance of highly trained swimmers, runners and shot-putters. The experimental procedure used a double blind administration procedure, where neither the experimenter nor the subject knew whether drug or placebo had been given until after the experiment. The majority of athletes, 67 % of the swimmers, 93 % of the runners and 85 % of the shot-putters, performed better under the influence of amphetamine, compared with placebo. (Smith and Beecher, 1959). In addition to these measures, the athletes were assessed on their own feelings about their mood and physical state (Smith and Beecher, 1960a). It was found that amphetamine increased feelings of boldness and elation, which the investigators felt might facilitate antisocial behaviour in certain circumstances. However, the increased boldness and elation were usually accompanied by feelings of greater friendliness which were socially positive. The circumstances in which these feelings would lead to socially constructive behaviour and antisocial behaviour in individuals with different personalities has not been studied. In addition to the feelings of increased mental activation, the athletes reported that they felt more physically active, as if they had more strength and endurance, and as if they had improved coordination. However, they either did not recognize that the amphetamine had improved their performance, or did not appreciate the extent of the improvement (Smith and Beecher, 1960b).

In conclusion, the data of this study are in agreement with the animal data. Amphetamine increased physical activity and enhanced motor performance, not only restoring performance lowered by fatigue but also raising performance to supramaximal levels in the rested person (Weiss and Laties, 1962). In addition the pattern of behaviour is very similar to that observed in manic patients. Using the line of reasoning outlined with reserpine, it seems reasonable to argue that since the behavioural effects of amphetamine in man are analogous to those of manic patients, and similar to those observed in animals, then the biochemical correlates of amphetamine which modified the animal's behaviour are relevant to mania. In other words, mania may result from increases in the functional levels of norepinephrine. However, as Schildkraut and Kety (1967) point out, it is not possible to confirm or reject the catecholamine hypothesis on the basis of currently available chemical data. Confirmation of the hypothesis can only result from 'direct demonstration of the biochemical abnormality in the naturally occurring illness' (p. 29). In the meantime it has proved useful to treat depressed states with agents increasing the functional amounts of norepinephrine and some of these will be considered next.

Therapeutic effectiveness of antidepressants

In this discussion the major doses of drugs in use will be compared. These are the psychomotor stimulants which were first used in the 1930's, and two other classes of drugs, the monoamine oxidase inhibitors and the tricyclic compounds which have been synthesized since 1955.

Psychomotor stimulants

These psychopharmacologically defined groups of drugs include cocaine, amphetamine, methamphetamine and methylphenidate, and seem to be structurally heterogeneous. Behaviourally they are also heterogeneous producing a variety of other behavioural effects beside their mood elevating properties. The most important of the compounds is amphetamine, which we discussed in the last section. As one would expect, it alleviates the symptoms of depression in some diagnostic categories (Roberts, 1959). However, after the euphoria there is usually a return to a more intense depressive state, although these symptoms can be eliminated temporarily by another dose of the drug, tolerance quickly develops for this compound, and so larger doses are required for mood elevation. The prolonged use of amphetamines usually leads to dependence and a psychotic state characterized by hallucinations and paranoia (see Chapter 8). As a result, amphetamine and the other so-called psychomotor stimulants are not useful for the therapy of the depressions. However, the selective effect of amphetamine on different sorts of depressives has led to its use as a predictor of the clinical response to imipramine, the tricyclic antidepressant (Schildkraut, 1967).

Monoamine oxidase inhibitors

The first effects of the monoamine oxidase inhibitor iproniazid (Marsalid) were observed in the early 1950's, when the drug was used in the treatment of tuberculosis. Not only did it have beneficial effects on the course of this disease, it also induced euphoria and hyperactivity. By 1957 the drug was in widespread use, although it had not been subjected to controlled clinical testing. Careful clinical trials show that there is an improvement in about 30–50 % of depressed patients after 2–4 weeks treatment, in comparison with about a 10 % remission rate for placebo (Paré and Sandler, 1959; Kiloh, Child and Latner, 1960). Since this research, a number of other monoamine oxidase inhibitors have been developed including pargyline (Eutonyl) and tranylcypromine (Parnate). The latter are nonhydrazine derivatives, and as a result do not produce the liver damage of the earlier hydrazine derivatives, although this was extremely uncommon (Paré, 1968). There seems to be a favourable response with pargyline in about 30 % of the depressed patients regardless of their diagnosis and psychiatric history (Kline, 1963) with the first improvement seen as little as 6 days after starting the drug.

Tranylcypromine is probably the most often prescribed monoamine oxidase inhibitor still in current use. In a review of the 54 studies in the 1950's Atkinson and Ditman (1965) found that the drug was used on depressive reactions in only 16 of these, representing a test on 927 patients. Eleven of these found the drug was beneficial and 5 obtained unfavourable results. When the drug effects on different depressive subclasses are examined, it was found that the drug was helpful in reactive and psychoneurotic depressives, and of less help in agitated and psychotic depressives. Several other investigators agree that patients who respond to monoamine oxidase inhibitors seem to fall into the category of reactive depressives (Sargant and Dally, 1962; Kiloh, Ball and Garside, 1962). The features of these patients seem to indicate the drug is effective when there is a good premorbid personality, lassitude, obsessionality and phobic anxiety with a tendency for the symptoms to increase throughout the day so that the patient has difficulty in getting to sleep (West and Dally, 1959).

Patients receiving a monoamine oxidase inhibitor have to be instructed very carefully about their interactions with other drugs and with certain foods. They must not take adrenalin, amphetamines, barbiturates, opiates, or thyroid preparations, because of dangerous interactions. There are also interactions with certain foods containing tyramine, including avocado, beer, cream, chicken livers, aromatic cheeses, game, broad beans and yeast extracts. The list is so long that most depressed patients find it difficult to remember. If tyramine containing foods are eaten, instead of the tyramine being destroyed by the monoamine oxidase in the liver, the tyramine tends to build up in the blood, producing headaches, cardiovascular crises and in some cases cerebral haemorrhage. In addition, dopa decarboxylase and dopamine beta-hydroxylase synthesize the false transmitters, octopamine and alpha-methyl-octopamine, from tyramine. These are taken up by adrenergic neurons and stored until released, when they compete with norepinephrine for the postsynaptic receptors producing hypotensive effects (Kopin, 1968).

Tricyclic antidepressants

The first tricyclic agent which was found to be an antidepressant was imipramine (Tofranil). It was developed because of its chemical relationship to the phenothiazines and has a similar triple ring structure. It was an ineffective tranquillizer, and its properties were not detected in the animal-running tests. However, Kuhn (1958) discovered that the agent was helpful in the treatment of certain types of depression, and it led to the synthesis of similar compounds like desmethylimipramine (desipramine) amitriptyline, nortriptyline and protriptyline. In general, imipramine is a mood elevating antidepressant, amitriptyline is an antianxiety-sedative antidepressant, while desipramine, nortriptyline and protriptyline are stimulant antidepressants. A detailed discussion of the structure–activity relationships of these drugs can be found in Biel (1970). Some clinical studies have suggested that amitriptyline is more useful for treating the agitated depressions, whereas the rest may be more useful in the endogenous depressives showing motor retardation.

Treatment with the tricyclic antidepressants is a long process. Some subjective improvement is only reported after a week, compared with an almost immediate change with amphetamine. In some patients it may be as much as 3 weeks, which constitutes a very definite problem with the suicidal patient, and in these cases it is often considered necessary to give emergency treatment with electro-convulsive shock before pharmacotherapy. After 1 or 2 weeks treatment hypomanic episodes are usually observed, and in the manic depressives a manic phase can be precipitated. In his original description Kuhn (1958) described the recovery pattern as very similar to that of spontaneous remission. The difference between patients with and without medication seems to be that the tricyclic antidepressants prevent relapses to which the depressive is prone (Lehmann, 1966). As a result, the spontaneous remission rate appears to be about 19 %, the placebo recovery rate as around 40 %, while impiramine produced an average of 66·7 % in a survey of 42 controlled studies (Lehmann, 1966). This figure is even greater, over 75 %, if the patients are endogenous depressives (Ball and Kiloh, 1959).

Types of depressive patients

The differences in the efficacy of drugs when depressives are subdivided into different diagnostic categories is of great theoretical importance, because it implies that there may be more than one biochemical type of depression. Some interesting evidence for this position has come from a large study of imipramine and thioridazine, a phenothiazine which has a similar sort of chemical structure, on a group of 68 schizophrenic and 77 depressed patients. These patients were newly admitted to hospital and were classified on the basis of a profile of independent symptoms rated on the seven points of the Brief Psychiatric Rating Scale. The drugs were administered using a 'triple-blind design' in the sense that the nature as well as the assignment of treatments was kept from the physicians and patients. Imipramine was not very effective with the schizophrenics, but rather surprisingly it was found that thioridazine was more effective with the depressives than imipramine (Overall, Hollister, Meyer, Kimbell and Shelton, 1964). The depressive group were then subjected to a computer analysis which revealed three depressive subgroups, anxious (agitated), retarded and hostile. In other classifications the anxious depressives would have been classified among the 'neurotic depressives' while the retarded would have been grouped in the 'endogenous' category. It was found that thioridazine was better for the treatment of agitated depressives, while imipramine was more effective for the retarded depressives (Overall, Hollister, Johnson and Pennington, 1965). As there were only 15 % retarded depressives and 50 % agitated depressives the phenothiazine was the most effective for the group as a whole. These results were confirmed by other studies which also showed that some depressives improved significantly with another phenothiazine, chlorpromazine (Fink, Klein and Kramer, 1965).

There seems to be general agreement also that imipramine is the drug of choice for endogenous depressives. In the original reports on this drug Kuhn (1958) noted that the main indication was without doubt simple endogenous depression. Every complication of the depression impaired the chances of success. It was not much use in manic depression tending to precipitate mania. With reactive depression there was not such clear cut success, and in cases where difficult neurotic problems predominated imipramine was unable to develop its effect until these symptoms could be dealt with. In a computer-based correlation of individual symptoms and drug improvement, this result was confirmed (Kiloh, Ball and Garside, 1962). Imipramine gave a good response when the age of the patient was 40 years or more; if the depression was qualitatively different from 'normal' depression and associated with early waking; if the duration of the illness was under 1 year; and if a weight loss of 3 kilos or more had occurred. These are some of the diagnostic signs used to put patients in the class of endogenous depressives. On the other hand, patients under 40 years in whom the depression was of sudden onset precipitated by external factors; who were irritable and unable to concentrate, who had difficulty falling asleep and were restless while asleep; and who showed neurotic symptoms. These patients were classified as 'neurotic' or 'reactive' depressives and tended to do badly with imipramine. As commented earlier, reactive and psychoneurotic depressives seem to be best treated by monoamine oxidase inhibitors, while agitated and psychotic depressions were not improved (Atkinson and Ditman, 1965).

These results have led to the suggestion that there are at least two independent categories of depressives, reactive and endogenous (Kiloh, Ball and Garside, 1962), and probably, from the above data of Overall et al. (1964), a third, agitated.

Genetic evidence for the two different categories comes from the study of Paré, Rees and Sainsbury (1962) in which they showed that first-degree relatives of depressed patients who became depressed themselves respond to monoamine oxidase inhibitors and tricyclic antidepressants in the same way. In all seven cases examined, the drug result in the relative correlated with that in the patient, suggesting that there are two genetically specific types of depression. Other evidence comes from the results of electroconvulsive shock therapy on endogenous and reactive depressives. Endogenous depressives respond better to shock therapy than reactive depressives. This suggests that the symptoms may be more than a mild variety of endogenous depression if they fail to respond or can be made worse by a treatment which is effective on the other condition (Kiloh, Ball and Garside, 1963).

These data suggest that the simple catecholamine hypothesis of mood must be made more specific with respect to depression. One way this might be done is to suggest that endogenous depressives synthesize sufficient norepinephrine, but that there is either inadequate release or an inactivation mechanism which is too effective. Thus, any drug like imipramine which increases the amount of transmitter available in the synapse will alleviate the disorder, but any treatment affecting mechanisms prior to release like monoamine oxidase would be ineffective. On the other hand, reactive depressives might not synthesize enough norepinephrine or be able to store it while the succeeding stages in transmission would be normal. Thus, any drug which increases the amount of transmitter available for release, such as monoamine oxidase inhibitors, would ameliorate the symptoms.

The remaining type of depression is agitated depression which seems to be distinguishable from the other two types by being susceptible to treatment by phenothiazines (Overall et al., 1965) and by phenothiazine-like tricyclic antidepressants such as amitriptyline (Biel, 1970). Phenothiazines were described as agents which reduce the functional amounts of norepinephrine reaching the synapse in Section A of this chapter and in Chapter 1. However, the catecholamine hypothesis of mood says that depressed states result from a diminished amount of functional norepinephrine. This contradiction can be resolved by considering the experiment of Segal and Mandell (1969) discussed in Section A, in which it was found that small increases in the functional norepinephrine produced increased activity, but higher levels decreased activity and the animals appeared sedated. If agitated depressives had excessive quantities of norepinephrine, so that there was depolarization block then the drug of choice would be an agent which reduced the functional levels of norepinephrine to the normal range. This suggestion and the preceding ones are mere speculation, and as we stated before, they are only directly testable by studying abnormal function in the depressed patient.

Nevertheless, these results suggest that the physician might adopt the following strategy when prescribing for depressed patients. First, he should enquire whether a close relative had been treated and choose the drug that was effective in that case as Paré (1968) recommends. Second, he should use the diagnostic signs to categorize the patient as either endogenous or neurotic and prescribe imipramine if the patient fits the endogenous category. If he is neurotic he should test with amphetamine. Improvement with amphetamine suggests a reactive depressive who could be treated with a monoamine oxidase inhibitor while an agitated depressive would fail to respond to amphetamine and a phenothiazine or phenothiazine-like tricyclic antidepressant such as amitriptyline. One interesting point to note is that imipramine seems to be more effective in men than women (Medical Research Council, 1965). This suggests that some mood disorders may differ in men and women and some types may be hormonally based, and this suggestion will be examined next.

C. SEX HORMONES AND MOOD

There are three depressive states which seem to be associated with hormonal changes during life. These are menstrual depression, postpartum depression and involutional melancholia. Considerable controversy exists about the contribution of the hormonal changes to these disorders particularly in the case of the involutional melancholia, where the disorder is associated with the stresses common to late middle age such as unfulfilled ambitions, recognition of missed opportunities and a realization of a decreased capacity to cope with life. In spite of these obvious psychological factors, there is some evidence for relating involutional depression to the climatic change in hormones, and much stronger evidence associating menstrual depression and postpartum depression with the decreases in circulating hormones. The first section will relate the menstrual changes in hormones to the adrenergic system, and then this information will be extrapolated to account for postpartum depression and involutional melancholia.

In the male the level of the androgens remains virtually constant throughout the life, but in the female hormone production is variable and two main classes, the oestrogens and progestogens, are secreted as well as small amounts of androgens from the ovary and adrenal cortex. The cyclical production of the female hormones results in the characteristic ovarian cycle of most adult females (Harris and Naftolin, 1970). In the human female during the first phase of this cycle the ovarian follicles and the oocytes are immature and the levels of hormones are low. Oestrogens begin to be secreted by the ovary under the influence of the follicle-stimulating hormone, FSH, from the pituitary. The production of FSH is controlled by feedback of oestrogens to the hypothalamus. Oestrogens also stimulate the release of the luteinizing hormone, LH, from the pituitary. A preovulatory surge of LH results in rupture of a follicle and the release of a mature oocyte about 24 hr after the surge. Following the ovulatory phase, granulose cells of the follicle change to form the corpus luteum which develops under the influence of prolactin. The corpus luteum releases progesterone and is the source of oestrogens which inhibit prolactin after about 14 days causing regression of the corpus luteum. This premenstrual phase is characterized by the reduced secretion of progesterone and oestrogens. The withdrawal of these two hormones brings on menstruation and simultaneously the low levels of oestrogen enable the secretion of FSH to initiate the new cycle.

One obvious way by which the hormones could activate the nervous system and change mood is by modifying the functional amounts of norepinephrine transmitter at the synapses in the brain. This sort of interaction seems to occur, and assays for norepinephrine in the hypothalamus of the rat have been made at different phases in the ovarian cycle (Stefano and Donoso, 1967). In the anterior and medial hypothalamus it was found that there was an increase in the levels of this transmitter during dioestrous, reaching a peak during the afternoon of proestrous just prior to ovulation. On the afternoon of oestrous the levels of norepinephrine fell to their lowest levels suggesting that within this 24 hr period when ovulation is occurring there was marked activity in the adrenergic neurons of the hypothalamus, resulting in the release of norepinephrine. The dramatic changes in norepinephrine at ovulation closely parallel the variations in release of LH, and it has been found that depletion of brain monoamines, including norepinephrine, blocks LH release and ovulation (Meyerson and Sawyer, 1968; Ratner and McCann, 1971) suggesting that monoaminergic pathways are necessary for triggering the ovulatory surge of LH. The changes in norepinephrine levels could be due to either FSH or oestrogen, and it is known that gonadectomy, which increases FSH, also increases the synthesis of norepinephrine by acting on the synthesizing enzyme, tyrosine hydroxylase (Beattie, Rodgers and Soyka, 1972). However, in the same study injections of oestrogen which inhibit FSH also increased the activity of tyrosine hydroxylase by 30 %, showing that oestrogen has an independent action on tyrosine hydrosylase activity. Comparison of LH, oestrogen and norepinephrine levels show the closest relationship is between the oestrogen and the transmitter.

Another factor which would contribute to the changing norepinephrine levels is the variation in monoamine oxidase during the oestrous cycle (Kobayashi, Kabayashi, Kato and Minaguchi, 1966). Monoamine oxidase is the inactivating enzyme which controls the levels of free norepinephrine within the neuron, so that when its activity is high the levels will be low. It was observed that there was a proestrous decrease in monoamine oxidase activity in the hypothalamus from a peak at 10 a.m. to its lowest level of the cycle at 6 p.m. (Kamberi and Kobayashi, 1970). Ovariectomy produced a large increase in monoamine oxidase activity, while injections of oestrogen lowered this activity (Kobayashi et al., 1966), i.e. oestrogen acts like an inhibitor of the enzyme. Similarly, castration raised the levels of enzyme activity in adult male rats, and injections of androgens reduced these again (Zeller, cited in Brover, Klauber, Kobayaski and Vogel, 1968).

From this information it seems that both testosterone and oestrogen act on the adrenergic transmitter system of the rat in a similar fashion, but the levels of oestrogen are changing cyclically, so there are cyclic changes in the amounts of transmitter in the female. The rat has been used successfully for testing drugs for their efficacy in the therapy of mania and depression (Russell and Warburton, 1973), and it has been found that drugs which raise the functional norepinephrine levels in the rat are effective antidepressants, while drugs which lower the functional levels of transmitter at the synapse are effective in treating mania (Schildkraut and Kety, 1967). It would seem to be worthwhile to examine the effects of gonadal hormone on activity, self-stimulation and avoidance, which we have seen are useful situations for evaluating antidepressants in the rat.

Behavioural effects of oestrogen

It is one of the classical findings of physiological psychology that gonadal hormones influence the locomotor activity of animals. In general, adult male rats are less active than adult females and castration of the adult rat further reduces spontaneous activity (Hoskins, 1925). In contrast to the male, the female rat shows cyclical changes in activity with up to threefold increases in activity at oestrous (Wang, 1923). These effects are due to the cyclical release of oestrogens and begin at puberty. They can be abolished by ovariectomy (Wang, 1923; Richter, 1927), and the effects restored by injections of oestrogens (Young and Fish, 1945). In fact, increased activity in the normal adult male rat can also be produced by injections of an oestrogen (Hoskins and Bevin, 1941). It is interesting that a much smaller change in activity is observed in a stabilimeter than in a running wheel (Eayrs, 1954) showing that it is *not* gross activity but locomotion specifically that is increased. Nevertheless, the same pattern holds in measures of ambulation in an open field for normal oestrous (Burke and Broadhurst, 1966) and for oestrous induced by injections of oestrogen and progesterone (Gray and Levine, 1964). This increased locomotion is consistent with the observation that on the evening of oestrous the female rat leaves her burrow and lays scent trails over a wide territory in order to lead males back to her. This form of behaviour is highly adaptive and contradicts the widely held notion of sexual passivity in the female rat. The changes in locomotion can be explained in terms of the effects of the hormones on the brain neurochemical systems. Biochemically we have seen that oestrogen makes more free norepinephrine available for release, potentially enhancing adrenergic neuron function. From this effect one would predict that oestrogen will act like drugs increasing the amount of functional norepinephrine, and this is precisely what has been found in the majority of studies.

In a study of self-stimulation performance, a sensitive test for adrenergic activation (Stein, 1962), Prescott (1966) found that the highest mean bar pressing rate in female rats occurred on the day of oestrous for five out of the six rats, while the sixth rat had the highest score on the day after. These changes, coincident with the appearance of vaginal cornification, cannot be explained in terms of increased random activity because responses on a non-contingent lever did not increase. Contradictory results have been obtained in a study where oestrous was induced by a single injection of oestrogens (Hodos and Valenstein, 1960). However, it has been found that repeated injections of oestrogen are necessary to restore high levels of wheel running in ovarectomized rats, which probably explains the disparate results (Prescott, 1966). In male rats the self-stimulation rate is also correlated with the levels of testosterone (Olds, 1958), even though the androgens appear to be much less potent monoamine oxidase inhibitors (Wurtman and Axelrod, 1963). In other conditioning situations females are superior to males in avoidance conditioning, both lever pressing (Nakamura and Anderson, 1962) and in a runway (Levine and Broadhurst, 1963), which correlates with the hormonal differences in monoamine oxidase inhibition. In castrated males there is even an inverted U-shaped dose-response relation between lever pressing performance and replacement doses of testosterone (Broverman et al., 1968) paralleling the inverted U-shaped dose-response curve obtained in studies of brain levels of norepinephrine and performance.

Mood changes during the menstrual cycle

Extrapolating from these rat data on norepinephrine and the oestrous cycle to humans, we would anticipate that clinical surveys would show cyclic changes in mood throughout the menstrual cycle, and that these would coincide with the levels of oestrogen reaching the adrenergic synapses in the CNS. A good example of such a survey is one conducted by Moos (Hamburg, Moos and Yalom, 1968) on various symptoms during the menstrual cycle. Some of these were known to be specifically related, e.g. crying, tension, breast swelling, while others were control symptoms, e.g. fuzzy vision, heart pounding, ringing in ears. It was found that there were adverse changes in mood, concentration and performance premenstrually and during menstruation. Moderate or severe changes seem to occur in 25 % of all women with some symptoms of negative affect in up to 90 % of the women questioned (Janowsky, Gorney and Mandell, 1967). In particular, depression was highest premenstrually when the oestrogen and progesterone levels are lowest (Hamburg, Moos and Yalom, 1968) while there seems to be a feeling of elation and well being around ovulation when the oestrogen levels are high (Benedek and Rubinstein, 1959). Correlative information on depression comes from a survey conducted by the Suicide Prevention Centre in Los Angeles (Mandell and Mandell, 1967). They asked telephone callers to give date of onset of menstrual period, expected date of next period and usual length of cycle. They subdivided the cycles into sevenths and recorded the frequency of calls in each seventh for the group. They found that the calls were highest during early menstruation, second highest premenstrually with an unexpected peak in the mid-cycle (4/7). The peak followed the lowest incidence at 3/7th, i.e. 12 days in a 28-day cycle.

In a companion study to the one on depression throughout the menstrual cycle, Hamburg et al. (1968) examined the cyclic variation in anxiety and sex arousal during the cycle. It was found that anxiety was highest during the menstrual flow, decreasing to a low point during the days around the time of ovulation at the mid-cycle, and then gradually increasing throughout the cycle until the onset of the menstrual flow. The self-rating of sexual arousal was the mirror image of this pattern, with the greatest arousal at mid-cycle. The same pattern of change in anxiety was obtained by Gottschalk (1969), using a verbal sample technique in which the subjects were asked to talk for 5 min about their personal experience on 3–7 days a week. The samples were scored for anxiety, by technicians, who were unfamiliar about the purposes of the study. Anxiety decreased around ovulation, but also in three women around menstruation. In a larger sample of 25 Ivey and Bardwick (1968) confirmed that anxiety was significantly lower at the presumptive ovulation time, but there was no premenstrual decrease. These mood fluctuations were also reflected in various sorts of performance. Studies of schoolgirls showed that the classwork performance of 27 % decreased in the week preceding menstruation (Dalton, 1960; Hamburg et al., 1968). In a study of examination performance Dalton (1968) found that in the Ordinary Level Examination, taken by girls aged 14–17 years, there was a drop of 5 % in the pass rate for girls taking the examination while menstruating. In the Advanced Level Examinations, taken by 16–19 year old girls, there was a difference in pass rate of 13 % between girls who took the examination just prior to menstruation and those who took the examination in the middle of a cycle.

Therapy of sex hormone depression

As we saw earlier, these changes are manifestations of a cyclic pattern of changes in concentration and mood (Hamburg et al., 1968). These authors found that there were fewer, 24 % compared with 36 %, reports of premenstrual negative affect in women taking oral contraceptives. This did seem to be explainable in terms of generalized group differences because there were no differences in the frequency of symptoms reported during the intermenstrual phase. There were also differences in the symptoms depending on the type of contraceptive used with the greatest frequency (Herzberg, Johnson and Brown, 1970), reported by women using sequential preparations (Hamburg et al., 1968). One explanation for the syndrome is the withdrawal of pro-gesterone (Hamburg et al., 1968; Dalton, 1969) and treatment with progesterone relieves the physical symptoms (Greene and Dalton, 1953; Dalton, 1959). However, Herzberg et al. (1970) found no difference in the incidence of de-pressive symptoms with contraceptive preparations containing different amounts of progesterone. In animal studies no effects of low doses of pro-gesterone on norepinephrine release and metabolism were found (Ladisich and Baumann, 1971), so at the present time there is no direct evidence for progesterone affecting mood by its action on the adrenergic pathways. The crucial factor may be the withdrawal of oestrogen, and consequently a fall in the functional norepinephrine. Certainly premenstrual depression can be alleviated by means of monoamine oxidase inhibitors like isocarboxazid and phenelzine.

Postpartum depression is also clearly correlated with a marked change in hormonal levels with the sharp decline in the levels of progesterone. Episodic crying was found in 67 % of all women tested in the first 10 days postpartum by Hamburg et al. (1968). This was treble the number who had cried in the 10 days prior to their prenatal interview. Other evidence of dysphoria, insomnia, restlessness and undue concern for the health of the infant. One more interesting observation was that only one out of the 39 women tested had ever received psychiatric treatment. These effects observed after the withdrawal of pro-gesterone are not directly explainable in terms of effects on the adrenergic system. One possible way out of this dilemma, proposed by Mandell (1970), could be that progesterone increases the corticosteroids secreted by the adrenal cortex, and as we will see later in Chapter 11, the corticosteroids act on the adrenergic system. Explanations of involutional melancholia are simply explained in terms of the abrupt decline of hormones in women at the cli-macteric, and the similar but slower decline in men with their consequent diminished stimulation of the adrenergic system.

4
Control of Attention

In classical physiological psychology it was always assumed that the state of behavioural arousal could be identified by the amount of electrocortical arousal. They are usually correlated but they can be dissociated by lesions, and drugs and are dissociated during normal sleep (see Chapter 6). In this chapter we will be discussing the effects of cholinergic drugs on behaviour together with the concomitant changes in electrocortical arousal. It will be concluded that the behavioural changes can be ascribed to modifications in the processes of stimulus selection or 'attention' due to the drug effects on cholinergic pathways ascending from the reticular formation.

Although this will be the conclusion, earlier work on the influence of cholinergic drugs on behaviour suggested that a cholinergic system was involved in response inhibition. Since this hypothesis is still frequently used to explain the effects of cholinergic drugs, evidence contradicting it will be outlined.

A. RESPONSE INHIBITION HYPOTHESIS

This hypothesis has been implicit in the work of many investigators but has rarely been stated explicitly. In its naïve form response inhibition has been used to describe the decrease in responding observed after injections of anticholinesterases, and response disinhibition applied to the increased responding often observed after injections of cholinolytics. For example, Herz (1968) discussed the effects of cholinolytics on passive avoidance behaviour and on discrimination behaviour and concluded that there is a muscarinic cholinergic system in the brain and that blockade of this system 'produces some sort of disinhibition which may be connected with drug-induced motor hyperactivity' (p. 73). In his conclusion he argues from the performance decrements produced by cholinesterase inhibition, that increases in central cholinergic activity result in greater response inhibition. This sort of hypothesis makes no attempt to distinguish between the effects of the drug on the brain and those on the peripheral nervous system. It ignores the fact that injections of a quarternary anticholinesterase, e.g. neostigmine that passes the blood–brain barrier poorly, disrupts performance by impairing neuromuscular transmission and not by acting on the CNS (see Chapter 5 for cholinergic involvement in muscle control).

A behaviourally more sophisticated hypothesis was proposed by Russell (1966a), when he suggested that there was a cholinergic system in the brain which was essential for the inhibitory control of competing responses. During training an increase in the probability of the reinforced response depends on the suppression of all the alternative responses that are possible in that experimental situation, but irrelevant for reinforcement. When reinforcement is withheld during experimental extinction or discrimination, the decrease in probability of a response results from the increase in competing responses. This model predicts that cholinolytics which block this system will accelerate the decrease in probability of the dominant response by impairing the suppression of competing responses, but the evidence (Hearst, 1959) does not support this prediction.

A revision of this hypothesis (Warburton, 1967) propounded the view that brain cholinergic system was important in the inhibitory control of both competing and dominant responses. A blockade of central neural function by cholinolytics and high doses of anticholinesterases would result in a disinhibition of all responses, but with proportionally greater increases in the lower probability responses, especially when the high probability responses reached their response ceiling. Once at its response ceiling the probability of the response will decrease as the competing responses are increased, giving the U-shaped dose-response curve frequently found with cholinolytics (see Stone, 1964; Warburton and Heise, 1972). Small doses of anticholinesterases which facilitated cholinergic synaptic transmission would inhance response withholding and improve extinction and discriminate performance as Russell (Russell, Watson and Frankenhaeuser, 1961; Banks and Russell, 1967) had found.

Another sort of response inhibition hypothesis has been proposed by Bignami and Rosić (1970), in which they suggested that the deficit caused by cholinolytics is on the motor side of the stimulus-response chain producing perseveration. This perseveration is manifested in terms of modes of responding similar to the 'set perseveration' of Mishkin (1964), which does not necessarily lead to hyperresponding, but may produce hyporesponding if the original set had a strong 'no go' tendency. This version would seem to predict uniform increases or decreases in responding, but not U-shaped dose response curves unless some sort of response ceiling assumption was introduced.

Since the various versions of the hypothesis were proposed, a number of studies have suggested that the modifications are occurring in the stimulus input rather than in the response control, in particular, those which have demonstrated state-dependent learning where the drug effects transfer to responding in the non-drug state. In studies of this phenomenon, a 2×2 factorial design has been used in which drugged and undrugged animals were trained. The two groups were then subdivided and retested with and without the training drug. Dissociation was demonstrated by the failure to show transfer when, either the saline–scopolamine group was compared with the saline–saline group, or the scopolamine–saline group was compared with the scopolamine–scopolamine group (Gruber, Stone and Reed, 1967; Oliverio, 1968; Berger and Stein, 1969). The crucial groups from our point of view are the transfers from scopolamine to saline and from the saline to saline. Although both groups are responding in a test situation after saline injections, the performance of the scopolamine trained animals is inferior to that of the saline trained. Thus scopolamine impaired the input of information to storage, and not response inhibition, because the animal's responding was measured only in the saline conditions.

B. STIMULUS CONTROL HYPOTHESIS

Any organism has many sorts of stimuli influencing it while in the experimental situation. From the external environment, stimuli impinge on the exteroceptors and these may be relevant, irrelevant or partially relevant for the responding of the organism, while after a response there are outcome stimuli which are determined by the experimenter. In addition, responses are influenced by various stimuli from interoceptors, e.g. blood chemical levels, response produced cues and by previous, 'remembered', stimulus sequences. Any single response will be determined by a subset of stimuli from these four sources and disruption of one of these sources of input would impair responding. Several hypotheses have been formulated in terms of selective impairment of one or other of these inputs, and in this section evidence will be presented which contradicts these specific suggestions.

Outcome Stimuli: Most behaviour patterns studied in the laboratory have programmed consequences such as reinforcement so that the acquisition of a behaviour sequence depends on the outcome of the trials, and stable performance is maintained as long as the outcomes remain constant. It follows that a drug impairing the input and use of outcome stimuli will disrupt behaviour. One hypothesis (Carlton, 1963) suggested that a brain cholinergic system was involved in the mediation of non-reinforcement effects and so any attenuation of cholinergic activity would impair performance in situations involving non-reinforcement. Many of the early experiments were clearly consistent with this hypothesis, but it was much more difficult to explain the changes in continuous avoidance responding (Heise and Boff, 1962; Stone, 1964) in these terms, and impossible to explain the increased responding in conditioned suppression situations (Vogel, Hughes and Carlton, 1967).

Prior Stimuli: Evidence from the clinical literature suggested that patients, injected with scopolamine, showed amnesia (Goodman and Gilman, 1965), and Meyers and Domino (1964), proposed the hypothesis that attenuation of cholinergic activity impaired recent memory for stimuli. However, this simple hypothesis was contradicted by many studies (e.g. Hearst, 1959; Laties and Weiss, 1966; Warburton, 1969) which showed that responding was disrupted even though a discriminative stimulus was actually present at the time of the response. Other studies have also shown that learning does occur when the animal has been injected with scopolamine, but that there appears to be an 'amnesic' effect where there is poor transfer to the non-drug state (e.g. Pazzagli and Pepeu, 1964; Bohdanecký and Jarvik, 1967; Carlton, 1969; Warburton and Groves, 1969), and recently these effects have been shown to be cases of state-dependent learning (Berger and Stein, 1969).

Internal and External Stimulus Control: Performance in behavioural situations depends not only on the external cues but also on an organism's use of cues from the internal environment. Laties and Weiss (1966) have stated that discrimination based on interoceptive cues are less precise because they require more trials to reach a criterion than comparable exteroceptively cued discriminations. They suggest further that cholinergic drugs disrupted interoceptive discriminations more than discriminations based on exteroceptive control, *because* the interoceptive control was less precise. The study on which they based their hypothesis was a fixed interval schedule with and without an external cue, a stimulus changing with time. It was found that greater shifts in response distribution were produced by scopolamine when no external cue was available. Similar results were obtained by Wagman and Maxey (1969) in a 'counting' schedule where internal control was based on response-produced cues. In a test of this hypothesis, Heise and Lilie (1970) tested drug effects in two related situations based on reinforcement of only those responses preceded by three no-response trials. In one case the external stimuli changed when reinforcement was available while in the second no external cues were available, and behaviour was controlled by the internal cues of both time and outcome of previous trial. It was observed that scopolamine impaired performance to about the same extent for groups trained in both situations while atropine had slightly more effect on internally controlled performance.

Similar results have been obtained in other experiments in our laboratory (Heise, Laughlin and Keller, 1970; Warburton, 1969; Warburton and Heise, 1972). In one of these experiments (Warburton and Heise, 1972), animals were trained on a double alternation where the first response of a sequence was controlled by an exteroceptive cue, a tone, while the remaining responses were controlled by interoceptive cues which were the outcomes of the previous trials. This situation enabled the effect of scopolamine on the two types of responding to be tested in the same animal. Doses disrupted both types of performance, and there was little evidence that the drug had a greater effect on interoceptively controlled performance.

The apparent differences in sensitivities of externally and internally cued responding found by Laties and Weiss (1966) can be explained in terms of stimuli used in the experiment. In the typical 'internal cue' situation the subjects base their response on a single temporal or response-produced stimulus. In contrast the experimenter programmes the external cue situation by adding extra external stimulus so that responding is effectively controlled by two stimuli. It is not surprising therefore that the two situations used by Laties and Weiss (1966) show differential sensitivity. In the study of Heise and Lilie (1970) it will be remembered that the internal responding was controlled by two cues, and no differences were obtained. In a study (Warburton, 1974) to be discussed later it was shown that sensitivity to disruption by cholinolytics is also reduced when two relevant external cues are available to the animal.

In summary, the studies in this section show that cholinergic blockade does not have specific effects on any kind of stimulus-controlled responding but seems to disrupt responsiveness to all stimuli irrespective of their origin, either past or present, internal or external.

C. STIMULUS SELECTION HYPOTHESIS

Variations in organisms' responsiveness to stimuli have frequently been attributed to changes in attention. As it is generally used in psychology, the term 'attention' has little explanatory value but is useful for classifying the cases where a response has been shown to be under the control of a specific stimulus or set of stimuli (Terrace, 1966). It implies that the organism is selecting in some way from the mass of stimuli impinging from the external and internal environment. From the evidence discussed it is hypothesized that a brain cholinergic system is mediating the selection of relevant stimuli in the environment, i.e. those cues which set the occasion for the experimenter-determined 'relevant' response, and those which set the occasion for withholding irrelevant responses. The next studies were designed to examine the sensitivity to interoceptive cues after injection with various cholinergic drugs.

Stimulus Detection: The discrimination situations used were differential reinforcement of low rate, DRL 15 sec with reset of the interval for responses during the last 10 sec (Warburton and Brown, 1971) and a similar discrete

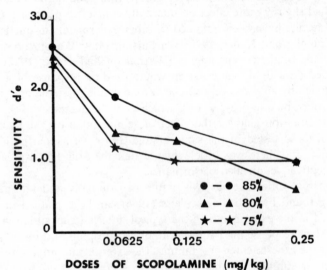

Figure 18. Change in the sensitivity index after injections of various doses of scopolamine for various deprivation levels. From Warburton, D. M., and Brown, K. (1971). *Nature,* **230,** 126–127. Reproduced by permission of Macmillan and Co. Ltd.

trial situation with a variable intertrial interval, mean 15 sec (Warburton, 1972; Warburton and Brown, 1972). Responding in these schedules was analysed by means of a theory of signal detectability analysis which yielded independent estimates of a subject's sensitivity to stimuli and his response bias, the non-sensory factors which also influence responding. The response-inhibition hypothesis mentioned earlier predicts that scopolamine would have impaired inhibition increasing both 'hits' and 'false alarms' and resulting in a decrease in the bias index while physostigmine would have the opposite effect. Figure 18 shows that scopolamine decreased the sensitivity index (Brown and Warburton, 1971; Warburton and Brown, 1971, 1972) while physostigmine increased sensitivity (Warburton and Brown, 1972). At the high doses of both drugs, 0·75 mg/kg scopolamine and 1·0 mg/kg physostigmine, there was an increase in the bias index which was consistent with motor impairment and/or a reduction in food intake but not with loss of response inhibition.

In perceptual theory, attention has been interpreted as a process of selection produced by increases in the signal-to-noise ratio in sensory systems (Treisman, 1964) and shifts in attention have been found to produce changes in sensitivity with no consistent shift in the response criterion (Broadbent and Gregory, 1963) which was exactly what was found in this study. On the basis of this notion it would be predicted that giving the animal two relevant cues to signal each trial would ameliorate the deficits produced by cholinergic blockade and the next experiment was designed to test this hypothesis.

Cue Redundancy: Three groups of animals were trained in a discrete trial situation with either a tone, a light or a tone–light combination as the trial stimulus. When they were at a 90 % criterion of responding, they were given a transfer test with both the tone, the light and the tone–light combination, with reinforcement given for responses only during the trials originally re-inforced during training. It was found that both tone alone and light alone showed transfer to the tone–light combination, but not to each other, while the tone–light group showed transfer on both types of single-cued trial, but there were individual differences in the amount of transfer to each cue. Later, the three group were tested with four doses of scopolamine to obtain a dose response curve for each group. It was found that the tone–light group were least disrupted by the drug (see Figure 19), and that the amount of disruption was directly correlated with the extent to which the organism relied on one cue as shown by the amount of transfer to each cue (Warburton, 1974). None of the response inhibition hypotheses would predict any differences in sensitivity to the drug with different numbers of relevant stimuli. However, these results and those of Laties and Weiss (1966) can easily be explained in terms of an impairment of stimulus selection, so that if the signal-to-noise ratio in all modalities was reduced by cholinolyptics, then the probability of selecting a relevant stimulus on a given trial would be p for light and q for the tone, but $p + q$ for the tone–light combination.

Stimulus Selection in Learning: In the two-stage theory of discrimination

72

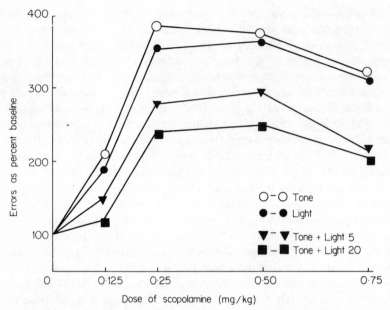

Figure 19. Dose response curve showing the effects of various doses of scopolamine on performance under the discriminative control of different stimulus combinations and amounts of training. From Warburton, D. M. (1974). *Quart. J. exp. Psychol.*, **26**, 395–404

learning proposed by Sutherland (1964) there are two separate processes: (a) learning to switch in the sensory analyser whose outputs discriminate

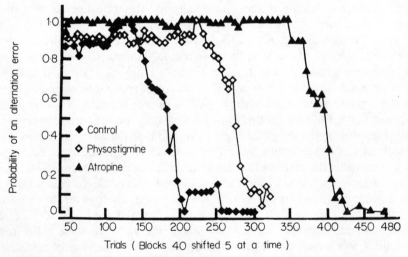

Figure 20. Acquisition performance of single alternation by three rats injected with either saline (control), physostigmine sulphate or atropine sulphate. From Warburton, D. M. (1969). *J. comp. physiol. Psych.*, **68**, 56–64

between relevant and irrelevant stimuli; and (b) learning which responses to link with those outputs. If learning is proceeding in two stages we would expect that if the second stage was dependent on the first, the learning curves would display discontinuity. Thus, for example, if the stimulus selection phase was difficult, a long presolution plateau would be expected prior to the discrimination learning phase when the errors decrease (Zeaman and House, 1963). Biochemical changes impairing the selection process would be expected to increase the length of the presolution plateau while leaving the discrimination learning rate unchanged. Data consistent with this prediction have been obtained in a study of cholinergic drugs on the acquisition of single alternation (Warburton, 1969; 1972). Use of a mathematical model (Warburton and Greeno, 1970) showed that a blockade of the cholinergic system only increased the length of the presolution plateau (see Figure 20), but left the discrimination learning rate parameter unchanged (Warburton, 1969). This was interpreted as an increase in the time when the organism was responding to irrelevant attributes of the situation or, in other words, failing to select and learn the relevant stimuli. In a related study (Warburton, 1972), the effects of small doses of physostigmine sulfate (0·05 mg/kg and 0·1 mg/kg) on single alternation were examined, and it was found that this drug facilitated acquisition by reducing the presolution plateau, while the rate of decrease in errors remained constant. An experiment in the same article tested the effect on the presolution plateau of shortening and lengthening the 'intensity' of the persisting outcome stimuli which control performance in this situation (Heise, Keller and Khavari and Laughlin, 1969). It was found that a decrease in the intertrial interval also reduced the presolution plateau but did not change the learning rate. It was concluded that physostigmine facilitated acquisition by increasing the probability of selecting the relevant stimulus, perhaps by increasing the signal-to-noise ratio in this modality. Facilitation of discrimination acquisition with the same doses of physostigmine were obtained by Whitehouse (1966), but his results cannot be analysed in order to pinpoint the improved process. The pattern of results with cholinergic blockade after cholinolytics is not what one would expect from the simple response inhibition hypothesis. It would be predicted that if the correct stimuli had been learned but the response could not be inhibited, then the rate of decrease in errors would be faster than control when the cholinergic system recovered from the blockade. This is clearly not the case even though the learning rate parameter is particularly sensitive to facilitation (Warburton, 1972).

Dissociated Learning: In the studies discussed so far it has been proposed that when cholinergic function in the brain is attenuated there is an impairment of discrimination in acquisition performance and well trained performance. In terms of understanding the behavioural mode of action of the cholinergic drugs, it is important to consider the studies which have examined the effects of scopolamine on the performance of animals trained under the influence of scopolamine, i.e. the experiments on state-dependent learning which were cited earlier as evidence against the response inhibition and simple amnesia hypotheses. These studies have shown that there is an impairment of transfer both from the drug state to the no-drug state and vice versa (see Figure 18). However, this impairment is asymmetrical with less transfer from the drugged to the no-drug state (D-ND) compared with non-drugged to drug (ND-D), and less transfer from the drugged to drug state (D-D) than from the non-drugged to the non drug (ND-ND) state (Gruber, Stone and Reed, 1967; Berger and Stein, 1969). This asymmetry contradicts the simple hypothesis that the drug was only part of the stimulus complex which became associated with the drug. It is also inconsistent with the simple hypothesis of amnesia produced by the drug because one would expect equal transfer from scopolamine to saline as scopolamine to scopolamine. The only hypothesis that fits the data is an impairment of both acquisition and the stimulus generalization between the two situations (Berger and Stein, 1969) and both these two deficits would result from an impairment of stimulus selection mechanisms.

As we saw in the last section, scopolamine impairs acquisition which would explain why there is poorer performance in the D-D condition than in the ND-ND condition. However, there must be some failure of generalization as well, otherwise the D-D score would be larger than the D-ND score. This generalization deficit can be explained in terms of the change in the pattern of stimuli produced by scopolamine.

D. NEURAL LOCATION OF STIMULUS SELECTION

Although there are several sites along the primary sensory pathways where cholinergic drugs could modify stimulus selection processes, there is only one site where changes in activity could modify the sensitivity to internal as well as external stimuli, and that is at the cortex. There appears to be a generalized arousal mechanism operating at the cortex, and it is proposed that this electrocortical arousal forms a background noise which is capable of masking of the less intense inputs to the cortex, and determines the final behavioural output. Electrocortical arousal appears to be controlled from the reticular formation (Thompson, 1967), since it is changed by cholinergic drugs and so studies of the ascending cholinergic pathways from the reticular formation to the cortex will be examined.

Studies of the Cholinergic Reticular Pathways: The projections of the cholinergic pathways having their origin in the brain stem have been analysed thoroughly by Shute and Lewis over the last 10 years, and their work summarized in *Brain* (Shute and Lewis, 1967; Lewis and Shute, 1967). From their studies it can be seen that there are at least three main systems—two ascending reticular systems and a hippocampal circuit. The ascending reticular systems are the dorsal and ventral tegmental pathways. The dorsal tegmental system projects to the thalamic regions including some, at least, of the primary sensory pathways and corresponds to the thalamic reticular activating system of Starzl,

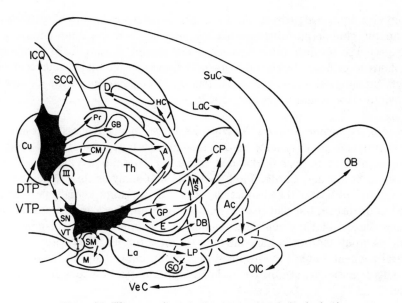

Figure 21. The ascending cholinergic pathways in the brain.

Abbreviations:		
A:	anterior thalamic nuclei	
Ac:	nucleus accumbens	
CM:	centromedian nucleus	
CP:	caudate nucleus and putamen	
Cu:	nucleus cuneiformis	
D:	dentate gyrus	
DB:	diagonal band	
DTP:	dorsal tegmental pathway	
E:	entopeduncular nucleus	
GB:	geniculate bodies	
GP:	globus pallidus	
HC:	hippocampus	
I:	interpeduncular nucleus	
ICQ:	inferior corpus quadrigeminum	
III:	oculomotor nucleus	
La:	lateral hypothalamic area	

LaC: lateral cortex
LP: lateral preoptic area
M: mamillary body
MS: medial septal nucleus
O: olfactory tubercle
OB: olfactory bulb
OIC: olfactory cortex
Pr: pretectal area
SCQ: superior corpus quadrigeminum
SM: supramamillary nucleus
SN: substantia nigra
SO: supraoptic nucleus
SuC: superior cortex
Th: thalamus
VeC: ventral (entorhinal) cortex
VT: ventral tegmental area
VTP: ventral tegmental pathway

From Shute, C. C. D., and Lewis, P. R. (1966). *Brit. med. Bull.,* **22,** 221–226. Reproduced by permission of the authors and the Medical Department, the British Council

Taylor and Magoun (1951), while the ventral tegmental system projects to the neocortex and to diencephalic nuclei. The hippocampal circuit also originates in the ventral tegmental area and projects to the hippocampus via the medial septal nuclei. From the hippocampus there are non-cholinergic afferent fibres returning to brain stem nuclei from which the three systems originated. These three systems are shown in Figure 21.

Indirect evidence for the relative importance of these pathways in explaining cholinergic drug effects comes from lesion and electrical stimulation studies. Thus it would be predicted that a lesion in a cholinergic system mediating selection would lead to behavioural deficits corresponding to the effects of attenuating cholinergic function while low intensity electrical stimulation would be expected to produce opposite effects. The results of using these techniques enable us to exclude the dorsal tegmental system as the major source of our effects because lesions in the dorsal tegmental area produce little change in behaviour (Hodos and Valenstein, 1962).

Ventral Tegmental Systems: In contrast to the 'ineffectiveness' of lesions in the dorsal tegmental system, lesions in the ventral tegmental area produce 'hyperactivity' in an open field and in extinction (Hodos and Valenstein, 1962; Le Moal, Stinus and Cardo, 1969). Schiff (1964) analyses this 'hyperactivity' further and observed that in fact the lesioned rats were actually less active when undisturbed, but that their responses to stimuli were exaggerated. These studies suggest that the ventral tegmental area is important in the control of stimulus input, but not in the control of response output. This evidence of the ventral tegmental area's involvement in stimulus control still does not enable us to distinguish between the cortical and hippocampal branches of the ventral system.

The hippocampal circuit ascends from the ventral tegmental area via the lateral preoptic, the diagonal band and the medial septal nuclei to the hippocampus. Although there is little evidence for the involvement of the lateral preoptic nuclei, there is strong support for the importance of the medial septal nuclei and hippocampal formation in mediating stimulus control. The effects of lesions in the hippocampal and septal areas closely parallel the changes obtained after attenuation of cholinergic function.

Single intraperitoneal injections of anticholinesterases, like physostigmine, produce increased hippocampal theta activity which is analogous to the increases resulting from sensory stimulation (Stumpf, 1965). Bures, Bohdanecký and Weiss (1962) injected doses of $0.5-1.0$ mg/kg i.p. of physostigmine and found that theta activity began at 3–4 min and was fully developed in 10–30 min. Animals, given the same doses, were placed in a passive avoidance situation and at 3–4 min there was facilitated passive avoidance, but after 10 min there was impaired performance. A lower dose (0.35 mg/kg) was found to block responding in active avoidance during marked theta activity (Erickson and Chalmers, 1966). In contrast to the intensification of theta activity produced by acute injections of anticholinesterases, doses of cholinolytics, like atropine and scopolamine, abolish theta activity completely, and a close correlation has been found between the abolition of theta activity and the disruption in the acquisition and stable performance of passive avoidance in rats (Bures et al., 1962).

In view of the similar origins of reticulo-hippocampal and reticulo-cortical pathways, it is not surprising that there is usually a strong correlation between the hippocampal activity and neocortical activity (Stumpf, 1965). In addition it has been thought for some time that desynchronization of the cortex was mediated by a cholinergic system (Funderburk and Case, 1951). It has been shown that atropine induced cortical synchronization and with higher doses there was an abolition of the response to sensory stimuli even though the animal was behaviourally awake, while the opposite effects were produced by physostigmine which induced a cortical activity characteristic of the aroused animal. These effects were demonstrated in cats (Bradley and Elkes, 1953; Funderburk and Case, 1951), dogs (Wikler, 1952), rabbits (Rinaldi and Himwich, 1955), monkeys (Funderburk and Case, 1951) and man (White, Rinaldi and Himwich, 1956; Ostfeld, Machne and Unna, 1960; Grob, Harvey, Langworthy and Lilienthal, 1947).

Hence attempts have been made to localize the cholinergic neurons involved in the electrocortical arousal responses. Rinaldi and Himwich (1955) infused acetylcholine into the carotid of a rabbit with a postcollicular postpontine transsection (cerveau isolé), which abolished the afferent input, and showed that a cortical arousal response was produced. This effect could be blocked by intraperitoneal injections of atropine. No effects were observed in an isolated cortex preparation, showing that the cholinergic arousal was dependent on afferents from subcortical structures and, from others studies, this region appears to be the midbrain reticular formation (e.g. Smirnov and Il'yutchenok, 1962; Kanai and Szerb, 1965). Thus stimulation of the tegmental nuclei induces electrocortical arousal and a large increase in the acetylcholine released at the cortex (Kanai and Szerb, 1965). From microelectrophoresis it appears that about 15–25 % of the cortical cells, mainly in the sensory cortex, are excited by direct application of ACh and this excitation is blocked by atropine and potentiated by physostigmine (Krnjević and Phillis, 1963a and b). The latter authors concluded that desynchronization was due to an increase in the random spontaneous activity rather than acting as a mediator from the specific sensory terminals. Recently Krnjević and his coworkers have confirmed the excitatory action of ACh on cortical neurons with intracellular recording (Krnjević, Pumain and Renaud, 1971). It appears that ACh lowers the resting membrane conductance of the cortical cells for potassium as well as the delayed potassium currents associated with the cortical action potential. The net result of these changes is a 'marked enhancement of the excitatory action of other inputs, with a particularly striking prolongation of evoked discharges'. (p. 265 Krnjević et al., 1971). It seems reasonable that any mechanism which prolongs the after-discharge of liminal evoked potentials will increase the probability of them being detected, and indeed Spehlmann (1969) observed that acetylcholine applied to the visual cortex prolonged both the discharge after mesencephalic reticular stimulation and the after-discharge of stimulating the visual afferent pathways, while atropine had the opposite effects on the evoked potential discharge. In addition, muscarinic cholinolytics increase the threshold of facilitation of neuronal discharge by electrical stimulation of the midbrain reticular formation (Il'yutchenok and Gilinskii, 1969).

The studies discussed so far have only demonstrated that cholinergic drugs modify the function of cortical neurons important in the response to sensory input and that activity of these neurons is controlled by the afferents from the tegmental nuclei in the reticular formation. The next question is whether this pathway is cholinergic as well. One bit of evidence is the heavy staining for cholinesterase along the ascending pathway from the ventral tegmental region to the cortex (Shute and Lewis, 1967). More critical evidence comes from a series of experiments on anaesthetized cats which examined the effects of cholinergic drugs injected into various brain regions and on electrical activity in the cortex (Endroczi, Hartmann and Lissák, 1963). Injections of carbamylcholine chloride and physostigmine into the dorsal and ventral tegmental areas resulted in neocortical desynchronization for up to $\frac{1}{2}$ hr, and also gave rise to marked hippocampal theta activity for 5–10 min. Similarly, injections into the medial septal region induced marked theta activity. Injections of acetylcholine directly into the dorsal hippocampus of the cat, which is equivalent to the ventral hippocampus of the rat, did not induce theta activity but did produce transient desynchronization in the cortex, i.e. the effects were qualitatively similar to cholinergic injections into the tegmental regions showing the relations between the hippocampus, septal area, tegmentum and cortex.

Cholinergic injections close to ascending reticular pathways

Early work on the effects of centrally injected cholinergic drugs on behaviour were carried out by Feldberg and Sherwood (1954). In a set of observational studies they found that injections into the lateral ventricles of either acetylcholine or anticholinesterases like DFP and physostigmine resulted in dramatic changes in alertness culminating in a catatonic state with high doses and no observable changes with low doses. Injections of atropine 'produce increased liveliness, but with signs that the cat's appraisal of its surroundings was impaired' (Feldberg and Sherwood, 1955). In the latter article the authors suggest that the catatonic effects may have been due to a depolarization block of the cholinergic synapses. From this proposal it would be expected that enhanced awareness of the environment would have occurred with the appropriate low dose of acetylcholine, but there was no evidence presented that support this conclusion. The problem of intraventricular studies like these and those of Warburton (1969) discussed earlier are that they are relatively nonspecific in the structures on which they are acting. One large structure close to the ventricles is the hippocampus, and it is not surprising that direct injection into this structure was made to follow up Feldberg and Sherwood's studies.

The pioneering work on direct chemical stimulation of the hippocampus was carried out by MacLean (1957). In these studies he introduced crystalline carbamylcholine chloride into the hippocampal formation of cats. After about 15 min high voltage spiking became practically continuous. During the height of this seizure the cat lapsed into a 'stuporous' state in which it did not respond to environmental stimuli and there was a generalized underreactivity to all forms of moderately strong stimuli. Intense stimulation produced activity which was violent and sudden in its onset and termination, and completely coordinated. When not stimulated it remained in a prolonged catatonic-like stance. These catatonic postures were similar to those reported by Feldberg and Sherwood (1954) after intraventricular injections close to the hippocampal formation. MacLean describes the behavioural changes as a 'loss of awareness during the seizure', and in support of this he cites a conditioning study in which he found that during the seizure the subjects did not respond to the conditioned stimulus, and sometimes were unresponsive to the unconditioned stimulus (MacLean, Flanigan, Flynn, Kim and Stevens, 1956). These experiments show that blockade of hippocampal function by inducing seizures abolished the typical responsiveness to stimuli. These findings were in marked contrast to the effects of lower doses of carbachol and of acetylcholine. In these cases the cats showed intensified responsiveness to stimulation in the form of enhanced pleasure and grooming reactions, which included loud purring to stroking, rolling over on back and rubbing the head and body against the experimenter and inanimate objects. There was also enhanced receptivity to genital stimulation.

In a similar fashion, Warburton and Russell (1969) followed up the study of the effects of intraventricular injection of cholinergic drugs on single alternation in rats (Warburton, 1969) by injecting the drugs into the hippocampus using the same behavioural test. In this study it was found that injections of 0·43 µg of carbachol improved well-trained single alternation responding in the sense of reducing the intertrial and no-go trial responding, while the reinforced responding remained unchanged and opposite effects were obtained with atropine (see Figure 22). The dose of physostigmine used had variable

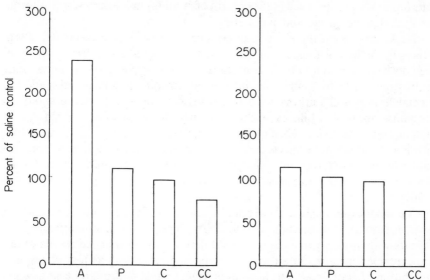

Figure 22. The effects on stable single alternation performance, measured in terms of errors (left) and intertrial responses (right), of injecting atropine sulphate (A), carbamylcholine chloride (CC), physostigmine (P) or saline (S) into the hippocampal formation. From Warburton, D. M., and Russell, R. W. (1969). *Life Sci.*, **8,** 617–627. Reproduced by permission of Pergamon Press

results in the sense of improving the animals with the poorer performances, and impairing the performance of those with better baselines. This fits in with the previously mentioned data obtained with intraperitoneal injections of physostigmine (Warburton, 1972; Warburton and Brown, 1972). It is also consistent with a study of the effects of a cholinesterase inhibitor, di-iso-propylfluorophosphate, on discrimination learning after minimal training and overtraining (Leibowitz, 1968). It was found that the number of discrimination errors was reduced when the injection was given after minimal training, but that errors were increased by the injection in the overtrained subjects. Thus we also have both facilitation and impairment with the same dose of the drug depending on the pre-drug performance of the subjects.

A related example of this type of variation has been observed in different strains of mice. It was found that C57BL mice had greater exploration scores measured by rearing, leaning against the wall and sniffing, than the DBA strain (van Abeelen, Gilissen, Hanssen and Lenders, 1972). Intrahippocampal injections of neostigmine reduced the scores for both strains. In contrast it was found that injections of scopolamine methyl bromide into the hippocampus increased the exploration scores of the DBA mice and reduced them in strain C57BL The same results were obtained with intraperitoneal injections in these strains (van Abeelan, Smits and Raaijmakers, 1971). Thus once again the effects of the drug were correlated with the base-line performance of the subjects, showing that the results obtained with intrahippocampal injections parallel the findings and suggesting very strongly that the cholinergic cells in the hippocampus involved in the control of stimulus processing were important in both discrimination and exploration.

As we have seen, there is strong evidence that the origin of the cholinergic fibres in the hippocampus is in the medial septal area, and injections into this region have been made by Grossman and his coworkers. In the initial study (Grossman, 1964b) found that intraseptal atropine minimized the effect of septal lesions and improved the performance of a lever pressing avoidance response in rats, while carbachol impaired lever pressing. Much greater impairment was found when carbachol injected rats were tested in a two-way avoidance situation. Subjects failed to jump during the conditioned stimulus, although they did escape immediately from shock. These results were confirmed by studies of cats (Hamilton and Grossman, 1969; Kelsey and Grossman, 1969).

A crucial experiment in these studies demonstrated that one-way avoidance was impaired by intraseptal atropine, like septal lesions, in contrast to the results with two-way avoidance. This difference shows that the results cannot be explained in terms of either a learning impairment or as loss of response inhibition. Let us consider how impairment of stimulus processing would explain this data. In his recent interpretation of avoidance behaviour in rats, Bolles (1971) has argued that a motor pattern can only become an avoidance response if it is based on a species-specific defence reaction. Flight from stimulus events and particularly geographical locations that elicit fear is a species-specific defence reaction of the rat, and so one-way avoidance is learned easily. Two-way avoidance in a shuttlebox introduces an element of conflict because the avoidance response involves movement towards stimuli previously associated with shock. However, scopolamine reduced this conflict by impairing the discrimination of the shock associated stimuli, whereas in one-way avoidance impaired discrimination will obviously lead to performance decrements by reducing the probality of detecting the warning stimuli, and also leading to increased intertrial responding.

In the only discrimination study on the ventral tegmental region (Warburton, 1972), a solution of carbamylcholine chloride and atropine methylbromide were injected bilaterally into, or close to, this area of the midbrain reticular formation, and the effects of these drugs examined using stable go–no go alternation performance as in the experiments on intrahippocampal injection. Unfortunately, the cannula itself produced lesion effects, resulting in an impairment of responding, so that the data in Figure 23 are presented as percentage of the saline injected baseline. The effects of atropine methlbromide were unequivocal with an increase in both the internally cued no-go responding and the externally cued ITI responding at the 4, 8 and 12 µg doses. The effects of carbachol were less clear cut; at the lowest dose, 0·10 µg, some animals showed a decrease in errors, i.e. no-go responding and ITI responding, but, at the two higher doses, 0·20 and 0·40 µg, there were decreases in the reinforced, exterocep-

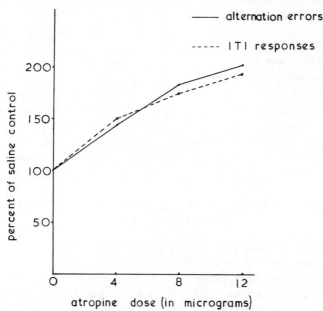

Figure 23. The effects on stable single alternation performance of injecting atropine sulphate close to the ventral tegmental region. From Warburton, D. M. (1972). In R. Boakes, and M. S. Halliday (Eds.) *Inhibition and Learning.* Reproduced by permission of Academic Press

tively cued, go responding as well. The interpretation of the atropine effects can be made simply in terms of an impairment of stimulus control by the interoceptive and the exteroceptive cues, and the improved performance with the lowest dose of carbachol is consistent with this finding. The impairment with higher doses could be due to a blockade of neural function by the carbachol, but this is unlikely in view of the effectiveness of these same doses in eliciting drinking (e.g. Russell, 1966) and in stimulation of the hippocampus, (Warburton and Russell, 1969).

Cortical filtering of stimuli

The results of the last section demonstrate that injections of cholinergic drugs into the hippocampus, septal area and ventral tegmental region parallel the behavioural effects of the same drugs injected intraperitoneally. Behavioural analysis showed that these effects could be best interpreted in terms of a modification of the normal processing of stimulus selection. Electrophysiological studies demonstrated that the behavioural changes after cholinergic drugs were correlated with changes in electrocortical arousal. Precisely the same changes in arousal resulted from intraperitoneal injections and direct injections into the hippocampal formation, septal area and ventral tegmental region. Specifically, injections of drugs enhancing cholinergic function produced neocortical desynchronization, while cholinolytics induced cortical synchrony. One way of relating these findings is to hypothesize that increases in activation of the ascending reticular pathways induces cortical desynchronization, which interacts with the sensory evoked potentials. Small increases in activation will enhance the probability of responding to the larger evoked potentials by masking the smaller ones. However, high levels of reticular activation result in occlusive interaction with the larger evoked potentials (see Bremer, 1961) impairing responding. This latter suggestion would explain the experiment where the ventral tegmental area was stimulated with large doses of carbachol and impaired performance was produced. In contrast, a reduction of activity in the ascending cholinergic pathways by cholinolytics will reduce the desynchronization and enable irrelevant stimuli to influence behaviour.

E. THE EFFECTS OF CHOLINERGIC COMPOUNDS ON HUMAN BEHAVIOUR

Throughout the ages man has been concerned about his perception of external events, and the factors which influence the selection of information from the world and from within himself. This selective aspect of perception has been termed 'attention' and historically it has been linked to various subjective concepts like of 'consciousness' and 'awareness'. Man's attempts to manipulate his own awareness of the world has led him to experiment with chemicals of many kinds. One group of compounds that has been used for thousands of years are the hallucinogens. The most widely used of these agents are the cholinolytic hallucinogens atropine, hyocyanine and scopolamine.

These agents are found naturally occurring in the botanical family, Solanaceae, and various genera grow profusely in Africa, North and South America, Asia, Australia and Europe. As a consequence, the powerful effects of the cholinolytic hallucinogens have been used by cultures on all these continents. Accidental poisoning by one of the solanaceous plants is a terrifying experience for the victim, and any onlookers. Typically, he has bulging eyes, flushed face, marked agitation verging on mania, disturbed speech and multimodality hallucinations, loss of attention and finally lapsing into sleep with terrifying dreams. It is not surprising that the use of the solanaceae have been used by the shamans of South America, the Aztec priests, the Navaho medicine men (Schultes, 1969), the Roman oracles, the sorcerers in the Middle Ages, the assassins in Persia (Lewin, 1923 reprinted in 1964) and even American teenagers in their search for 'mind expanding' agents (Goldsmith, Frank and Ungerleider, 1968). Lewin (1923, reprinted in 1964) has gone so far as to suggest that the deployment of these agents has changed the course of history on a number of occasions, and it has been said that it was used in the wooing of Caesar by Cleopatra! Some examples of behaviour after voluntary or accidental ingestion of solanaceous material will be given in Chapter 7.

Studies of the effects of cholinolytics in man

Early studies of cholinolytics were initiated because of the possible use of these compounds as antidotes for anticholinesterase chemical warefare agents, the 'nerve gases'. In one of these studies it was found that doses of 1·95–9·75 mg intramuscularly produced a flattening of the electroencephalogram with increased synchronization, and at the same time the subjects showed loss of attention, (White, Rinaldi and Himwich, 1956). In an early study Miles (1955) found that 2 mg of atropine intramuscularly impaired performance on a simple reaction time where the subject had to react either a tone or a light, but enhanced performance in the comparable choice reaction time. Hypotheses which might account for this effect would be that either atropine increased 'switching of attention' or it increased 'focus of attention'. These notions were tested in experiments by Calloway and by Ostfeld to be discussed next.

Calloway and Band (1958) injected volunteers with 2 mg of atropine intramuscularly and gave them a battery of objective and subjective tests with each subject serving as his own control in some tests. In the Stroop test the subject was given a number of cards with the names of colours printed on them in black and asked to read them as quickly as possible. He was then given a series of cards with a number of coloured spots on them and asked to identify them. A third set of cards had colour names on them printed in colours different from the name, e.g. RED printed in green, and he is required to name the colour ignoring the printed words. The score was the difference between the times for the second and third list. The effects of atropine were to increase the mean time required to read the third list by a small amount, and a Wilcoxon test showed an increase in scores in the atropine group for the twenty subjects ($p = 0·051$).

In the Luchins test, which consists of problems of how to obtain a given volume of water with three different measures, it was found that atropine improved performance when the mode of solution was changed. In other words, subjects receiving atropine discovered the new short method of solving the problems faster than the control group. Calloway and Band (1958) claim that atropine improved performance by broadening attention, so that the subject attended to aspects of the task which were not essential for the original task, but helped in discovering the simpler method. Broadened attention will impair performance on problems where irrelevant stimulus information will interfere with the solution, like the Stroop test. This evidence, while not impressive, is consistent with the hypothesis of changes in attention induced by this drug.

A more convincing study using both behavioural tests and electroencephalographic responses was performed by Ostfeld, Machne and Unna (1960). The dose of 10 mg of atropine orally impaired recent memory and attention span was reduced to 15 sec or less, so that remembering three objects and repeating them and recalling a series of numbers could not be performed. However, all subjects could answer simple questions and perform tasks which did not require prolonged attention or memory span.

The onset and duration of these behavioural changes were correlated with the occurrence of EEG synchronization (see Figure 19). In addition the duration of the cortical response to flashes of light was reduced to about half the pre-drug magnitude by $\frac{1}{2}$ hr after the dose was ingested. In a companion study of scopolamine (Ostfeld and Aruguete, 1962) found a significant impairment in the Stroop test, supporting Calloway and Band's finding with atropine, reading memory, associative memory in card object recall at both 10 sec and 60 sec. In addition, subjects were impaired on a buzzer press test where the subject was required to hold down a button for 1 min. The number of times the buzzer stopped was a measure of the subject's 'attentional set', and it was found that scopolamine increased this score significantly. Ostfeld and Aruguete (1962) concluded that the results of these studies showed that cholinolytics interfere with the capacity to maintain an attentive set rather than an inability to focus attention *per se*.

The effects of anticholinesterases in man

The first anticholinesterases were synthesized about 120 years ago, but the pharmacological properties of these synthetic compounds did not become clear until about 1930, in spite of their extreme toxicity (Holmstedt, 1963). In the next 10 years considerable research was devoted to the discovery, study and manufacture of organophosphorus anticholinesterases for military use in Germany and later in Britain, the United States and Soviet Union. The pharmacological effects of these compounds turned out to be cholinesterase inhibition similar to physostigmine. Most of these agents developed for military use, such as GB (Sarin) and VX (probably ethyl S-dimethylaminoethylmethylmethylphosphothiolate), are lethal within minutes in doses of 2–10 mg for men, and their effects on behaviour are academic. More recently a number have been used, e.g. parathion and malathion, as insecticides, and acute poisoning by these compounds is characterised by giddiness, withdrawal, anxiety, uneasiness, restlessness and anorexia. If exposure was moderate, these symptoms were followed by a headache, insomnia or sleep with excessive dreaming. At higher doses there was difficulty in concentration, confusion, memory disturbances, and occasionally disorientation, (Grob and Harvey, 1958). The behavioural effects of the organophosphorus compounds were correlated with changes in EEG, such as increased frequency and amplitude, as well as irregular rhythm (Grob, Harvey, Langworthy and Lilienthal, 1947). These observations give some direct evidence in support of the hypothesis that there is a cholinergic system in man involved in attention. Experiments in the Soviet Union showed that small doses of anticholinesterases produced an 'inability to concentrate attention' in arithmetic tests, and the hallucinations produced by cholinolytics (Michelson, 1961) suggesting that small increases in acetylcholine might facilitate performance, but no direct evidence from studies of anticholinesterases have supported this possibility.

One compound which is known to release brain acetylcholine is nicotine (Armitage and Hall, 1967; Holmstedt and Lundgren, 1967). Injections of nicotine produce electrocortical arousal in the cat (Armitage, Hall and Morrison, 1968) and smoking produces similar effects in man (Hauser, Schwartz, Ross and Bickford, 1958). It is significant that many smokers claim that their concentration and efficiency are improved after smoking a cigarette. More concrete evidence comes from a visual reaction time test. Frankenhaeuser, Myrsten, Post and Johansson (1971) found that cigarettes spaced 20 min apart prevented the increase in reaction time that normally occurs with testing for 80 min under monotonous conditions. In tests in a simulated driving situation, performance was maintained at a higher level than controls by cigarettes (Heimstra, Bancroft and DeKoch, 1967).

These data were sufficiently encouraging for us to test subjects in a prolonged signal detection task which paralleled the rat studies (Brown and Warburton, 1971; Warburton and Brown, 1971; 1972) discussed in Section B. Subjects were required to detect pauses in the sweep of a second hand of a clock. It was found that detection performance was improved by smoking, and that there was less deterioration in detection over the test period, compared with controls (Wesnes and Warburton, unpublished). This result is consistent with the previous studies with rats that showed small increases in brain acetylcholine levels improved stimulus detection performance. It must be pointed out, however, that nicotine also acts on both the norepinephrine and serotonin systems in the brain, and also increases plasma epinephrine (Frankenhaeuser, Myrsten, Johansson and Post, 1970), and the possible influence of these on performance must be borne in mind. Nevertheless, these data taken in conjuction with the other evidence suggest very strongly that an acetylcholine system in the human brain is important in attention. The possible importance of this system will be discussed further in Chapter 7 on hallucinations.

5
Motor Control

Observable behaviour consists of contraction of skeletal muscles in appropriate temporal and spatial patterns. In this chapter we will outline some of the neurochemical systems involved in programming these muscular contractions. In the first part the excitation of muscles by the motoneurons will be outlined and in the second some aspects of central control will be considered. The sections on neuromuscular function and spinal control have been included because of their importance as confounding factors in studies of the involvement of the cholinergic system in attention, mentioned in Chapter 4, and also as we shall see in the interpretation of cholinergic drugs on the central cholinergic motor system.

A. PERIPHERAL MUSCULAR CONTROL

The progress of our understanding of skeletal neuromuscular transmission is one of the exciting stories of modern neurophysiology. The key papers in the development of ideas in this field have been reviewed in Eccles (1964) and summaries of the research can be found in all neurophysiology texts and most physiological psychology texts. Thus it will not be necessary to go into great detail about the morphology of the process and the major emphasis will be placed on its pharmacology.

The cellular unit of muscular contraction is the muscle fibre. This threadlike fibre can be as long as 30 cm but on contraction it will shorten to about 57 % of its resting length. Single fibres never contract, instead small groups of them contract together because they are innervated by the terminals on branches from the same motoneuron. The number of fibres innervated by one axon depends on the fineness of the movement controlled (see Basmajian, 1972). The cell body of the motoneurons are located in the grey matter of the anterior horn of the spinal cord. If one of these generates an action potential then a group of muscle fibres will contract and it has proved possible to record the muscular activity electrically (see Basmajian, 1972). In general it is necessary to activate two or three of these motor units to give a visible movement, but the contraction of one motor unit can be recorded electromyographically. The junction between the nerve and muscle was shown to have special properties by Bernard (1857), in his work on curare. He demonstrated that curare interfered with muscular contraction triggered by the motoneuron but not when the muscle was stimulated directly. In the first decade of this century Langley found that drops of weak nicotine solutions initiated a contraction only when it was applied close to the nerve terminals (Langley, 1905). The concept of neuromuscular transmission as a chemical process was finally demonstrated by Dale when he showed by perfusion that acetylcholine was released by the nerve terminals on the muscle (Dale, Feldberg and Vogt, 1936).

At the beginning of the 1950's techniques of microelectrical recording had advanced so that it was possible to record the arrival of the action potential at the terminal and the generation of a muscle impulse. Kuffler (1949) demonstrated that there was nearly 1 msec delay between the two events, which confirmed that there was a transmitter substance released and it seems logical to suppose that it is acetylcholine. Studies of the junction by Katz, Fatt and del Castillo have shown that about 10^{-18} moles of acetylcholine are required to evoke a muscle action potential (Fatt, 1959). In the absence of a presynaptic potential there is still release of small packets (or quanta) of acetylcholine from the motor nerve terminal producing spontaneous miniature potentials and it is believed that the presynaptic impulse, by depolarising the motor nerve terminal, increases the probability of quantal release by a factor of several hundred thousand (Katz, 1958a). This transmitter diffuses across the junction and interacts with the receptors which are localized under the terminals on the external surface of the membrane (Katz, 1958b; Fatt, 1959). As a result chemical agents may diffuse into the junction and modify the mechanism of receptor interaction. Soon after the interaction the acetylcholine is hydrolysed by acetylcholinesterase which appears to be located nearby, but not actually under, the terminal. The post synaptic potential lasts about 10 msec and triggers muscle potentials which produces a minute contraction. If there is a high frequency of impulses down the motoneuron then there will be a series of summating contractions that will increase muscle tension. There is coordination of antagonistic muscles so that increased tension in one muscle is generally accompanied by relaxation in the opposing muscle and so there is a change in position of some body structure. It is clear from what we have said that neuromuscular function and so movement can be modified by chemical agents that act on the cholinergic system.

Increased cholinergic function at the neuromuscular junction

The major technique for increasing the functional amount of acetylcholine at the neuromuscular synapse is by preventing inactivation of the transmitter by acetylcholinesterase. It is known that this enzyme is located on the postsynaptic membrane close to the receptor sites (Barnnett, 1962) so that it is readily available for hydrolysis. Numerous drugs interfere with the hydrolysis of acetylcholine and we listed some of the major ones in Chapter 4, and it was noted in that discussion that one of the major sources of confounding in the interpretation of the central effect of anticholinesterases was their effect on the neuromuscular junction. The method of distinguishing the central and peripheral effects is to compare acetylcholinesterase agents which do and do not pass the blood-brain barrier or to use a peripheral blocking agent like a quaternary cholinolytic. Thus it has been observed that motor behaviour in cats is severely disrupted by neostigmine, a quaternary acetylcholinesterase that does not pass the blood-brain barrier, and as a result there was an impairment of shuttle box avoidance (Funderburk and Case, 1947). These effects could be reversed by quaternary cholinolytics as well as tertiary cholinolytics showing that it was a peripheral effect. A review of situations demonstrating the toxicity of anticholinesterase agents in animals is given in Bignami and Gatti (1966).

In humans the toxicity of anticholinesterase compounds has been investigated extensively for chemical warfare. These agents, or nerve gases, are organophosphorous esters and were developed by Schrader in Germany as insecticides. The major agents developed were tabun (known as GA in the United States) and sarin (GB) which were at least 20 times as toxic as phosgene used in World War I. Soman (GE) was also developed in Germany, but was not manufactured as its toxicity was only comparable to sarin. Since 1950 a new range of anticholinesterase agents, the V compounds, were developed that were more lethal—12–10 mg depending on the site of application—because they can be absorbed through the skin due to their low volatility. Their chemical formula is secret but the World Health Organization suggests that VX may be ethyl.S-diamethylamino-ethylmethyl-phosphonothialate (Meselson, 1970). This compound achieved notoriety because of an accident at the Dugway Proving Grounds when freak weather conditions over the testing zone resulted in VX being carried away from the testing grounds and killing sheep in Utah (Boffey, 1968). One of the first signs of poisoning in animals by all these anticholinesterases like VX is the muscular impairment. The surface skeletal muscles fasciculate, i.e. violent twitching in resting muscles. The normal muscular response is converted to a tetanic response due to the prolongation of the post-synaptic potential by the anticholinesterase. In resting muscles there are also fibrillations which appear as a faint, rapid, uncoordinated flutter in the whole muscle that results from the enhancement of the miniature potentials. As a consequence of this muscular impairment the animal shows jerky movements and a staggering gait; it was these early symptoms which led to an investigation of the sheep poisoned in Utah. The general effects are similar in all species but it is interesting that there are quantitative differences in the muscles impaired, that different muscles are affected in different species and that different compounds modify different muscle groups. In general the effects of low doses are confined to the face and neck and at higher doses the muscular weakness extends to the extremities. At lethal doses the respiratory and heart muscles become affected and all reflexes are abolished (Grob, 1956).

The symptoms of nerve gas poisoning have been emphasized to show the consequences of impaired neuromuscular function, and also because a number of related agents are in use as insecticides. These include parathion, malathion, tetraethyl pyrophosphate (TEPP), hexaethyl tetraphosphate (HETP) and exposure may occur during spraying, during the harvesting, or as a result of ingesting sprayed vegetable produce which has been insufficiently washed (Grob, Garlick, Merrill and Freimuth, 1949). Antidotes for this sort of poisoning are a combination of atropine and pyridine 2-aldoxime methiodide (PAM, PZ-AM (Proloxime)). The oxime acts by reactivating the phosphorylated enzyme and is a rare example of a drug engineered specifically for a purpose

The anticholinesterases have been used medically in the control of myasthenia gravis. In this disorder all movements result in muscular weakness and there is effectively paralysis. After rest, movement can be resumed at its original intensity so that muscle strength is greatest in the morning, but fatigue reoccurs quickly. It is now known that the disorder is the result of decreased functional acetylcholine at the neuromuscular junction, probably due to inadequate synthesis of the transmitter presynaptically (Desmedt, 1958). Anticholinesterases, by slowing up the inactivation of the acetylcholine, make more available for interaction with the postsynaptic receptors. In a comparison (Harvey, Lilienthal, Grob, Jones and Talbot, 1947) of the effects of neostigmine, which does not pass the blood-brain barrier, and di-isopropylfluorophosphate, which does enter the brain, showed that both improved the disorder by increasing motor power, as measured by grip strength, but that neostigmine was the more useful therapeutic agent because it did not produce the unpleasant central effects discussed in Chapter 4 and Chapter 7.

Decreased cholinergic function at the neuromuscular junction

In Chapter 1 it was mentioned that there were two main types of postsynaptic receptor at cholinergic synapses and that these were muscarinic and nicotinic receptors. Atropine and scopolamine are antimuscarinic compounds but it has been known since Langley's experiments in 1905 that the neuromuscular junction is nicotinic and it is not affected by atropine but by curare.

Curare is one of the best known of the non-depolarising neuromuscular blockers. It was brought to Europe from Guiana by Sir Walter Raleigh in 1595, and it was used throughout South American by the Indians as an arrow poison. When introduced subcutaneously, it causes a loss of proprioceptive tonus, then a loss of the voluntary muscle tonus, and finally loss of the respiratory movements. The first muscles affected are the short muscles of the toes, ears and eyes; then the limbs and finally the thoracic, diaphragm and abdominal muscles. At first there is a loss of capacity for sustained effort, then there are smaller contractions until these disappear completely. Death occurs from respiratory paralysis, but this can be blocked immediately by physostigmine doses. Curare-like drugs have played an important part in the development of learning theory, because they were used to see whether it was necessary for responses to occur in order for conditioning to take place, and it has been shown that learning without responding does occur using d-tubocurarine (e.g. Black, 1958). Botulinum toxin formed by the anaerobic bacteria, Clostridium botulinium is another naturally occurring neuromuscular blocker. It is worth mentioning in this section because of its mode of action. It produces muscular paralysis after a latency of about 12 hr by drastically reducing the probability of releasing quanta of acetylcholine. It does not affect the size of the miniature end plate potentials, showing that it is not blocking postsynaptic receptors (Thesleff, 1960). The contraction produced by direct stimulation of the muscle is unchanged. Physostigmine is not an antidote for any of the effects. Much the same pattern of muscular impairment as curare has been noted. It has no psychological or pharmacological use, although its use for military purposes has been investigated (Polley, Vick, Cinchta, Fischetti, Macchitelli and Montarelli, 1965).

B. SPINAL MOTOR CONTROL

The integration of information necessary for the co-ordination of muscles occurs in the spinal cord, and a full discussion of various sorts of information can be found in Ruch, Patton, Woodbury and Towe (1965). It will be centred on the most intensively investigated synapses in the central nervous system, which are the terminals on the Renshaw cells in the spinal cord. These are interneurons in the anterior horn of the spinal cord and investigations by Eccles have shown that they are activated by collaterals from the motoneurons (Eccles, Fatt and Koketsu, 1954) and the Renshaw cell itself makes inhibitory synaptic contact with motoneurons in the spinal cord which activated it. Dale (1935) proposed that nerve cells liberate the same transmitter at the synapses of all of its collaterals and this would be expected if the cell is maintaining any sort of metabolic unity. Since the terminals of the motoneurons release acetylcholine at the synapses on the muscles, it would be expected that the recurrent collaterals to the Renshaw cells would be cholinergic as well. Initial pharmacological evidence by Eccles et al. (1954) supported this expectation (Curtis and Eccles, 1958) by showing that it was excited by acetylcholine and carbachol introduced by microelectrophoresis close to the cell. The function of the Renshaw cell appears to be negative feedback control of the firing in the alpha motoneurone (Eccles, 1957) so that the higher the frequency of firing in the motoneurone the greater the firing in collateral to the Renshaw interneuron. Since it is specific to one segment with no connection to other motoneurons it does not coordinate function but merely prevents convulsive contraction in the muscles. Thus blockade of cholinergic function will tend to induce convulsive contraction in the muscles. More significant for Dale's principle, the Renshaw cell was excited by nicotine in the same way as the neuromuscular junction. The nicotinic properties of the receptors were confirmed by the blockade by dihydro-β-erythroidine, which is specific to this sort of receptor, and it is also blocked by d-tubocurarine. More recently it has been argued that there may also be muscarinic synapses ending on the Renshaw cell whose axons have different origins (Curtis and Ryall, 1966).

Many investigations have been undertaken to discover the other transmitters involved in other aspects of spinal function and Phillis (1970) has summarized these experiments with other putative transmitter substances. It appears that serotonin but not norepinephrine is an inhibitory transmitter on the Renshaw cells. Both norepinephrine and serotonin are localized in the descending nerve fibres from the brain stem, and seem to be mediating the central control of motor function. The adrenergic pathway seems to augment the monosynaptic extensor reflex and depress the flexor while the descending serotonergic fibres seem to increase the excitability in both flexor and extensor motoneurons. The pathways originate in the reticular formation and play an essential part in the control of movement by control of activity in the gamma system (Granit, Holmgren and Merton, 1955). If there is a constant rate of gamma discharge the muscle will maintain a fixed length independent of tension, producing rigidity. From the evidence cited above, the serotonergic pathways may mediate this function so that activity in this pathway will facilitate the myotatic reflexes of the extensor muscles resulting in hypotonus in these muscles. The reticular control of movement was investigated by Magoun (Magoun, 1950) before the discovery of the ascending reticular arousal systems discussed in Chapter 4. Magoun, (1950) stimulated regions in the ventromedial part of the medulla and found inhibition of cortically initiated muscle responses and jerk reflexes. The inhibition was usually bilateral, but with weak stimulation it was ipsilateral. It had rapid onset after switching the stimulation on and responses returned rapidly after current offset. They also described a lateral facilitatory region which had opposite effects (Magoun, 1950). Lindsley, Shreiner and Magoun (1949) mapped out the brain systems involved, in showing that they extended from the frontal cortex to the reticular formation and received inputs from the basal ganglia (see Figure 24). From the reticular formation muscular control was influenced by changing the balance of excita-

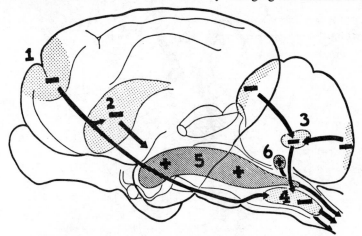

Figure 24. Diagram of the descending excitatory and inhibitory inputs from the frontal cortex (1), the basal ganglia (2), cerebellum (3, 6) and reticular formation (4, 5) to the motor systems of the spinal cord. From Lindsley, D. B., Shreiner, L. H., and Magoun, H. W. (1949). *J. Neurophysiol.*, **12**, 197–205. Reproduced by permission of C. C. Thomas

tion or inhibition on the alpha motoneurons indirectly via the gamma moto-neurons controlling contraction of the intrafusal fibres of the muscle spindles (Ruch, Patton, Woodbury and Towe, 1965). In the next section we will consider the inputs to this descending system from basal ganglia and the neurochemical systems that appear to be involved.

C. CENTRAL MOTOR CONTROL

The control of motor muscular activity is organized in different levels of complexity. In the grey matter of the spinal cord the basic reflexes necessary for coordinated movements are controlled. Many functions like walking are reflexive but cannot function autonomously, and normal control depends on the high levels in the C.N.S. The next highest level in the C.N.S. is the descending reticular formation, and it appears that some of the axons descend from the medulla to the motoneurons in the spinal cord to inhibit the function of the antigravity muscles, extensors, and excite the flexors. More laterally there are axons descending from the medulla which excite the extensors and inhibit the flexors. Some of these functions relate to the reflexive control of balance. The

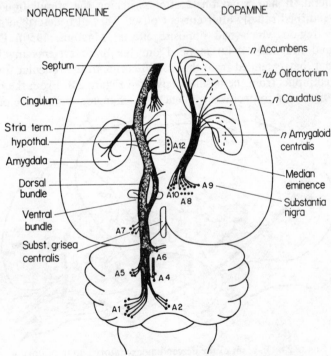

Figure 25. The projections of the catecholamine neurons in the rat brain showing the dopamine neurons ascending from the substantia nigra to the caudate nucleus. From Ungerstedt, U. (1971) *Acta. physiol. scand., suppl.*, **367**, 1–48. Reproduced by permission of the Scandinavian Society for Physiology

motor cortex and cerebellum appear to be involved in the control of voluntary and complex automatic responses, respectively. One part of the motor system has a less obvious function in control, and that is the basal ganglia. This group of structures lie above the thalamus and below the cortex in the forebrain, and consist of the caudate nucleus, putamen and globus pallidus (see Figure 25). The caudate and putamen receive efferents from the cortex and their efferents feed to the globus pallidus. The latter connects with nuclei in the thalamus and the descending reticular formation, especially the substantia nigra and the red nucleus, and these structures with the basal ganglia are frequently referred to as the extrapyramidal motor system (Jung and Hassler, 1959). The early attempts to study the function of the basal ganglia in animals were disappointing because nothing seemed to happen during stimulation. However, when the cortex was stimulated it was observed that the concomitant stimulation of the caudate and globus pallidus inhibited the motor reactions. Stimulation of the caudate abolished the movements, while pallidal stimulation arrested movements in midcourse (Ruch, Patton, Woodbury and Towe, 1965). Lesion studies by Rosvold (1968) have shown that the caudate is part of a motor modulation system involving the prefrontal cortex, and that there seem to be two separate functions. The orbital frontal cortex and ventrolateral caudate appear to control whether or not a response will occur, while the anterodorsal caudate and lateral frontal cortex guide the direction of a response, particularly eye–hand coordination.

Motor disorders

Much of our information on the functions of the extrapyramidal motor systems has come from neurological observations in man, rather than from animal experimentation. The problem of this approach has been that lesions in this region are produced by diffuse pathological changes. Two main classes of motor abnormality have been distinguished: the hyperkinetic-dystonic syndrome and the hypokinetic-rigid syndrome. The following description of these two major groups is taken from Jung and Hassler (1959). The hyperkinetic syndrome is characterised by an excess of motor activity. For example, the choreic syndrome is seen in the irregular occurrence of involuntary jerks or well coordinated fragments of movements. These episodes are of short duration, but can be enhanced by behavioural excitement or even voluntary movements. This disorder is believed to be the result of destruction of cells in the caudate and putamen. In contrast hypokinesis or Parkinson's disease is characterized by an absence of spontaneous, reactive and automatic movements. There is clear poverty of all movement with sometimes a distinctive posture and shuffling gait. There is often tremor when the patient is at rest, but less obvious when the patient is making some voluntary movement. Often when no tremor is visible it can be discovered when he is asked to draw or write. The onset is usually unilateral and most patients can carry on their normal work for some time, except for skilled craftsmen and musicians. Mental alertness remains unimpaired for many years, and many Parkinsonian patients outlive their generations. Thus the disorder is predominantly a motor disability. The tremor can be enhanced by stress and disappears during sleep. The defect is believed to be due to a loss of nerve cells in the substantia nigra (Jung and Hassler, 1959), but other structures may be involved, such as the mesencephalic tegmentum (Ruch, Patton, Woodbury and Towe, 1965). One method of decreasing the rigidity and tremor was a surgical lesion in the region of the globus pallidus (Cooper, 1965) but more recently several sorts of drug therapy have proved effective in alleviating hypokinesia (see Section D). In addition clinical studies have enabled the formulation of hypotheses about the biochemical control of movement, and some information has come from drug-induced motor disorders as well.

Drug-induced Parkinsonism

There are a number of chemicals that can induce Parkinsonism, such as carbon monoxide poisoning, but in this section only the recently observed motor sequelae to chemotherapy of the psychoses will be mentioned. These have been observed in the first groups of patients treated with reserpine and chlorpromazine (Steck, 1954). The 'tranquillizing' effect observed with these drugs were not infrequently accompanied by classical Parkinsonian symptoms of akinesia and hypokinesia, with poverty and slowness of movements and the typical mask-like face. In some patients there may also be hypotonia and the development of this depends on the drug dose and the susceptibility of the patient. Susceptibility to the Parkinsonian side effects increases with age, but there is a marked increase in the occurrence of the signs in those patients over 50 years (Ayd, 1960). Ayd also observed a higher occurrence of the signs in women (47·8 %) than men (29·6 %) with the same oral dose, but no adjustment was made in these estimates for the higher dose per unit weight in women. Steck (1954) argued that dose was a crucial factor in the development of the extrapyramidal symptoms, and that it appeared as if the effects were cumulative. Freyhan, (1956) argued against this notion, but if his 48 cases are examined it can be seen that the probability of the occurrence of Parkinsonian symptoms increase with each month on the drug, although 40 % occur in the first month. Similarly, the probability of occurrence increased as the dose gets larger.

The effects observed with chlorpromazine have been observed with the newer phenothiazines and butyrophenones, like haloperidol, and the development of more potent psychoactive agents has been parallelled by an increasing incidence of unwanted extrapyramidal effects. In particular, the piperazine derivatives, like thioproperazine and fluphenazine, are most likely to produce Parkinsonian signs, and these are the most useful drugs in manic excitement and other agitated states. It is uncommon for these side effects to persist long after medication has been discontinued, so they are linked directly to biochemical changes in the brain and not to drug-induced lesions in the extrapyramidal system. These biochemical effects can be reversed by a number of drugs, including amphetamines and monoamine oxidase inhibitors, tricyclic antidepressants and cholinolytics (Delay and Deniker, 1968). Obviously it would be expected that *l*-dopa would reverse the symptoms, and one has been published (see p. 101).

In summary, the drug-included Parkinsonism seems to be identical with the naturally occurring disorder, and it seems likely that the drugs are merely precipitating the latent disorder by blocking the same pathways associated with the basal ganglia. From the drugs listed it would seem that catecholamine pathways and cholinergic pathways may be involved.

D. PHARMACOTHERAPY OF MOTOR DISORDERS

In this section evidence will be reviewed which suggests that the motor disorders discussed in the last section are the consequence of a disturbance in the balance of activity in an excitatory cholinergic system. The evidence from pharmacotherapy is a little one-sided because the incidence of Parkinsonism is about five times the incidence of Huntington's Chorea.

Parkinson's Disease

The first piece of evidence is that there is a good correlation between the levels of dopamine in the striatum and the amount of degeneration observed in the substantia nigra during autopsies of Parkinsonian patients (Hornykiewicz, 1966). These levels were reduced by about 80 % in the caudate nucleus and putamen, with similar decreases in concentrations of its major metabolite, homovanillic acid. In a recent study, Lloyd and Hornykiewicz (1970) demonstrated that the *l*-dopa decarboxylase activity, the synthesizing enzyme, was greatly reduced in the caudate nucleus and putamen of patients who had suffered from Parkinson's disease, probably as a result of the loss of neurons. Since this enzyme is not rate limiting, it is not the central cause of the disorder and this is supported by the finding that *l*-dopa can be used to alleviate the characteristic hypokinesia (Hornykiewicz, 1966). The problem with this compound as a therapeutic agent is the unpleasant unwanted effects of nausea, gastric distress and blood pressure changes that accompany its use in effective doses. In fact, doses of 12–16 gm or more of *dl*-dopa were needed to produce amelioration of symptoms (Cotzias, Van Woert and Schiffer, 1967). In this study seven out of eight patients recovered with this dosage, while marked improvement was obtained in only one out of eight patients given less than 12 gm, with slight improvement in two more. The use of *l*-dopa, the biologically active isomer, which is converted to dopamine, enabled the doses to be lowered. In a study of the first 60 patients admitted to hospital (Yahr, Duvoisin, Schear, Barrett and Hoehn, 1969), doses of the drug were increased from around 1 gm per day until intolerable side effects developed, or until a daily dose of 8 gm was reached. They were scored daily on rigidity, tremor, speech, facial expression, bradykinesia, posture, gait, dexterity and postural stability on five point scales in addition to handwriting, films and tape recordings of speech. The results were expressed in terms of an improvement score. Thirty-five % of the subjects had shown 50 % or greater improvement in the trial period of 4–8 weeks, and in the follow-up period, 59 % had improved to this extent. The average reduction of rigidity was 72 %, with a 66 % reduction of tremor and less of a loss, 57 %, in akinesia. Replacement of *l*-dopa by placebo capsules resulted in a gradual deterioration over a period of about a week. Restoration of the full dose led to a restoration of the therapeutic effect over the next few days. In comparison with other forms of Parkinsonian therapy, including cholinolytics, used by some of the patients, *l*-dopa was more than three times as effective as the previous remedies. However, some patients exhibited more improvement by combining *l*-dopa with one of the cholinolytics. Biochemical studies in conjunction with these studies demonstrated a marked increase in cerebrospinal homovanillic acid, the metabolite of dopamine, to normal levels or above, suggesting an increase in the synthesis of this transmitter in the brain. In a more sophisticated study of improvements in motor performance, Velasco and Velasco (1973) measured the linear movement velocity by the limbs, rate of alternating finger movements reaction time, amplitude of arm swing and tremor. Lineal movement velocity, the time taken to press two switches separated by a metre with the same finger, and arm swing were significantly improved by *l*-dopa showing that the drug relieved rigidity. In only two cases was reaction time reduced by *l*-dopa, and fine alternating finger movements were not changed, suggesting that *l*-dopa is having a limited effect on akinesia. Tremor also remained unchanged, although the authors admit that the technique might have been insensitive to change, and that periods of treatment may have been too short. In addition, the doses used in this study were lower than those found to be most effective, i.e. greater than 8 gm of *l*-dopa (Yahr et al., 1969).

Intensification of the Parkinsonian symptoms can be achieved by agents which reduce dopamine function, like reserpine, chlorpromazine and alpha-methyltyrosine, which blocks synthesis at the rate-limiting step. From this evidence it looks as if impairment of the dopaminergic system results in the predominance of some other motor control producing rigidity. Some hint of the nature of this system can be found in the older drugs that were used clinically to reduce rigidity. These were the cholinolytic drugs like atropine and scopolamine which block cholinergic neurons. The best results were obtained with a mixture of a cholinolytic and dl-amphetamine (Walshe, 1952). It is now known that both *d*- and *l*-amphetamine are potent inhibitors of dopamine uptake presynaptically (Taylor and Synder, 1970), and this will increase the functional dopamine at synapses. This study is consistent with the idea that there are two antagonistic dopaminergic and cholinergic systems, so that a drug which increases functional dopamine, like amphetamine or *l*-dopa, will add to the antiparkinsonian effect of a cholinolytic drug which decreases the amount of functional acetylcholine. The evidence shows clearly that acetylcholine is the excitatory system while dopamine is the inhibitory one.

There is also fragmentary evidence for a serotonergic inhibitory input to the motor control system. It will also be remembered that in the last section it was noted that tremor disappeared during sleep. In view of the involvement of serotonin (5–HT) in sleep discussed in Chapter 6, it would seem reasonable that there might be an involvement of the serotonin system in motor control. There is some evidence of a decrease of serotonin in the striatum of Parkinsonain patients (Bernheimer, Birkmayer and Hornykiewicz, 1961) suggesting that there may be impaired synthesis of this substance. In line with this notion is the observation that Parkinsonian tremor is improved by 5-hydroxytryptophan, the precursor of 5-HT (Hornykiewicz, Markham, Clark and Fleming, 1970). It should also be noted that reserpine depletes 5-HT as well as dopamine and norepinepherine, which would make this drug very effective as a tremorigenic agent. Although serotonin is implicated in motor control, it is not clear what its precise role is. It may be that it is only coordinating motor function with sleep, or perhaps it is the transmitter in the descending pathways in the spinal chord, mentioned in the last section. A third possibility is raised by pharmacotherapy of chorea.

From the point of view of trying to understand the neurochemical mechanisms involved in normal motor control, it is significant that *l*-dopa therapy itself results in the development of abnormal movements. These have been reviewed thoroughly by Barbeau and MacDowell (1970), but a few of the significant points will be included here. These movements range from generalized choreoathetoid movements to specific dystonia. They can occur during involuntary movements and at rest, and during movement they are not confined to the muscles involved in the activity. Occasionally they may appear only during concentration such as reading, mental arithmetic and controlled movements. They do not seem to be confined to the afflicted side of the body. In general the dyskinesia occurs in patients who have experienced marked relief of the Parkinsonian symptoms, and a reduction in the dose of *l*-dopa to prevent the dyskinesia usually results in decreased benefit to the Parkinsonism. This suggests that it is the balance between two systems which is important, and both hyperactivity and hypoactivity can produce motor dysfunction.

Huntington's chorea

As we stated earlier, this illness is much less common than Parkinson's disease, and so the data is much more limited, but the information is consistent with our hypothesis. In particular, the choreiform movements and the accompanying psychotic behaviour are controlled by phenothiazines including chlorpromazine, and with the amine depleting agents like reserpine (Sim, 1963). In contrast, *l*-dopa and monoamine oxidase inhibitors tend to accentuate the choreiform movements which supports the hypothesis (Hornykiewicz et al., 1970). However, the levels of dopamine and homovanillic acid in the striatum and pallidum of patients who have died from Huntington's Chorea are within normal limits, suggesting there are no abnormalities in the synthesis and inactivation of dopamine (Hornykiewicz, 1966). Administration of 5-hydroxy-tryptophan, the precursor of serotonin that improved Parkinsonian symptoms, exacerbated the motor symptoms of Huntington's Chorea (Lee, Markham, and Clark, 1968), and these authors argue that serotonin was acting on the globus pallidus which has been released from striatal control. It has been suggested that an important factor in the appearance of chorea is an imbalance in the striatal dopamine and serotonin activity with a shift in favour of dopamine (see Lee, Markham and Clark, 1969). However, it is clear from the above data that injections of both serotonin and dopamine precursors aggravate Huntington's chorea and this argues for a cholinergic–non-cholinergic imbalance in chorea with the shift in favour of the non-cholinergic systems, dopamine and serotonin. Thus, Huntington's chorea is in a sense the opposite disorder to Parkinson's disease. If Parkinsonism is due to a deficiency of functional dopamine in the nigro-striatal system modulating a cholinergic pathway, then chorea could be related to an increased effect of dopamine at synapses in the same pathway (Klawans, 1970), or alternatively, hypofunction in the cholinergic pathway. The assays of brain dopamine indicate that it is not hypersecretion of dopamine and Klawans (1970) has argued that the dyskinesia may result from dopamine receptor hypersensitivity. This would fit in with the finding that homovanillic acid, the major catabolic product of dopamine in choreiform patients is normal. No evidence of a disorder in serotonin function has been found, but again the disorder could be explained in terms of increased receptor sensitivity at the serotonin terminals.

Studies of acetylcholine function have also disclosed no obvious evidence that would be expected from a hyposynthesis of acetylcholine (Duvoisin, 1967), although receptor changes have not been examined. The indirect involvement of the striatal cholinergic system in chorea has demonstrated by a comprehensive study by Klawans and Rubovits (1972), in which they tested various anticholinesterases and cholinolytics on hand stability and drawing. Physostigmine improved hand stability and motor control in drawing. The simultaneous administration of scopolamine methylbromide which does not pass the blood-brain barrier did not block the improvement. Edrophonium, an anticholinesterase that did not pass the blood-brain barrier, had no effect. Benzatropine, a centrally acting cholinolytic, worsened the chorea, and this resulted in marked deterioration in drawing ability. Physostigmine reversed the effect of the benzatropine confirming the action of the drugs in the central nervous system. Since the disorder has been associated with nigro-striatal degeneration with atrophy of the caudate and putamen and less often globus pallidus (Jung and Hassler, 1959), it is presumed that this drug effect is occurring in the basal ganglia.

Neurochemical systems

The first suggestion of a brain transmitter system being involved in central motor control came from Carlsson (1959) when he proposed that dopamine is essential for the normal functioning of the extrapyramidal system. This suggestion was based on his finding that the high concentrations of dopamine were present in the caudate nucleus and putamen, in contrast to the very low levels of norepinephrine (Vogt, 1954). In the substantia nigra and pallidum, the dopamine levels are about ten times the levels of norepinephrine. Further evidence for the location of dopamine neurons in the basal ganglia is the presence of the synthesizing enzyme, dopa decarboxylase, in the caudate nucleus (Holtz, 1959) which appears to have high activity and forms dopamine rapidly (Hornykiewicz, 1966). It appears from studies on the subcellular localization of dopamine in the caudate that most of this transmitter is not bound to protein like other transmitters, but is occurring in a free form in the presynaptic cytoplasm prior to release (Laverty, Michaelson, Sharman and Whittaker, 1963). After release it is inactivated successively by MAO and COMT to form homovanillic acid, whose anatomical distribution parallels the distribution of dopamine. From this discussion it is obvious that the biochemistry of this system is similar to that of the norepinephrine systems discussed in Chapter 3.

Neurons containing dopamine appear to have their soma in the pars compacta of the substantia nigra (Dahlström and Fuxe, 1964). It has proved difficult to locate the terminals of these dopamine neurons, but it has generally been accepted that some synapse in the corpus striatum and that most of the dopamine containing axons connect the substantia nigra with the striatum (Dahlström and Fuxe, 1964). This has been confirmed by Ungerstedt (1971) using anterograde or retrograde degeneration after lesions in rats to find the origin and termination of the bundles and Figure 25 summarizes current information. The neurons seem to originate in the zona compacta and ventral tegmental region, ascend through the lateral hypothalamus and ultimately flow out in the globus pallidus and enter the nucleus caudatus putamen. In a complementary experiment in the cat, neurochemical assays showed that the loss of 'dopamine' fluorescence observed by Ungerstedt was parallelled by severe losses in tyrosine hydroxylase, dopa decarboxylase and dopamine in the caudate nucleus after lesions in the nigrostriatal pathway (Moore, Bhatragar and Heller, 1971) and the amount of loss was proportional to the amount of the pathway destroyed. There was no evidence in this study that gross motor activity or reflexes were affected. However, lesions in the nigrostriatal region in monkeys that produced decreases in brain dopamine resulted in hypokinesia with tremor rigidity and akinesia on the side opposite the lesion, similar to human Parkinsonism (Poirier and Sourkes, 1965). It can be concluded from investigations that animal models of motor disorder may throw some light on human motor control. The problem of this work on animal models lies in the similarity of the dopamine and norepinephrine biochemistry, which makes interpretation of the results of intraperitoneal injections difficult.

Studies of the dopamine inhibitory system

Induction of hyperfunction in the central dopamine neurons can be achieved by several drugs, and drug combinations with the most obvious being injecting *l*-dopa This drug was injected into monkeys some of whom had lesions in the ventral tegmental region (Goldstein, Battista, Ohmoto, Anagnoste and Fuxe, 1973). They injected *l*-dopa in combination with *l*-alpha-methyldopa hydrazine (MK 486), which inhibits dopa decarboxylase peripherally, and prevents the drug-increasing dopamine outside the brain. The effects of this were a series of abnormal movements including aggressiveness, chattering, irritability, hypersensitivity and increased drinking. In monkeys who had developed a tremor as a result of a unilateral tegmental lesion, the drug combination relieved the tremor, but induced stereotyped facial movements, hyperkinesia, chorea-like movements, increased grooming, purposeless repetitive hand movements, unusual sitting and walking postures, as well as the behaviour observed in the intact animals given this treatment. Injections of Trivastol, a piperazine deriva-tive which stimulates dopamine receptors, had the same effects as *l*-dopa. A combination of Trivastol with 5-hydroxytryptophan, the precursor of serotonin, reduced the aggressiveness and irritability, but had no effect on the stereotyped facial movements and hyperkinesia. Fuscaric acid, which inhibits norepine-phrine synthesis, had no effect on the Trivastol's effects, showing that the abnormal movements were not a consequence of adrenergic activation. These data show that tremor reduction and the genesis of various abnormal move-ments had a common origin in increases in functional dopamine. This confirms the findings of human therapy with *l*-dopa that dykinesia appeared in every patient who showed improvement in Parkinsonian symptoms, suggesting that tremor results from hypofunction in a dopaminergic pathway, while dyskinesia and hypotonia are a consequence of hyperfunction.

Increase of dopamine function in brain

From the brief biochemical discussion earlier it would be of interest to examine the effect of a monoamine oxidase inhibitor. These compounds would prevent the breakdown of dopamine and so increase the functional levels of this transmitter. In human patients these compounds can produce a choreiform syndrome, characterized by persistent involuntary rhythmic or jerky muscular movements. Hyperkinetic 'side effects' result sometimes from other drugs increasing central dopamine levels including amphetamine and phenothiazines with piperazine side chains (Hornykiewicz, Markham, Clark and Fleming, 1970). These compounds have decreased antiadrenergic properties (Shepherd, Lader and Rodnight, 1963), from which we can infer that they may have weaker antidopamine activity too. The hyperkinesia is most common among children compared with hypokinesia, which occurs with older patients (Ayd, 1960).

Direct injection of dopamine, dopa and d-amphetamine into the caudate nucleus of cats (Cools and Van Rossum, 1970) resulted in a decrease in moving, walking and standing, and nearly all the time the cats were lying and turning the head contralateral to the injection side, with abnormal leg flexion. Increased doses changed the movements from contralateral to homolateral. These effects could be blocked by haloperidol, and it appeared that this was competitive inhibition, probably at the postsynaptic receptor, because larger doses of dopamine would counteract the effect.

Reduction of dopamine function in brain

One of the drugs used has been reserpine, which we have already discussed in terms of norepinephrine depletion in Chapter 3. In addition, dopamine is depleted from the brain, including the basal ganglia (Zbinden, 1960). In human patients given reserpine, one of the most common unwanted effects is the Parkinsonian side effects of tremor, rigidity and akinesia. These effects could be counteracted by subsequent injections of *l*-dopa (Carlsson, Lindquist and Magnussen, 1957). In rats similar effects are observed, and it has been demonstrated that the muscular rigidity is parallelled by an increase in alpha motoneuron activity, and a decrease in gamma fibre activity, and that this effect of reserpine was also abolished by *l*-dopa (Roos and Steg, 1964).

In a survey of human patients it was found that about 40 % of those patients given phenothiazines, like chlorpromazine, and butyrophenones, like haloperidol, suffered from Parkinsonian effects (Ayd, 1960). This sensitivity to phenothiazines is highly correlated with sensitivity of reserpine, and patients over fifty were particularly susceptible. It has been found that *l*-dopa antagonises the effect in some patients (McGeer, Boulding, Gibson and Foulkes, 1961). Chlorpromazine is known to block norepinephrine neurons (see Chapter 3), and it is thought it may block the dopamine receptors as well. *L*-dopa would counteract this effect by increasing the amount of dopamine synthesized, and probably the amount released. From this we can conclude that reduction of dopamine function in the brain impairs muscular control, so that rigidity and hypokinesia is observed in man and rats.

The cholinergic–dopaminergic interaction at the caudate

In Chapter 4 some of the cholinergic pathways (Shute and Lewis, 1967; Lewis and Shute, 1967) ascending from the reticular formation were discussed. One of the systems shown in Figure 21, eliminated from consideration as a pathway controlling attentional effects, was the fibres passing from the ventral tegmental region to the basal ganglia, and from the basal ganglia to the lateral cortex. Shute and Lewis (1967) found that in normal rat brains the caudate-putamen region was so rich in acetylcholinesterase that it was difficult to distinguish fibres from cell bodies. However, lesions in the ventral tegmental pathway from the substantia nigra (pars compacta) resulted in a loss of enzyme in the caudate-putamen, and an accumulation of the enzyme caudal to the lesion, showing that the enzyme was being synthesized in the ventral tegmental region, and being transported by axoplasmic flow to the synapses in the caudate. The striatal radiation to the lateral cortex appears to originate from the globus pallidus which is also innervated by the ventral tegmental pathway. Shute and Lewis give no evidence for a direct cholinergic projection from the caudate to the cortex.

In a perfusion study of the caudate nucleus in cats, McLennan (1964) irrigated the area with a push–pull cannula, and activated the nucleus from different brain structures, and assayed the perfusate for acetylcholine and dopamine. He found increases in the amount of ACh released after stimulating the nucleus ventralis anterior of the thalamus, and increased release of dopamine after stimulation of the nucleus centromedianus of the thalamus. This shows that acetylcholine is present in the region as well as the inactivating enzyme. The next step in a neuropharmacological analysis is to try to evaluate the neural function of the identified compound. In a microelectrophoresis study of the caudate (Bloom, Costa and Salmoiraghi, 1965) it was found that 75 % of the caudate neurons tested were excited by acetylcholine, with several-fold increases in firing rate. Tests in the same region with the catecholamines dopamine and norepinephrine, showed that they depressed the firing rate of caudate neurons with the greatest number being affected by norepinephrine (81 %), while dopamine was less effective. Both catecholamines reduced or abolished the facilitatory responses to acetylcholine.

In further tests of caudate neurons, McLennan found about that the same proportion (60 %) of dopamine depressed neurons (McLennan and York, 1967). They demonstrated that dopamine depressed the firing rate of 80 % of cells excited by stimulation of the nucleus centromedianus, cells which release dopamine. In a related study, McLennan and York (1966) found that cholinergic stimulation of the cells which were excited by stimulation of the nucleus ventralis anterior increased their firing rate, while cells whose firing rates were decreased by thalamic stimulation were depressed by acetylcholine and cholinomimetics. Both sorts of effect were blocked by atropine. In an extension of the latter studies Connor (1970) examined the effects of dopamine on caudate neurons in cats. First he stimulated cell bodies in the substantia nigra and classified the caudate neurons in terms of their changes in firing rate. It was found that the iontophoretic application of dopamine depressed the firing rate of all those neurons depressed by nigral stimulation. Neurons facilitated by nigral stimulation had mixed responses. The depressant effects of dopamine could be blocked by concurrent iontophoresis of alpha-methyl-dopamine, a dopaminergic blocker.

At a neuropharmacological level there is clearly an excitatory cholinergic pathway terminating in the caudate. Parallel to it appears to be an inhibitory dopaminergic pathway. In addition there may be another dopamine pathway from the thalamus, and possibly an adrenergic pathway of unknown origin. The next obvious step in the analysis of the interaction is to evaluate the effects of cholinergic drugs on the caudate nucleus, and the effect of dopamine on the responses to cholinergic drugs.

Injections of acetylcholine and carbachol into the caudate of cats (Connor, Rossi and Baker, 1966; 1967) and rats (Dill, Nickey and Little, 1968) produced tremor. This consisted of fine tremor in the facial and neck muscles and coarser tremor of the hind-limbs and fore-limbs with a frequency which was essentially the same as that in Parkinson patients. Dill et al. (1968) noted that series of effects occurred with the extent of progress through the series, depending on the dose. At low doses there was contralateral hind-limb tremor, hyperextension of the trunk, fore-limb tremor, bilateral forelimb tremor. The gradation of effect for tremor is similar to the scale of effect found for the blockade of abnormal movements by intracaudal dopamine. These tremors of the limbs were in a plane parallel to the body. Injection of carbachol into other parts of the brain, including the dorsal hippocampal formation and mesencephalic reticular formation (no coordinates specified) had no tremor effect. Intraperitonal injections of atropine blocked the tremor induced by carbachol, whereas atropine methylbromide, which did pass the blood-brain barrier, had no effect, showing that carbachol was acting centrally (Connor et al., 1966).

The interaction between the cholinergic and dopaminergic systems was explored by injecting dopamine and other amines into the caudate about 20 min after an intracaudal injection of carbachol, during the period of maximal tremor (Connor, Rossi and Baker, 1967). Dopamine inhibited tremor with a latency of 6 to 9 min, and the antagonism could be achieved by injecting dopamine into either caudate nucleus, ipsilateral or contralateral to the site of the carbachol injection. Dopamine was also effective in blocking involuntary movements occurring in animals not injected with carbachol, but the dose required was about one-quarter that needed to antagonise carbachol tremor. Unfortunately, for a simple interpretation of the data the catecholamines, *l*-isoproterenol, *l*-epinephrine, and *l*-norepinephrine also suppressed tremor, with the first two abolishing tremor in less than 2 min. Noncatechol phenethylamines had no effect. This order of potency suggested that the catecholamines were acting on beta-adrenergic receptors, and this was supported by the antagonism to *l*-epinephrine by beta-adrenergic blockers like propanolol, while phentolamine and chlorpromazine, alpha-adrenergic blockers were not antagonistic. This finding links the behavioural data to the microelectrophoretic study of Bloom et al. (1965) which had found that dopamine was less effective than norepinephrine in depressing spontaneous and acetylcholine-induced firing of the caudate neurons. The most parsimonious interpretation is that although the receptors are beta-adrenergic, the actual transmitter is dopamine, but because of the structure of the receptors the other two catecholamines are 'supermimetics'. Alternatively there may be an adrenergic system involved and Hornykiewicz (1973) has hypothesized that a noradrenergic mechanism determines the sensitivity of the motor effector system through which the dopamine system acts.

Synthesis

In the earlier sections we discussed how the muscles contracted as the result of a cholinergic pathway from the spinal chord to the muscles. The alpha motoneuron which activated the muscles branched, feeding back to the inhibitory Renshaw cell which prevented convulsive contractions in the muscles of a particular segment. The movement of the muscles by the motor cortex was stabilized by at least two mechanisms (Ward, 1968). The stretch of the intrafusal fibres of the muscle spindle in a muscle excites motoneurons of synergistic muscles, and inhibits the antagonists of that muscle. The interneuron mediating the inhibitory control has inputs from higher levels in the central nervous system. In addition, the reticular formation, basal ganglia and frontal cortex appear to be part of a system controlling the gamma motoneurons which cause the muscle spindles to contract, making the intrafusal fibres more sensitive to the muscle stretching and therefore more likely to excite synergists. It is thought that this descending pathway to this inhibitory system may be serotonergic. In the brain one input is from a complex servo-network including the basal ganglia. At the caudate there is convergence of an excitatory cholinergic input and an inhibitory dopaminergic pathway. One candidate for the cholinergic input is the ascending cholinergic pathway to the caudate from the ventral tegmental area, which activates the descending inhibitory system so that during high electrocortical arousal, caused, for example, by novel stimuli there would be muscular rigidity causing the freezing behaviour. In normal circumstances this cholinergic system is modulated by a dopaminergic pathway which originates in the substantia nigra in the tegmentum. The activity in the caudate nucleus controls the globus pallidus that receives a serotonin input and activates the inhibitory alpha system and inhibits the descending gamma system. Steg (1964) showed that reserpine which depletes dopamine abolishes the gamma motoneuron activity, and that l-dopa injections restored it. The abolition of gamma activity is accompanied by increased alpha motoneuron activity in producing rigidity and tremor. It is interesting that injections of 5-hydroxy-tryptophan, the precursor of serotonin, will reverse the rigidity perhaps by restoring activity in the descending input to the gamma systems. This limited summary of the control of movement indicates the complexity of the servo-mechanisms involved. In his discussion of this complexity Ward (1968) states in his conclusions on the function of the basal ganglia that 'insight will grow as these systems are studied as systems rather than as isolated pathways or nuclei, and it is thus less profitable to attempt to describe function in terms of "motor" and sensory components' (p. 111). This viewpoint is in complete accord with the neurochemical network approach outlined in the first chapter.

6
Sleep and Dreams

Sleep plays such an important part in our lives that it is hardly surprising that its mechanisms have intrigued scientists for many years and pharmaceutical firms have spent millions on the discovery of sleep-inducing agents. Barbituric acid was synthesized over 100 years ago and the derivatives of barbital and phenobarbital were used 60 years ago as hypnotic agents. The mechanism of action of these agents and the other 2,500 barbiturates has been the subject of study ever since. The discovery of hypnotic agents lead naturally to the hypothesis that sleep resulted from the build up of a 'hypnotoxin' in the blood-stream as the result of bodily activity and that at a certain level of fatigue sleep is produced. Evidence for this sort of factor came from studies like those of *Monnier and Hosli* (1964) in which dialysed blood from one sleeping rabbit was injected into a free-moving rabbit. They found that sleep could be induced in the rabbit by repetitive electrical stimulation of the intralaminary thalamus. Blood was extracted from the sleeping animal, dialysed to remove the heavy particles, and then injected intravenously into an awake animal. The recipient fell asleep after 10–15 min, but this could be reversed immediately by tactile or acoustic stimuli, as in normal sleep. This study showed that there was some chemical system controlling sleep and offered the possibility of discovering the neurochemical mechanisms involved in sleep. The story of this dicovery will form the major part of this chapter, but first it is essential to consider the neurophysiological developments which were a prelude to this work.

A. PASSIVE THEORY OF SLEEP

The development of techniques for recording the electrical activity of the brain from the scalp enabled scientists to obtain independent evidence of brain function. These early workers were impressed by rhythmic activity present in the brain during sleep, and the organization imposed on this activity by sensory events. They argued that consciousness was a direct result of sensory input breaking up this synchronized activity (Adrian and Matthews, 1934). Thus sleep was a state in which sensation had 'lost its usual control of central neural firing' (Hebb, 1949).

This theory gained support from the famous study of Moruzzi and Magoun (1949), which showed that electrical stimulation of the reticular formation is an encephalé isolé preparation which showed an EEG synchronization resulted in desynchronization. This suggested that EEG activation and behavioural arousal were the result of a tonic flow of ascending impulses from the reticular formation to the cortex. From this dynamic theory of arousal it was logical to suggest that sleep could be explained parsimoniously as a consequence of a diminution of activity in this arousal system (Bremer, 1954). Similarly, it was plausible to argue that barbiturates were hypnotic agents because they depressed the tonic activity in the ascending reticular pathways. As a result there is a weakening of the after-discharge following the primary sensory response (Bremer, 1954). However, this process of deactivation was thought to be initiated in normal circumstances by synaptic fatigue at the reticular, diencephalic or cortical levels.

In the paper preceding Bremer's article from which the above discussion was taken, W. R. Hess (1954a) described experiments in which electrical stimulation of portions of the diencephalon induced drowsiness and a state resembling natural sleep in all respects as far as the behaviour was concerned. This was important evidence for abandoning the passive theory of sleep that was accepted by most of the major neurophysiologists in the mid-1950's (see Moruzzi, 1964).

Stronger evidence came from the discovery of two electrophysiologically separate states of sleep. R. Hess, the son of W. R. Hess, remarked in the discussion of his father's paper that the electrocortical activity was not always characteristic of sleep because sometimes the electrical activity was found to have the same pattern as waking (R. Hess, 1954). This comment anticipated the reports of Dement and Kleitman (1957) which demonstrated that there were two distinct phases of sleep. In normal subjects the first sign of falling asleep (Stage 1) was a slowing of the alpha rhythm, the electrical pattern of relaxed wakefulness, which becomes fragmented to be replaced by diffuse theta activity (4–6 cps) of Stage 2. As sleep becomes deeper in Stage 3, these waves become slower, 1–6 cps, with a larger voltage. In addition there are brief bursts of 12–14 cps waves, called 'spindles', together with K-complexes which are sensory-evoked potentials occurring at the onset and sometimes at the offset of the stimulus. Oswald (1962) has suggested that it represents a partial arousal response because the intensity and duration of the K-complex varies with the meaningfulness of the stimulus. Stage 4 is the deepest phase of nondreaming sleep and is called slow wave sleep. There is some muscle tension especially in the neck, but the eyeballs are still. After about 90 min the characteristics of the EEG change dramatically indicating the onset of paradoxical sleep.

Paradoxical sleep shows a pattern of activity characteristic of electrocortical arousal with low voltage (100 μV) fast waves (14–30 cps). The muscles are completely relaxed (atonia) except for the frequent, brief body movements and the eye muscles which produce eyeball movements. If awakened the person usually reports dreaming, but paradoxically the person is much harder to arouse than when he is in slow wave sleep despite the electrocortical arousal. The paradoxical sleep, also called rapid eye movement sleep, is always preceded by phasic activity, the ponto-geniculo-occipital spikes (spike activity) which occur 30–45 sec earlier. These 'deep sleep' spikes are almost identical to those recorded from the same structures during visual attention. Their rate is fairly constant during paradoxical sleep but their basis is not known. During para-doxical sleep it has been observed that there is spontaneous electrical activity resembling the effects of visual input in the awake animals, ponto-geniculo-occipital spikes, which may be triggering the eye movements and dreams as if there was visual input.

These facts have been some of the major discoveries of electrophysiology in the last 20 years. They have opened up a whole new era in sleep research, and laid the foundation of all the recent advances to be discussed in this chapter. This research has been helped by the discovery that all mammals have both sorts of sleep, and these occur in alternate phases. Thus sleep can be investigated in the common laboratory species like the rat and the cat, even though the length of the slow-wave sleep and paradoxical sleep phases differ from species to species. In addition it has turned out that cortical signs of slow-wave sleep, paradoxical sleep and spike activity are separable pharmacologically, indicating that there are probably different neurochemical systems controlling them.

B. NEUROCHEMICAL CONTROL OF SLOW-WAVE SLEEP

It follows from the introductory discussion that there must be at least two neurochemical systems involved in the control of the two states of sleep. Once it became clear that these two states depended on distinct regions of the brain then the stage was set for Jouvet, Koella and their coworkers to discover the important systems involved in short-wave sleep. One obvious approach would have been to inject the major neurohumours into the body and observe the effects on behaviour. Unfortunately, these compounds do not pass the blood-brain barrier and so cannot affect the nerve cells.

In order to overcome this difficulty, animals were given doses of precursors like 5-hydroxytryptophan, a precursor of serotonin that passes the blood brain barrier. It was observed that, after a short while, a state like slow-wave sleep was induced by this compound. This state lasts for 5 to 6 hr although the animal may be aroused briefly by loud noises. During this phase there is a suppression of paradoxical sleep and spike activity and after recovery there is a rebound increase in paradoxical sleep (Jouvet, 1967). It seems obvious from this one piece of evidence that slow-wave sleep is the result of serotonin release in the brain, whereas paradoxical sleep is not dependent on the serotonin system.

A second set of experiments were based on the use of drug combinations in order to obtain a selective effect. The first experiments were carried out by Matsumoto and Jouvet (1964) using cats with chronically implanted electrodes to record the electrocortical activity, eye movements, neck muscle tension, heart rate and respiration for the several days of the experiment. They found that an injection of 0·50 mg/kg of reserpine, intraperitoneally, suppressed slow-wave sleep totally for more than 18 hr, and it did not return to normal until 48 hr later. There was a total abolition of paradoxical sleep for 3 or 4 days after injection, demonstrating the depletion of norepinephrine and serotonin impaired normal sleep, although the animals were in a 'state of tranquillity'. The separation of the amine effects was obtained by means of precursor injections to selectively restore the amine levels. The injection of 5-hydroxytryptophan 4 hr and 14 hr after reserpine prevented the depletion of serotonin and restored all the signs of slow-wave sleep. In contrast paradoxical sleep remained abolished and did not reappear until the third day. This dissociation gave more strong evidence that slow-wave sleep might depend on serotonin pathways in the brain.

A follow-up to this study was aided by the coincidental discovery, at about the same time, of parachlorphenylalanine which selectively blocks serotonin synthesis. This led to a spate of studies of the effects of this drug on slow-wave sleep in rats (Mouret, Bobillier, Jouvet, 1968; Torda, 1967), cats (Delorme, Froment and Jouvet, 1966; Koella, Feldstein and Czicman, 1968), monkeys (Weitzman, Rapport, McGregor and Jacoby, 1968).

The study on monkeys (Weitzman et al., 1968) is discussed in more detail because it has discussed correlative measures of the biochemistry in seven regions of the brain and because of the ease with which the results can be extrapolated to man. The electrocortical activity and eye movement recordings were made from rhesus monkeys during a sleep period which lasted from 10.00 p.m. to 6.00 a.m. After baseline recordings for 4 days when the proportion of the 8 hr spent in paradoxical and slow-wave sleep was recorded, the animals were injected with doses of the inhibitor. It was found that there was a decrease in the amount slept, but that paradoxical sleep was unchanged so that a relatively greater proportion of time was spent in paradoxical sleep. The measure of serotonin levels showed a decrease of from 31–46 % in the brain regions although no information was obtained on the turnover rates or the relative effects on the different storage pools. It is critical that the greatest amount and thus the greatest depletion occurred in the region of the midbrain raphé, shown to be important in slow-wave sleep by the Jouvet lesion studies to be described next in this section.

The brilliant studies using histo-fluorescence techniques demonstrated that the raphé system along the midline of the pons and medulla has high concentration of serotonin (Dahlström and Fuxe, 1964). Lesions in this region which destroyed over 90 % of the neurones produced total insomnia with neither sorts of sleep, but there was no effect on sleep with only 15 % destruction and between 15 % and 90 % there was an inverse correlation between the amount of damage and the percentage of sleep (Jouvet and Renault, 1966; Jouvet, Bobillier, Pujol and Renault, 1966). In the first few days after lesions of the rostal portion of the raphé, a complete disappearance of slow-wave sleep was observed, but paradoxical sleep still occurred alternating with wakeful periods. Later paradoxical sleep only appeared when slow-wave sleep occurred for more than 15 % of the time (Jouvet et al., 1966).

C. NEUROCHEMICAL CONTROL OF DREAMING SLEEP

It can be seen from the findings of Jouvet that there may be a functional relation between the two sorts of sleep so that paradoxical sleep (PS) only occurs when slow-wave sleep has reached a certain threshold of 15 % (Jouvet et al., 1966), and the rate of occurrence of paradoxical sleep, after depletion of serotonin with parachlorphenylalanine is dependent on the endogenous levels of serotonin, and the quantity of slow-wave sleep (Mouret et al., 1968). As a result Jouvet (1967) proposed that some deaminated metabolite of serotonin might be responsible for triggering paradoxical sleep. This notion would be consistent with the increased frequency of paradoxical sleep during the last one-third of sleep, when the metabolites of serotonin would presumably be at their highest. Whatever the triggering system, it is certain that the neurochemical control system is not serotonin.

If we consider the electrical signs of paradoxical sleep we can get a clue about the likely biochemical candidate for the control system. The characteristics are fast low-voltage activity at the cortex, and theta rhythm in the hippocampus as well as an absence of muscle tension, except in the eye muscles. The tonic excitability in the brain is also characteristic of electrocortical arousal, and is believed to be controlled by the ascending cholinergic reticular pathways (see Chapter 4). It seems reasonable to suggest that cholinergic systems are involved in the production of PS and all the evidence is consistent with this finding. Jouvet (1962) found that paradoxical sleep in cats was blocked by atropine and enhanced by physostigmine, but there was no effect on behavioural slow-wave sleep. In a further study (Jouvet, 1967), atropine was injected at the onset of the recovery sleep after paradoxical sleep deprivation. The animals showed slow waves at the cortex, behavioural sleep, but no rebound cortical fast activity until the drug had worn off, although there was evidence of the phasic spike activity. This suggests very strongly that there is a cholinergic system in the brain directly controlling paradoxical sleep. The possible neuro-anatomical locus of this system was investigated in a series of studies by Hernandez-Peon (see reviews in Hernandez-Peon, 1963; 1965).

Hernandez-Peon has demonstrated that behavioural and electrographic manifestations of paradoxical sleep were produced in the cat shortly after local application of minute crystals of acetylcholine alone, acetylcholine plus physostigmine, or carbachol when applied only in a highly circumscribed pathway extending to the rostral forebrain from the pontine and bulbar tegmentum, perhaps identical with the ventral tegmental pathway discussed in Chapter 4. The sensitive loci were located along this pathway and included the preoptic region, the lateral and postero-medial hypothalamus at the level of the medial forebrain bundle, the ventromedial part of the midbrain teg-mentum (the ventral tegmental area) and the medial ponto-mesencephalic tegmentum (Hernandez-Peon, Chavez-Ibarra, Morgane and Timo-Iaria, 1963). Between 20 sec and 4 minutes after introducing a few crystals of cholinergic agents into the pathway, the cat fell asleep after the usual preparatory be-haviour of the cat, i.e. closing of the eyes, dropping the head, and curling up the body. Paradoxical sleep occurred with the following electrical changes: the desynchronized pattern of wakefulness was replaced by spindles and high-voltage slow-waves which in turn were quickly replaced by a fast low-voltage activity. The amplitude of the electromyogram of the neck muscles diminished during the 'synchronized' stage. High-amplitude rapid eye movements appeared periodically in bursts during the 'desynchronized' stage of sleep.

Although sometimes the cat was easily aroused by acoustic or nociceptive sensory stimulations, in some experiments sleep was so deep that the cat was aroused only by a strong electrical stimulation of the mesencephalic reticular formation, falling again asleep when the stimulus was turned off. The duration of sleep induced by a single application of cholinergic substances along the hypogenic pathway varied from 30 min to 4 hr. Awaking from cholinergic sleep was either sudden or else the animal passed throughout several periods of wakefulness alternated with short periods of sleep before it remained fully awake.

Since the experiments with carbachol stimulation supported the hypothesis of cholinergic involvement in paradoxical sleep, Hernandez-Peon tested the effects of local application of cholinolytics like atropine (Velluti and Hernandez-Peon, 1963). In a group of cats two cannulae were implanted with a rostral cannula in the preoptic region, and the caudal cannula in the interpeduncular nucleus of the tegmental area. In a first series of experiments, atropine was placed in the hypogenic points tested the day before, and it was found that the application of acetylcholine in the atropinized loci no longer induced sleep. In other experiments, the hypnogenic effects of acetylcholine were first tested in the rostral cannula. The next day, atropine was locally applied in the caudal cannula, and after a minimal period of 30 to 60 min, acetylcholine was applied again in the rostral cannula. Blocking of cholinergic sleep was obtained by application of atropine either to the interpeduncular nucleus or to the medial ponto-mesencephalic tegmentum.

In a further investigation of sites in the caudal mesencephalon and pontine reticular formation (George, Haslett and Jenden, 1964) found that a minute solution of carbachol injected into the region of the nucleus reticularis pontis oralis and the nucleus reticularis pontis caudalis induced all the signs of paradoxical sleep in cats; the nictitating membranes were relaxed, the pupils were myotic, continuous rapid movements of the eyeballs and the hippocampal activity was synchronized. In contrast to the results of Hernandez-Peon, slow-wave sleep with spindling was never obtained at any stage, either during induction or recovery. This difference would be due to the differences between using crystals and drug solutions. This evidence from animal studies is supported by the studies of the effects of anticholinesterases on human behaviour, discussed in Chapter 4. One of the typical findings of these studies (e.g. Grob et al., 1947) was that the subjects in these studies reported excessive dreaming. This study predated the discovery of paradoxical sleep, so that we do not have any other supporting electrophysiological data.

These results give clear support for cholinergic involvement in paradoxical sleep, and this would explain several phenomena of paradoxical sleep. In

Chapter 4 the case was made that our awareness of external stimuli, internal stimuli including ideas and associations, and learning can only occur during electrocortical arousal produced by cholinergic activation. From our own experience we know that dreams are based mainly on our past experiences, but that the dreams may also incorporate external stimuli, and this has been shown in the sleep laboratory (Berger, 1963). This period of sleep is also the only period when we can recall anything at all, even though the information is only held for about 10 sec in some short term store (Mandell, Mandell and Jacobson, 1965).

The idea of direct cholinergic involvement differs in emphasis from an early theory of Jouvet (1969) who places more importance on the norepinephrine system in its direct control. In favour of this hypothesis, Jouvet (1967; 1969) cites a number of studies that are not as clear cut as they seem. Firstly he quotes the experiment of Matsumoto and Jouvet (1964) which demonstrated that after depletion of serotonin and norepinephrine by reserpine and the abolition of slow-wave sleep and paradoxical sleep, DOPA, the precursor of norepinephrine, increased spike activity even further, and after about 30 min slow-wave sleep appeared briefly and was followed by paradoxical sleep. On the other hand, selective 50 % depletion of norepinephrine by alpha-methylparatyrosine did not suppress paradoxical sleep (Marantz and Rechtschaffen, 1967). Indirect evidence from a study of norepinephrine turnover showed that rats which have rebound increases of paradoxical sleep following selective deprivation, also have marked increases in the uptake and turnover of norepinephrine in the brain (Pujol, Mouret, Jouvet and Glowinski, 1968), including the brain stem. However, *increases* in the intraneuronal norepinephrine with four monoamine oxidase inhibitors, including pargyline, selectively *suppressed* paradoxical sleep (Jouvet, Vimont and Delorme, 1965). This would suggest that reduction of functional norepinephrine is necessary for the appearance of paradoxical sleep. The finding that people are hardest to arouse during paradoxical sleep would be predicted from the notions that a norepinephrine system is involved in behavioural arousal and that paradoxical sleep is correlated with inactivity in the ascending NE pathways. Nevertheless, there seems to be some sort of relationship between the phasic spike activity and the noradrenergic systems of the pontine portion of the brain stem ventral and caudal to the locus coeruleus. It is thought by Jouvet (1967) that the hallucinatory visual scenery of dreaming may be produced by the initiation of the spikes activity due to the release of monoamines at this site. Further evidence on these issues will be presented in the next section on the function of sleep.

This discussion of mechanisms and drugs affecting them, while giving interesting examples of the methodology of studies in the neurochemical control of behaviour, has not considered the function of such a complex control system in maintaining normal behaviour. This omission will be remedied to some extent by a more speculative section on the significance of sleep in the light of the behavioural and biochemical effects of sleep deprivation.

D. SIGNIFICANCE OF SLEEP PHENOMENA

The obvious answer to the question 'What is the function of sleep?' is that this period provides a time for the restoration of body tissues after the day's activity. Studies of incapacitated subjects have shown that they do not necessarily sleep less than normal subjects, so that it does not seem to be anything simple like dissipation of the products of muscular activity. One way to study the importance of sleep is to examine the consequence of sleep deprivation.

In the last 20 years after World War II numerous studies have measured the behavioural and biochemical consequences of sleep loss in man and animals. In a typical human subject there was a progressive deterioration in all types of performance due to lapses in attention and mental confusion. Some subjects confused their thoughts with reality to the point of hallucinating, together with mood changes until a temporary psychosis appeared (Lubin, 1967). Animal studies have shown that sleep loss can produce cell changes in the central nervous system (Kleitman, 1963), although we do not know if these are permanent. In humans, it has been demonstrated that there is depletion of the body's reserves of the catalyst, adenosine triphosphate (ATP), which are essential for the release of energy from food (Luby, Grisell, Frohman, Lees, Cohen & Gottlieb, 1961). The importance of ATP can be gauged from the estimate that man produces and uses 70 kg of this compound, his own bodyweight, every day (Karlson, 1963). It has been demonstrated that ATP production is essential for the mobilization of the amino acids required for protein synthesis, so that an ATP deficiency will result in a failure of cell renewal which might explain the brain lesions. The synthesis of ATP depends on glucose in some tissues (Mahler & Cordes, 1966) and it is known that somatotropin, the growth hormone which is important in enhancing protein synthesis and raising blood sugar levels (Karlson, 1963), is released into the bloodstream during sleep Stages 3 and 4, i.e. in slow-wave sleep, but not during paradoxical sleep or during relaxed wakefulness (Sassin, Parker, Mace, Gotlin, Johnson and Rossman, 1969). As slow-wave sleep occurs mainly during the first third of the night the peak plasma levels were during the first 90 min, but even during the last third of sleep the subsidiary peaks tended to coincide with slow-wave sleep. Although the total amount of slow-wave sleep was similar for all subjects in the study there was some variation in its distribution so that subjects with more frequent periods of slow-wave sleep had larger initial peak levels and more frequent subsidiary peaks. This study points very clearly to the involvement of slow-wave sleep in tissue building.

This brings us to a consideration of paradoxical sleep and its importance to normal behaviour. Before considering paradoxical sleep deprivation studies it will be of interest to discuss some speculations on the function of this phase of sleep which are closely linked to those discussed above.

In their paper on the ontogenetic development of the sleep-dream cycle, Roffwarg, Muzio and Dement (1966) argue that paradoxical sleep including the spike activity plays a role in the structural maturation and maintenance of the CNS by providing excitation of the sensory and motor areas. They speculate that the paradoxical sleep with its dreams may provide the mechanism for hallucinatory repetition of accumulated experience in the infant. From this speculation it is not a too great leap to the notion that this phase of information storage in both infants and adults, perhaps by means of protein synthesis. In Chapter 10 the possible role of protein synthesis in learning will be discussed in detail, but at this stage it is only necessary to point out the evidence for increased chemical synthesis in the brain during paradoxical sleep. One bit of evidence is that oxygen use is greatest during paradoxical sleep and least during Stage 4 of slow-wave sleep (Brebbia and Altshuler, 1965). In addition, recordings of brain temperature by implanting thermistors in the cortex and hypothalamus of cats show a marked rise in temperature at the onset of paradoxical sleep (Rechtschaffen, Cornwall and Zimmerman, 1965). Other parts of the body show no evidence of enhanced synthesis at this time. This speculation can be linked with the release of growth hormone during slow-wave sleep. Studies of the action of growth hormone show that there is a minimum lag period of 50–60 minutes before the enhancement of protein synthesis can be detected, and it is several hours before the optimum stimulation of synthesis is found, at least in rat liver (Talwar, Gupta, Pandian, Sharma, Rao, Jaikhani, Sen, Sopori, Basu and Jha, (1971). From this we can conclude it would be essential, for full efficiency, that the release of growth hormone precede the time of protein synthesis by an hour or more. There is some evidence, as we mentioned earlier, that paradoxical sleep will not occur until after a certain amount of slow-wave sleep, about 90 min in the normal human, and that there seems to be a biochemical 'lock' preventing its occurrence (Jouvet, 1967). The memory disturbances of the sleep-deprived individual could be a direct consequence of the loss of this storage mechanism.

Studies on paradoxical sleep deprivation have led to other lines of speculation that do not at first sight seem to be related to the tissue building and memory consolidation notions outlined above. These studies were designed to determine whether the hallucination seen in sleep deprivation may be the result of paradoxical sleep loss. In order to investigate this possibility human subjects' sleep was interrupted whenever desynchronization appeared and it was found that deprivation was followed by a compensatory rebound of paradoxical sleep when the subject was allowed to sleep undisturbed (Dement, 1965). The effect of deprivation was cumulative so that a greater rebound was found after longer deprivation, although there was never complete compensation. This effect did not dissipate with time in the absence of additional paradoxical sleep, suggesting that there is a build-up of some chemical in the CNS which triggered this phase of sleep in some way. Studies in cats by Dement (Dement, Ferguson, Cohen and Barchas, 1969) showed that cerebrospinal fluid withdrawn from the ventricles of a paradoxical sleep-deprived cat induced paradoxical sleep when injected into the ventricles of a cat who had been sleeping normally. In contrast, cats injected with cerebrospinal fluid from a normally treated cat showed paradoxical sleep suppression as the result of the control injection, so that the effect of the fluid from the PS deprived cat is even more striking.

The behaviour of cats deprived of this sort of sleep was dramatic, with the appearance of compulsive mounting behaviour in a number of the male cats. In male rats a similar pattern was observed with 80 % of the deprived rats mounting an oestrous female, while only 14 % of the control rats did so. Other observations were increase in the speed of eating by paradoxical sleep-deprived subjects (Dement, Henry, Cohen, and Ferguson, 1967), the amount eaten (Bowers, Hartmann and Freeman, 1966), and in aggressive behaviour (Bowers et al., 1966; Morden, Conner, Mitchell, Dement and Levine, 1968). Dement et al. (1969) interprets these changes as generalized hyperexcitability, which results in an increase in drive-oriented behaviour as a consequence of the accumulation of some 'paradoxical sleep chemical'. These findings are reminiscent of the Kluver-Bucy syndrome produced by limbic lesions which cause amygdala dysfunction (Schreiner and Kling, 1953). The commonly accepted explanation of these changes is that the cats can no longer discriminate a strange environment from a home territory. As we showed in Chapter 4 a decrease in functional acetylcholine produces discrimination impairments, and it has been argued earlier in this chapter that the cholinergic system is involved in paradoxical sleep, so that any evidence on the amount of functional acetylcholine in the brain, particularly limbic system, would be important.

One study has demonstrated a significant 35 % decrease in acetylcholine levels in the telencephalon, consisting of cortex, limbic system and basal ganglia, but not in the diencephalon, and brain stem (Hernandez-Peon, 1965). This fall would account for hallucinations and confusion in paradoxical sleep-deprived subjects, although the reasons for this depletion are not understood. More recently studies by Hernandez-Peon (1969) showed that cats deprived of sleep for 10–12 days showed a significant depletion of brain norepinephrine in the tegmental region, a large increase in dopamine with small increase in serotonin. The dopamine increase was interpreted as mobilization of the precursor for synthesis of the norepinephrine used up during wakefulness. The increase in serotonin would be expected from the involvement of this transmitter in slow-wave sleep. Incidentally, this depletion of norepinephrine would explain the greater uptake of tritiated norepinephrine in the turnover studies of Pujol et al. (1968) mentioned in Section C.

One curious finding of the animal studies by Dement et al. (1967) was that rats that were previously deprived of paradoxical sleep were more sensitive to a second period of deprivation of the same duration as the first in terms of the rebound paradoxical sleep. It turned out that repeated short periods of deprivation were more effective than one long period. Dement proposed that repeated sleep loss might be a factor in the development of an acute psychotic episode, since this symptom is reported frequently in case histories. In particular, Dement notes that narcolepsy has been shown to be a sudden occurrence of paradoxical sleep, often at inappropriate times, and these patients report chronic sleep disturbance.

In these studies it was not clear whether the important part of paradoxical sleep deprivation was the prevention of paradoxical sleep or the reduction in the phasic spike activity. In a thorough study, Dement et al. (1969) disturbed the cat every time a spike appeared (up to 200 times an hour in some animals). The large reduction in phasic activity resulting from this procedure substantially increased the rebound paradoxical sleep periods. Obviously, it is the amount of spike activity deprivation which determines the compensatory paradoxical sleep length.

When considering the function of the spikes, it is significant that actively psychotic schizophrenics do not show compensatory increases in paradoxical sleep when deprived, whereas the same subjects do show rebound increases when tested during recovery from the illness (Zarcone, Gulevich, Pivik and Dement, 1968). These subjects showed no obvious differences from normal subjects in terms of the frequency and amount of paradoxical sleep recorded during an undisturbed night. Dement et al. (1969) hypothesized that the phasic events must be occurring during the waking state in actively psychotic schizophrenics, and thus the chemical associated with them does not accumulate to produce rebound paradoxical sleep. Further, they suggest that this sort of discharge of phasic activity during wakefulness may account for the hallucinatory episodes of the active psychosis. Consistent with this hypothesis are the many reports of human subjects deprived of sleep who have hallucinations (Luby et al., 1961). In addition, there is at least one study which has demonstrated that rapid eye movements and decreases in muscle tonus, two of the characteristics of paradoxical sleep, did occur in awake normal subjects at times when they had daydreams with active personal involvement. Other sorts of thinking were not associated with rapid eye movements, and decreased muscle tonus (Othmer, Hayden and Segelbaum, 1969). This suggests that some features of paradoxical sleep do occur during waking, but the frequency of these may be increased during some schizophrenic breakdown and after deprivation of paradoxical sleep, and that in the latter cases the subjects cannot discriminate between fantasy and reality. From this it is apparent that paradoxical sleep and spike activity is serving some important function, and that there are neurochemical mechanisms set up to ensure that, on the average, the same proportion of each day is devoted to these forms of activity.

The functional significance of these sleep phases is still a matter for speculation. It does not follow necessarily that because a certain type of dysfunction results from deprivation that these sleep phases are directly involved in preventing this sort of impairment, although this would be one explanation. Thus, Dement et al. (1969) have postulated that the ponto-geniculo-occipital spike activity is related to the brain mechanisms controlling instinctual behaviour, and that the occurrence of phasic spikes prevents these types of behaviour occurring during the waking state. Thus the high frequency of paradoxical sleep in childhood results from the large amount of spike activity, and represents the development of CNS control over drive-oriented behaviour. Since this control is the result of social learning, this notion can be related to the previous hypothesis that paradoxical sleep might be involved in learning.

E. HYPNOTIC DRUGS

Hypnotic agents are the major psychoactive drugs in use today, ranging from the prescribed sedative drugs to the 'nightcap' drink. In spite of their widespread use we know relatively little about their action and after-effects. Ideally a hypnotic agent should induce sleep which has the normal quantities of the two types of sleep and spike activity, and which leaves the patient refreshed and alert the next morning. As well as these properties, an ideal agent should have no withdrawal effects when it is discontinued. At the present time there are no drugs which satisfy these criteria, and in fact the barbiturates, the most frequently prescribed hypontics, fail on all three counts.

An injection of a barbiturate induces low-voltage fast activity at low doses, but as the dose is increased slow waves appear and the patient becomes unconscious (Brazier, 1963). At a light level of anaesthesia, sensory stimuli induce the K-complexes seen in Stage 3 of slow-wave sleep, but in deep surgical anaesthesia K-complexes are not seen, and this state seems to be equivalent to Stage 4 of slow-wave sleep. However, the cortical evoked response to sensory stimuli produced under barbiturates differs from normal sleep. In an anaesthetised subject the evoked response, recorded from the scalp and averaged to eliminate random background activity, shows an initial positive response followed by a negative response, and then a second positive and negative wave. The initial positive and negative waves are enhanced, but the later components are not seen in subjects anaesthetized with barbiturates. It is thought that it is the secondary response components which determine awareness and abolition of these waves give the barbiturates their anaesthetic properties (see Chapter 4).

Another difference from normal sleep is the suppression of paradoxical sleep. Studies were reviewed by Oswald (1968) which showed that barbiturates delayed its onset, reduced the proportion of it per night, decreased the frequency of eye movements during paradoxical sleep, but did not appear to affect the periodicity. After withdrawal of the drug compensatory rebound of paradoxical sleep occurs with decreased delay of onset and an increase in the vividness of dreams and frequency of nightmares, just as other paradoxical studies have found (see Section C). One common after-effect of barbiturate-induced sleep is the 'hangover', which has been attributed to the effects of paradoxical sleep suppression, although another contributory factor is the unmetabolized drug remaining in the morning. Withdrawal from barbiturates (see Chapter 8) shows similar symptoms to paradoxical sleep deprivation in humans, including irritability and hallucinations (Seevers and Deneau, 1963).

The effects of the drug on sleep can be explained in terms of the drug's action on the neurochemical mechanism underlying slow-wave sleep. It turns out that an anaesthetic dose of barbiturate, like 50 kg of pentobarbital doubled the levels of serotonin in the rat (Anderson and Bonnycastle, 1960) and there was a direct correlation between the progress of narcosis and sleep. Thus the anaesthetic effects were greatest at the time of peak biochemical effect, and continued as the serotonin was released, with recovery when the levels returned to normal. In an earlier section it was shown that slow-wave sleep was increased and paradoxical sleep was suppressed by treatments that increased the available serotonin (Jouvet, 1967) and barbiturates seem to have a similar effect.

Another chemical agent which suppresses paradoxical sleep is alcohol. A dose of 1 gm/kg administered in the evening decreased the percentage of paradoxical sleep in the first half of the night, but there was some compensation in the second half (Yules, Freedman and Chandler, 1966), whereas double the dose suppressed it altogether (Knowles, Laverty and Kuechler, 1968). In the study of Yules et al. it was found that although paradoxical sleep was partly suppressed on the first night there was less and less effect on four successive nights. On the first night without alcohol, rebound paradoxical sleep was observed. It has been suggested that prolonged PS suppression might be the cause of the irritability and hallucinations of the alcoholic (Kalant, Leblanc and Gibbins, 1971), but as these authors point out 'This area must still be regarded as one for extensive investigation rather than as a basis of a strong plausible hypothesis' (p. 260).

From the effects of barbiturates on serotonin and paradoxical sleep suppression it could be argued that alcohol was also acting on the serotonin systems to increase the functional amount of transmitter. Evidence consistent with this was obtained by Bonnycastle, Bonnycastle and Anderson, (1962) but this has not been supported by most experimenters (Wallgren and Barry, 1971). Instead, a different mechanism has more support, and that is the reduction of functional acetylcholine which we have proposed as one transmitter important in paradoxical sleep. The release of acetylcholine from rat brain tissue was inhibited by alcohol (Kalant, Israel and Mahon, 1967) leading to an accumulation of acetylcholine in the brain after a single dose (Hosein and Koh, 1966). The consequences of this inhibition will be exactly like those of a cholinolytic like atropine, scopolamine or benactazine which have all been used to induce sleep. Thus it seems possible that although alcohol and barbiturates suppress PS the underlying mechanism is different for the two drugs.

A third group of drugs which suppress paradoxical sleep are the antidepressant agents like amphetamine, the monoamine oxidase inhibitors and the tricyclic antidepressants which were discussed in Chapter 3. Amphetamine was described as valueless because of the adverse effects associated with its use and in particular the hallucinations, confusion and irritability. As in the case of barbiturates and alcohol, these reactions have been attributed to the loss of paradoxical sleep produced by amphetamine, but in fact amphetamine addicts show normal amounts of sleep (Oswald and Thacore, 1963) showing that tolerance develops and this was confirmed by rebound paradoxical sleep appearing during withdrawal. Nevertheless, there is considerable paradoxical sleep suppression when amphetamine is first given, with delay in onset and decrease in frequency (Rechtschaffen and Maron, 1964).

It has already been argued in an earlier section that norepinephrine could not be the activating transmitter for paradoxical sleep, rather these studies suggest that norepinephrine may be a transmitter for a pathway inhibiting paradoxical sleep, so that dreaming and behavioural arousal are incompatible. In support of this, sleep studies with other drugs increasing functional norepinephrine show that monoamine oxidase inhibitors, like tranylcypromine and nialimide, suppress paradoxical sleep, while tricyclic antidepressants like imipramine and amitriptyline reduced paradoxical sleep (Oswald, 1968). In contrast, drugs reducing functional norepinephrine, like chlorpromazine and reserpine, increase the proportion of paradoxical sleep with an earlier onset to the periods (Hartmann, 1966; Lester and Guerrero-Figueroa, 1966).

7
Schizophrenias and Experimental Psychoses

This chapter has been included because of the marked increase in biological studies of schizophrenic behaviour in the last two decades. Unlike the previous chapters, there is not a well-established body of knowledge, although there is strong evidence for some form of methylation disorder in schizophrenia. Much of the work has been based on the notion of producing 'experimental psychoses' with drugs, known as 'psychotomimetic' or 'hallucinogenic' drugs. Strictly speaking, both of these terms are inaccurate because drugs rarely produce effects indistinguishable from true hallucinations where the subject is convinced of the existence of the object. Nevertheless, they have proved useful as tools for producing 'model psychoses' (Russell, 1960) in which symptoms analogous to those found in psychoses can be studied in the laboratory. It has often been assumed naïvely that a drug might produce all the symptoms of the schizophrenic state, but no drug of this sort has been discovered. However, this does not mean that all the studies on 'hallucinogenic' drugs have been useless. On the contrary, they have given experimenters insights into the neurochemical mechanisms underlying disturbances in perception. It remains to be demonstrated that these disturbances are homologous with those occurring in the schizophrenias. One important difference that must be borne in mind is that the subjects in laboratory studies were aware that they were receiving a drug which could produce perceptual disturbances, whereas the schizophrenic has no information at all.

The 'hallucinogens' are loose a classification of drugs whose *predominant* symptoms are hallucinations. One consequence of this mode of action is that only human subjects can be used to study these main effects although neurophysiological and biochemical evidence may provide some explanation of the main effect. Behavioural studies of animals have provided rather uninteresting information; in general the animal displays random behaviour with the secondary effects of the drug, like excitation, superimposed on it. Human studies have been devised to analyse the subjective effects using questionnaires designed to disentangle the sensorimotor, perceptual and mood phenomena (Linton and Langs, 1962; Levine and Ludwig, 1965; Isbell and Gorodetzky, 1966). In order to minimize extraneous stimuli, experiments are usually conducted in a plain room with subdued lighting and the patient is left undisturbed as much as possible.

A. THE LYSERGIC ACID DIETHYLAMIDE (LSD) MODEL

As most people know, LSD-25 was synthesized from the fungus, ergot (Claviceps purpurea) in 1938. It was first tasted by Albert Hofmann the research chemist in 1943, and he found that it produced visual hallucinations with small doses (Hofmann, 1958). These findings have been substantiated by many subjects since then, but it has been found that an individual's experiences depend critically on the experimental environment, his preceding experiences, his personality and his culture, and it seems as if the LSD state is an extremely suggestible one and that the experiences during the drug can be programmed to a greater or lesser extent depending on the subject.

Perceptual changes

Although these basic effects may vary there are a number of perceptual effects common to the subjective experiences, and these are hallucinations, dream-like imagery, abstract imagery and perceptual distortion (Jacobsen, 1968). These will be considered in inverse order representing decreasing contact with reality.

Perceptual distortion occurs in all modalities although most subjects experience visual changes. Thus objects in the primary colours seem unusually intense and saturated, and colour after-images are prolonged while sounds may be intensified until they become unpleasant and the senses of touch, taste and smell seem more acute. There are often switches in sensory quality so that a wine may taste sour or sweet alternately, small objects may suddenly enlarge and large objects contract, and bright objects suddenly become dull. Synaesthesia often occurs so that sensations in one modality carry with them impressions belonging to another modality, e.g. coloured hearing, which often produces a fusion of percept, concept and affect. Subjective time is distorted so that time periods are overestimated. In an experimental study of this phenomenon (Aronson, Silverstein and Klee, 1959) drugged subjects were asked to estimate intervals ranging from 15 to 240 sec. In comparison with their undrugged performance the subjects overestimated the intervals, suggesting that LSD makes time seem to pass more slowly. Particularly disturbing were distortions of the body image in size and shape (Masters and Huston, 1966; Bercel, Travis, Obinger and Dreikurs, 1956). Typical comments of subjects injected with LSD include 'My arms and legs feel detached' (Bercel et al., 1956); 'My body feels as if it has been anaesthetized, and the flesh has become rubbery' (Masters and Huston, 1966); 'My body feels as if it is melting into the background (Bercel et al., 1956). There is a report of one woman who could not distinguish her body from the chair in which she was sitting and became very agitated when she thought 'she could not get back into herself' (Frosch, Robbins, and Stern, 1965). In some cases these distortions of body image are part of the visual distortions, for subjects frequently report that their hands change size enormously as they bring them nearer to their face. These effects are found mainly with higher doses, over 100 µg, and are some of the reliable changes observed with increasing dose (Abramson, Jarvik, Kaufman, Kornetsky, Levine and Wagner, 1955).

An extension of the perceptual distortions in the visual modality are the abstract coloured images that have frequently been reported. These consist of a coloured moving display of patterns shooting past when the eyes are closed or the subject is in a darkened room. These seem to be an elaboration of the phosphenes, flashes of colour seen by normal subjects when their eyes are closed (Knoll, Kugler, Höfer and Lawder, 1963). In the drugged subject these are much more vivid and elaborate, forming diamonds, latticework, and carpetlike patterns (Ditman, Moss, Forgy, Zunin, Lynch and Funk, 1969). These abstract images disappear when the eyes are opened or the light is switched on, and are replaced by the distortions of the surrounding world described previously.

Dream-like imagery consists of a stream of images and associations consisting very often of memories of scenes rather than events. They are often childhood memories or resemble such memories, and very often they have the complete affective states that accompanied the original events or those that would be appropriate for a fantasized event. It is this property of LSD in particular which has been claimed as an important adjunct to psychotherapy in the treatment of chronic neuroses (e.g. Sandison, 1954; Eisner and Cohen, 1958; Abramson, 1960), because it reveals material for the therapist to interpret. In a typical session the patient is prepared for a dramatic event, but it is also explained that help will be available if the experience becomes unpleasant. The patient is asked to record his images and feelings on paper when he is able to write. These records are then available for later interpretation by the therapist according to his theoretical orientation. Some therapists seem to have the skill to make the experience a highly meaningful and therapeutic one. They claim that LSD produces an upsurge of unconscious material to consciousness and that repressed memories are relived with remarkable clarity with a profound improvement in prognosis (Sandison, 1954). Most therapists regard the LSD session as an extension of the conventional therapeutic values of recall, reliving, insight and emotional release (Savage, Jackson and Terrill, 1962). As in the case of most psychodynamically oriented therapy it has proved difficult to estimate the success of LSD therapy especially when the case reports are shrouded in pseudomysticism.

Hallucinations are almost exclusively visual and like the dream images are not under the control of the subject or an experimenter. It is believed that they are triggered by the individual's affective changes rather than the other way round (Jacobsen, 1968). In many cases they develop from the dream images and seem to be based on childhood experiences. It seems as if hallucinations are a development of dream images and result from doses over 150 μg. They can be perceived when the eyes are closed, but become blurred by the intensification of the perception described earlier. They are usually visual experiences rather than auditory, olfactory or gustatory, in common with the perceptual distortions. The subject is usually not 'convinced' by the visions and so they

122

are more properly termed pseudohallucinations. It is only in rare cases where the subject experiences disturbances in thinking and feeling that reports are made of true hallucinations, i.e. objects and persons that are perceived with the eyes open and which the subject believes are real (Cohen, 1970).

Mood changes

While the perceptual effects are consistent from person to person, the mood changes are variable but nearly always present. They differ from person to person, from occasion to occasion in the same person, and often vary during the same drug experience (Katz, Waskow, Olsson, 1968). At one end of the continuum is extreme euphoria manifested in ecstasy and exaltation which some subjects interpret in religious–mystical terms (Masters and Huston, 1966) feeling that the experience has given them tremendous insight into the meaning of life. At the opposite end of the continuum some subjects experience severe depression, anxiety and paranoia which seem to be a result of the perceptual distortions and the mood before taking the dose. If the subject is feeling relaxed before the session he is more likely to experience euphoria during the session. An inexperienced subject is more likely to be apprehensive before taking the drug and it is found that these subjects tend to report anxiety more than experienced users (Smith and Rose, 1968).

The perceptual changes and the duration of the experience interact with the anxiety and result in a panic reaction. The subject feels that he has lost control completely and that he will never return to reality (Becker, 1967). At this stage the subject may seek medical assistance to recover from the 'bad trip'. When examined by the doctor he may be confused, depressed, anxious and suspicious with motor restlessness, grimacing, bizarre behaviour and inappropriate affect (Stern, 1966). It is significant for the notion of chemical psychoses that diagnosis is extremely difficult unless the subject admits that he has ingested LSD (Stern, 1966). The anxiety induced by the drug experience outlasts the drug's presence in the body by many days, but hardly ever recurs later than 2 or 3 weeks after the session. It must be emphasized that although these reactions can occur with higher doses of the drug, they are most likely in the illicit drug session. Cohen (1970) reviewed the outcome of administering hallucinogens to over 5,000 subjects and estimated that only 0·08 % experimental subjects and 0·18 % of hospital patients given hallucinogens had psychotic reactions lasting for more than 24 hr. In non-medical situations the percentages are undoubtedly higher, judging by the number of admissions to hospital.

A recurrence of part of the LSD experience or 'flashback' may happen days or weeks, even months, after the drug was ingested (Stern, 1966); in particular the hallucinatory experiences and anxiety seem to be common effects which reappear. These symptoms are particularly disturbing to the LSD user because they are not associated with taking the drug, and often give rise to feelings of depersonalization and panic reactions. The reasons for the recurrence are not understood but they are more likely to occur when the person is under stress (Smart and Bateman, 1967). For example, students who have taken LSD weeks before an examination may have a 'flashback' during the examination period. While these symptoms are seen most often in subjects that have taken LSD frequently, I have seen a number of patients that claim they had only taken LSD on one occasion, and in one case the dose was known to be less than 100 μg. In general, recurrence of the symptoms is only likely within the first 3 weeks of ingestion, which suggests that the LSD has produced a slowly reversible biochemical change that can be exacerbated by the chemicals activated during stress, producing a psychosis.

However, there do seem to be some cases where the biochemical change is not reversible and a schizophrenic-like psychosis develops (Stern, 1966). Particularly serious are the cases where the patients have been given the drug without them being aware of it. The panic reactions are particularly intense and prolonged, particularly in children (Cohen and Ditman, 1963; Millman, 1967). However, the majority of prolonged reactions are seen in subjects who have a previous psychiatric history suggesting that the drug has been taken as a form of self-medication (Glickman and Blumenfield, 1967). Certainly, over two-thirds of the LSD 'psychotics' admitted to several hospitals in the United States have been found to have had treatment for psychiatric disorders previously (Frosch et al., 1965; Ungerleider, Fisher and Fuller, 1966). The latter authors conclude that the incidence of serious temporary complications following LSD ingestion is not infrequent and that it is surprising that such a profound psychological experience has permanent adverse effects so rarely. From this brief survey it is clear that any LSD user is taking a risk of precipitating a psychotic reaction, but the risk is minimized when the drug is taken under medical supervision. If the source of LSD is illicit, then the dangers are compounded by impurities and other additives, like amphetamines (Cohen, 1970).

At this stage the effects of LSD and the schizophrenias will be compared. The perceptual effects are dissimilar, because in the schizophrenias the hallucinations are predominantly auditory, while LSD induces mainly visual changes. Both conditions show disturbances of affect, but schizophrenias have blunted affect while LSD usually produces euphoria. The common factor of the two states appears to be the anxiety reaction, which may be a secondary symptom resulting from the stress of the LSD and schizophrenic experience rather than being part of the primary disturbance. What is clear is that one consequence of the LSD experience may be the precipitation of a schizophrenic-like psychosis in susceptible individuals. As Cohen and Ditman (1963) concluded 'It is possible that LSD disrupts psychic homeostatic mechanisms and permits reinforcement of latent delusional or paranoid ideas' in individuals who are predisposed towards psychosis. In summary, LSD does not provide a good model of any of the schizophrenias, although the dramatic nature of the drug experience itself may precipitate a psychosis like schizophrenia.

Neurochemical basis for LSD hallucinations

The perceptual changes that we discussed in the last section were mainly visual, although auditory, gustatory and olfactory changes have been found as well. As a consequence, research has concentrated on the neural changes in the visual pathways. There are a vast number of papers that have examined the various steps in the primary sensory pathways, and it is only possible to refer the reader to a review of these by Brawley and Duffield (1972) for more information. Briefly, the evidence discussed in this article indicates that while LSD affects the spontaneous discharge of the retina, the locus of the effect is unknown, and it may be due to centrifugal fibres coming directly or indirectly from the reticular formation. At the lateral geniculate nucleus the effects of low doses are unclear, but high doses suppress spontaneous and evoked potentials. The apparent lack of effect of low doses may be an artifact due to the use in these studies of barbiturate anaesthetics which block the effects of LSD (Key, 1965). It is conceivable that LSD may act on geniculate synapses, perhaps by decreasing inhibition at these synapses. At the present time it is thought that serotonin may be one of the inhibitory transmitters at this nucleus (Curtis and Davis, 1963), and this notion would link up with the serotonergic blocking action of LSD which we will discuss later in this section.

Research by Bradley and his group (Bradley and Hance, 1957; Bradley and Key, 1958; Key, 1965) suggests that LSD has its effects on the reticular pathways making them more sensitive to the input along the collateral pathways. Thus, Bradley and Key (1959) found that LSD did not raise the level of activity in the brain stem in the absence of sensory input. There are both supporting and contradictory evidence from other studies (Brawley and Duffield, 1972). Iontophoretic studies of the lateral bulbar reticular formation, i.e. not the raphé nuclei, show that LSD antagonized serotonin excitation of cells (Boakes, Bradley, Briggs and Dray, 1970). There is some evidence (see Brawley and Duffield, 1972) to suggest that serotonin may be mediating inhibition of electrocortical activity (itself mediated by the cholinergic system, see Chapter 4), and so LSD causes electrocortical activation by blocking these neurons.

In two successive paragraphs it has been stated that LSD might have serotonin blocking properties. Consistent with this hypothesis LSD blocks slow-wave sleep (Oswald, 1968) that seems to be mediated by the serotonin pathway originating in the raphé nuclei of the reticular formation (see Chapter 6). The first direct evidence that LSD might block serotonin's pharmacological activity comes from LSD's antagonism of serotonin's action on the uterus (Gaddum, 1957). Studies of the effect on brain serotonin showed that LSD produced a rise in the level of serotonin in the brain and concomitantly a fall in the major metabolite of serotonin, 5-hydroxyindoleacetic acid (Anden, Corrodi, Fuxe and Hökfelt, 1968; Tonge and Leonard, 1969) suggesting that there was decreased release of serotonin. When the synthesis of serotonin was blocked by parachlorphenylalanine, the rate of depletion of serotonin was very significantly decreased (Tonge and Leonard, 1969) which confirms this hypothesis.

Both Andén et al. (1968) and Tonge and Leonard (1969) showed that LSD acts on serotonin receptors as well, and Anden et al. suggest that a parsimonious explanation might be that the serotonin was slowing serotonin release by inducing "a negative feedback mechanism on the presynaptic 5-HT neurones via a neurone chain" (Andén et al., 1968, p. 5). There is no precedent for such a mechanism in any neural system and it must be regarded as highly speculative. It is simpler at this stage of our knowledge to think of LSD having two opposite effects on serotonin neurons with the blocking action predominating. The consequence of this blockade seems to be removal of inhibition and so activation of the ascending cholinergic pathways producing electrocortical arousal. As we mentioned in Chapter 4, hyperactivity in these cholinergic pathways produced by inhibiting acetylcholinesterase does result in hallucinations. However, discussion of this possibility will be postponed until Section B of this chapter. A second possible contributory mechanism for the effects of LSD, suggested by Andén et al. (1968), is the release of norepinephrine. The release of norepinephrine by LSD and other hallucinogens was confirmed by Leonard and Tonge (1969), but this effect was small and probably explains the occurrence of euphoria (see Chapter 3) but not the hallucinations, although some compounds related to norepinephrine have hallucinogenic properties (see Section C).

B. ELECTROCORTICAL AROUSAL AND HALLUCINATIONS

In chapter 4 there was some discussion of the fact that modifications of eletrocortical arousal produced changes in the selection of stimuli, and thus the awareness of them; in particular decreases in electrocortical arousal were correlated with the incorrect detection of stimuli by the drugged organism and another way of putting this is that it is hallucinating.

Decreased cholinergic activity

During studies of electrocortical patterns during fantasies, illusions and hallucinations, Fink (1960) observed that these phenomena were correlated with cortical desynchronization. A group of effective agents for producing these changes were the cholinolytic drugs. An intensive investigation has been made of a group of these compounds, the piperidyl glycolates, by Abood and Biel (1962). They evaluated them with small groups of trained observers, consisting of psychologists, nurses, medical students and psychiatrists with respect to a number of subjective experiences, including the extent of confusion, disorientation, hallucinations, illusory experiences and mood changes. As one would expect from the result of experiments described in Chapter 4, there was spatial and temporal disorientation with general misinterpretation of the environment, and impairment of spatial and temporal awareness. Although the hallucinations were predominantly visual, there were frequent hallucinations in the auditory, tactile, olfactory and gustatory modalities. The visual hallucinations were of clearly defined objects, like people, or objects and often they could not be distinguished from reality, so that subjects reached for a hallucinated drink, made the appropriate drinking movements and commented on its taste and smell. Real objects were sometimes distorted and very strong illusions were experienced. Subjects also heared voices and carried on conversations with them, frequently 'recognizing' the voice as a familiar one. On other occasions, the auditory hallucinations were of music, single instruments and whole orchestras. Not surprisingly, these vivid experiences resulted in confusion, bewilderment and apprehension, and panic and anxiety reactions occurred in susceptible individuals. As well as these emotional changes, the subjects displayed negativism, paranoia and sudden angry outbursts and did not experience euphoria. There were also feelings of depersonalization and alienation, rather like LSD, but to a lesser extent. With lower doses the subjects could describe the slight effects coherently, but with larger doses the subjects talked aimlessly, losing their train of thought many times and in some subjects there was complete mutism. Typically, the subjects had amnesia for most of these experiences (see Chapter 9 for a discussion of the phenomenon of state dissociation with cholinolytics) although occasional subjects had good recall of the hallucinations.

These effects were seen in all the subjects tested by Abood and Biel (1962), and it is striking that such strong depersonalization and disorientation effects were seen with experienced trained subjects. Thus it is scarcely surprising that these reactions to the drug are even more dramatic in inexperienced individuals, especially those ingesting the drug unknowingly, e.g. by eating parts of the group of plants, the Solanaceae, which contain atropine, scopolamine and other cholinolytics. For example, Lewin (1923; reprinted 1964) cites a case of 17th century villagers who ate a plate of lentils containing thornapple (Datura stramonium) seeds, and showed the typical symptoms. One man thought he was a wheelwright and wanted to drill holes in wood. He grabbed one piece of the wood with a hole burned in it, held it to his mouth, made swallowing movements, shouting 'Now I am almost properly drunk. Oh, this drink is good indeed!' Another went to the blacksmith's forge and shouted for help to catch all the enormous numbers of fish that he saw there. The next day none of them recalled their strange behaviour, and they could not be convinced that they had acted so strangely.

Another case of thornapple poisoning occurred in Jamestown, Virginia at the beginning of the 18th century, when a group of soldiers boiled some thornapple plants, now called Jimson Weed after the incident. Those who ate it developed severe symptoms, if we are to believe the contemporary account in the 'History and Present State of Virginia' written by Robert Beverly in 1705. 'One would blow a feather in the Air; another would dart straws at it with much Fury; and another stark naked was sitting up in a Corner, like a Monkey grinning and making Mows at them; a Fourth would fondly kiss and paw his Companions, and snear in their Faces, with a Countenance more antik than any in a Dutch Droll. In this frantik Condition they were confined, lest they in their Folly should destroy themselves . . .'.

One of the strangest cases of accidental poisoning occurred in the Tennessee mountains (Roueché, 1965) when a family were taken to hospital after eating a meal of soup, spaghetti and meat sauce, tomatoes, corn bread and milk. One of the men was waving his hands and talking gibberish. He was peering into space and grabbing at the air and was so uncontrolled that he had to be restrained in a cot. He kept reaching for imaginary door knobs as if he wanted to get out of the room. Sometimes he appeared to be fighting off a swarm of insects and on other occasions he was calm and pointed up at the ceiling and spoke about the beautiful flowers. However, most of the time he was incoherent. The symptoms were similar to those of poisoning by thornapple and the connection between the meal and the poisoning was that the tomatoes had been grafted on to thornapple to produce a frost-free plant, and as a result the tomatoes had 6·36 mg of scopolamine in each tomato.

Scopolamine poisoning has also been observed as the result of suicide attempts using a 'harmless' sleeping preparation of scopolamine called 'Sominex'. The central nervous symptoms of six patients in Wyoming (Beach, Fitzgerald, Holmes, Phibbs and Stuckenhoff, 1964) were those described earlier: they were stuporous, irritable, semi-coherent, hallucinating and disoriented (Hoffman and Gay, 1959). Similar symptoms were found after ingestion of an asthma remedy containing a mixture of atropine and scopolamine (Goldsmith, Frank and Ungerleider, 1968).

Increased cholinergic activity

The usual method for increasing the activity in central cholinergic neurones is by using doses of an anticholinesterase.

The behavioural effects of acetylcholinesterase inhibition have been investigated after both intentional exposure experimentally, and after accidental poisoning with insecticides. In a study of the nerve gas Sarin mentioned in Chapter 4, Grob and Harvey (1958) administered 22 μg orally and produced mild symptoms consisting of anxiety, restlessness, emotional lability, and insomnia with excessive dreaming and nightmares. At the higher dose when eight additional micrograms were given within 8 hr there was withdrawal and mild or moderate dysphoria. The mildest symptoms were accompanied by a diminution in the voltage of the electroencephalogram and the moderate symptoms were accompanied by abnormal high voltage slow waves, two-and-a-half to six per second, and 70 to 150 mV. Estimates of the inhibition of plasma and red blood cell cholinesterase showed that 22 μg produced around 35 % inhibition of the blood enzyme. In a more extensive study of a compound, EA-1701 (perhaps GE) similar to Sarin, on 93 military volunteers (Bowers, Goodman and Sim, 1964) it was found that psychological changes did not occur until the blood cholinesterase fell to below 40 % of control. This critical value is the same as that given in Chapters 1 and 4 as the critical brain enzyme level for behavioural changes; below this level loss in discrimination performance was observed in animals, and it was suggested that there were attentional deficits. In the Bowers et al. (1964) study, subjects mentioned an inability to concentrate and this was manifested in the subjects finding difficulty in expressing their thoughts. Questions were met with long pauses and conversations were often broken off in mid-course similar to the 'thought blocking' seen in schizophrenics. The subjects were perplexed but not disturbed by the inability to maintain the train of thought. Inappropriate and illogical thought processes were not observed. Hallucinations, delusions and perceptual distortion were not reported, even in subjects with 10 % blood cholinesterase levels. There were, as in the Grob and Harvey (1958) study, signs of anxiety and depression.

Clinical studies of the toxic effects of anticholinesterase insecticides closely parallel these findings. Thus, Parathion in lethal and near-lethal doses resulted in anxiety, confusion, sometimes disorientation, difficulty in concentrating and conversation (Grob, Garlick and Harvey, 1950). The post-mortem on two of these patients showed that the brain cholinesterase levels were 22–39 % in the cortex, and brain stem, while the plasma and red blood cell cholinesterase was between zero and 20 % of normal. Similarly Gershon and Shaw (1961) reported on sixteen cases of chronic exposure to these insecticides in agricultural workers, and in every case concentration and memory were sufficiently disturbed to interfere seriously with work and reading ability. Of these, five were diagnosed as schizophrenics and seven as depressives. One case report was of a scientific field officer checking the efficacy of the sprays. After 4 years exposure he was admitted to hospital with an acute psychosis. He had auditory hallucinations and schizophrenic ideas of religious influence; he thought that he was being called to Rome to replace the Pope. He spoke of sensations of walking on a cloud and of not being in contact with other people and the events around him. He had difficulty thinking and acting and spent a lot of the time sitting around the house 'like a zombie'. He often did things which he did not intend to do, and which he thought stupid afterwards. Prior to exposure, he had been healthy and had taken part normally in sporting and social activities.

In other cases of Gershon and Shaw (1961), auditory hallucinations were present, and one patient thought that voices were saying that someone intended to shoot him. He developed delusions of persecution, and thought people were staring at him in the street and that there were plots to remove him from his job. This prolonged exposure to anticholinesterases in unsuspecting subjects can result in a psychotic pattern resembling schizophrenic behaviour with auditory hallucinations. However, hallucinations were not usually seen in informed subjects exposed to anticholinesterases, although strong evidence for concentration and thought disorders were obtained.

From these two subsections it can be seen that insufficient functional acetylcholine and increased acetylcholine can produce hallucinations. This paradox has been resolved by Feldberg (1964) who invoked the concept of depolarization block described in Chapter 1, Section C, and used as an explanation in Chapters 3 and 4. Feldberg argued that excess acetylcholine blocks central cholinergic transmission, having the same effect as the synaptic block produced by cholinolytics. In support of this notion the critical enzyme activity level for producing hallucinations with anticholinesterases was 70 %, which was the critical activity level for producing a depolarization block in the cholinergic system. In Chapter 4 we saw that in animals doses of anticholinesterase inhibiting the enzyme more than this amount had the same effects on behaviour as cholinolytics. Thus, decreasing the function of the ascending cholinergic pathways seems to disrupt the selection of stimuli and hallucinations can result. The next question to be considered is whether some types of schizophrenia could result from impaired function of the cholinergic system.

Cholinergic hypothesis of schizophrenia

At the same time as the Grob and Harvey studies, Feldberg and Sherwood in Britain were also studying the intracerebral effects of acetylcholine and cholinergic drugs in cats and man. In cats (Feldberg and Sherwood, 1954; 1955) intraventricular injections of acetylcholine and an anticholinesterase induced a stuporous condition. In this state the cats could be put in abnormal postures which they retained for many seconds or minutes. They were capable of normal movements and could jump in the usual coordinated manner. Feldberg and Sherwood claimed that this condition could also be obtained by putting lesions in the diencephalon containing fibres connecting the dorsal and ventral tegmental nuclei and the hypothalamus (see Sherwood, 1958). This suggested that an increase in the cholinergic activity in the lower parts of the brain produced a modification of general awareness, and that this might be the basis of the disorder in some schizophrenics.

One prediction from this hypothesis would be that treatment of some schizophrenics with an anticholinesterase should exacerbate their symptoms. Tests of di-isopropylfluorophosphate showed that in 6 out of 17 schizophrenic patients the psychosis was activated with the reappearance of the florid symptoms which had been present at the start of the illness (Rowntree, Nevin and Wilson, 1950). These consisted of auditory hallucinations, ideas of reference, thought disorder and bizarre behaviour, and these changes persisted for some months after the injection. This was not a form of toxic delirium because there was no disorientation, the brain electrical activity was not characteristic of delirium and the symptoms were characteristic of the patient's earlier symptomatology.

A second prediction would be that agents reducing brain cholinergic activity would ameliorate the symptoms of some schizophrenics. Evidence for this prediction comes from the clinical studies of Sherwood (1952). He injected cholinesterase and the cholinergic blockers pentamethonium iodide intraventricularly into chronic catatonic patients. As an example, Case 1 from Sherwood (1952) will be described. This patient had been in a deep catatonic stupor for about 5 years and was mute, inaccessible and required every attention, including feeding. An injection of cholinesterase was given into the ventricles and within 1 hr she became cooperative, asked for a bed-pan, took a cup of tea and a biscuit and became more accessible. From the day after the injection she became more alert and although mute she responded to instruction, fed herself and was not incontinent. Two weeks after the injection she was out of bed, dressed herself, read and did needlework. Three weeks later she started to reply to questions. Seven weeks after injection she was earning pocket money for work done on the ward. Nevertheless, the tone in her muscles had begun to increase and another dose was given and her marked improvement continued. Six weeks after the second injection she went on the hospital outing to the seaside, after a hairdo and considerable preparation. The following day she lapsed into her pre-injection state of catatonic stupor. Unfortunately, there were no further supplies of cholinesterase available, but injections of penta-methonium iodide produced some improvement but they were not as effective as the cholinesterase injections. Of fifteen chronic patients treated with cholines-terase, six returned to normal at one time or another, and in one case this lasted for 4 years and 2 years in another without further injections.

A second set of studies which support the prediction that a reduction in brain cholinergic activity will be therapeutic for some schizophrenics comes from clinical studies of atropine therapy (reviewed by Forrer, 1956 and Forrer and Miller, 1958). It has been found that treatment of chronic schizophrenics with doses of atropine high enough to produce a coma resulted in a marked improvement. There was a very marked increase in accessibility, better contact with the surroundings and less affective dulling (Forrer and Miller, 1958). Schizophrenics showing high anxiety levels seem to be the most improved (Forrer, 1956; Grisell and Bynum, 1956). Forrer (1956) evaluated the effects of treatment of 87 schizophrenics with mixed diagnoses, of these 75 % were improved. It should be noted that some patients not diagnosed as schozophre-nics were also improved by this form of therapy which casts doubt on either the diagnostic criterion or the specificity of the therapy.

A last prediction from the hypothesis would be that the electrocortical activity of some schizophrenics will be similar to that of normal subjects dosed with an anticholinesterase. It has been found that some schizophrenics had slow voltage arhythmic activity with greater amounts of alpha activity (8–12 Hz) and that sensory stimulation does not block this activity (Farrell and Sherwood, 1956). Hyperventilation produced high-voltage slow activity (Szatmari and Schneider, 1955). After injection with an anticholinesterase there were slow waves of high voltage which were intensified by hyperventilation (Grob, Harvey, Langworthy and Lilienthal, 1947). Nevertheless, there were some schizophrenics who show low-voltage fast activity or no abnormality (Szatmari and Schneider, 1955). Farrell and Sherwood (1956) found that the improvement of the patient's behaviour after injection of cholinesterase was correlated with changes in the alpha activity with the frequency becoming more uniform, i.e. 'more normal', and alpha disappearing with sensory stimulation.

These predictions are consistent with the hypothesis, but there is the problem that in the study of Rowntree et al. (1950) one patient diagnosed as a schizophrenic was markedly improved by the anticholinesterase with a diminution of anxiety, while most deteriorated after this treatment. This finding was supported by a study (Pfeiffer and Jenney, 1957) in which physostigmine, an anticholinesterase, and arecoline, a muscarinic cholinomimetic, were tested in 23 catatonic schizophrenic patients. Of these, 16 patients showed lucid intervals for up to 20 min after injection. This suggests that some types of schizophrenia may result from insufficient functional acetylcholine as well, which would be functionally equivalent to a depolarization block.

C. TRANSMETHYLATION HYPOTHESES OF THE SCHIZOPHRENIAS

Adrenochrome hypothesis

At the beginning of the 1950's a young Englishman working in Canada was impressed by the strength of the analogy between mescaline's hallucinogenic properties and those symptoms of schizophrenia, and the similarity of the chemical structure of mescalin and epinephrine (Osmond and Smythies, 1952; Hoffer, Osmond and Smythies, 1954). Osmond and Smythies argued that schizophrenia might result from the production of a substance similar to mescaline and epinephrine, resulting from the pathological breakdown of epinephrine. This substance was called M substance, and their first paper (Osmond and Smythies, 1952) included a hypothesis by Harley-Mason which suggested that the final step in epinephrine synthesis was the transmethylation of norepinephrine with the methyl group being donated by methionine or choline. A pathological disturbance of transmethylation could lead to methylation of the phenolic hydroxyl groups instead of the amino group, producing hallucinogenic dimethoxyphenylethylamines. The transmethylation step in epinephrine synthesis has been confirmed (Blaschko, 1959), and as we saw in Chapter 3 0-methylation by catechol-0-methyltransferase is an important step in catecholamine metabolism.

Hoffer and Osmond tentatively identified the M substance as adrenolutin. Adrenochrome is the compound which gives a red colour to old epinephrine, which has been oxidized, and further conversion produces adrenolutin. Normally epinephrine is not oxidized, and the small quantity of adrenochrome produced is quickly metabolized to 5,6-dihydroxy-N-methyl indole with small quantities of adrenolutin (3,5,6-trihydroxy-N-methyl indole). In schizophrenics the synthesis of the dihydroxy indole is blocked so that adrenolutin is produced (Hoffer and Osmond, 1960). Production of adrenochrome will occur if the normal transmethylation of epinephrine is blocked. As a result of the excess production of adrenochrome, larger quantities of adrenolutin will result.

There is clinical evidence that doses of old stocks of epinephrine, especially when it was discoloured, induced psychotic disturbances. Accordingly, various psychopharmacological tests of adrenochrome and adrenolutin were carried out (Hoffer, Osmond and Smythies, 1954; reviewed in Hoffer and Osmond, 1968). Fourteen subjects received both placebo and adrenolutin, and the number reporting various sorts of changes will be listed. Nine out of fourteen experienced a change in mood during adrenolutin, with four feeling depressed and one feeling euphoric. Nine subjects felt less anxious after adrenolutin and one experienced more anxiety. Nine experienced more thought disorder under adrenolutin than placebo and six reported some sort of perceptual changes, although none had hallucinations. Eight subjects got a headache from the adrenolutin. Only three subjects showed the majority pattern of changed personality, thought disturbances, mood change, less anxiety, perceptual change and headache, while five exhibited the pattern of changed personality, thought disturbance, mood change and lowered anxiety. Clearly, there is not a consistent pattern of disturbance produced by adrenolutin. Hallucinations are not reported but Hoffer and Osmond (1968) emphasize that the presence of thought disorder with decreased anxiety and decreased insight is more characteristic of the schizophrenias than hallucinations.

Hoffer has claimed that there are abnormally high quantities of adrenochrome in the blood of schizophrenics (Hoffer and Payza, 1960), but this has not been confirmed by Szara, Axelrod and Perlin (1958), using a technique which was highly sensitive to adrenochrome and testing normals, acute and chronic schizophrenics. Hoffer and Osmond (1968) do not comment on, or even mention, this study in their 165-page chapter on adrenochrome in their book, where they mention twelve reports on adrenochrome and schizophrenia, of which eight showed significant differences between schizophrenic and normal comparison groups. The theory makes six interesting predictions on the basis of excess adrenolutin, and these have been substantiated to some extent (Hoffer and Osmond, 1966; Hoffer and Osmond, 1968). They are disturbed glucose metabolism, inhibition of glutamine acid decarboxylase, inhibition of acetylcholinesterase, antithyroid, antimitotic and antihistaminic, as well as a number of unusual clinical changes found in schizophrenics such as hair pigmentation.

As well as proposing a relationship between schizophrenia and adrenochrome and its derivatives, Hoffer and Osmond (1966, 1968) proposed that the actions of hallucinogens like LSD might be mediated via adrenochrome. Hoffer (1958) demonstrated in vitro that LSD increased the conversion of epinephrine to adrenolutin. It was also found that LSD in alcoholic patients increased the activity of epinephrine oxidase, which catalyses the conversion. In addition, LSD potentiated the psychological effects of adrenochrome doses (Hoffer and Osmond, 1960). However, Hoffer and Osmond (1968) point out that the changes produced by adrenochrome and adrenolutin are primarily on thought and mood, in contrast to the visual changes produced by LSD.

In conclusion, adrenochrome and its derivatives produce a number of the symptoms of schizophrenia, but the mood and perceptual changes are totally dissimilar. Thus, this hypothesis cannot be considered as a strong candidate for the biochemical disorder underlying some of the schizophrenias.

Catecholamine hypothesis

Another catecholaminergic hypothesis has the opposite assumption, that excessive methylation is a factor in the biochemical disorder underlying schizophrenia, in this case excessive methylation of epinephrine and dopamine which was the original suggestion of Harley-Mason in the paper by Osmond and Smythies (1952). The origin of this hypothesis was work on mescaline, which is an alkaloid extracted from the peyote cactus. This compound was first studied by Lewin (1923, reprinted 1964) at the turn of the century, and he established that this agent produced perceptual changes including visual hallucinations and euphoria and other effects similar to the hallucinogens discovered much later. Mescaline is a phenethylamine, 3,4,5-trimethoxy-phenethylamine, related to epinephrine and the amphetamines. A study of the structure activity relationship of mescaline (Smythies, Bradley, Johnston, Benington, Morin and Clark, 1967) showed that the psychomimetic properties of mescaline depended on all three methoxy groups in the 3,4,5-configuration. Tests of 3,4-dimethoxyphenethylamine and mescaline in humans show that while mescaline had its usual spectacular effects, 3,4-dimethoxyphenethylamine produced no discernible effects although it was excreted by schizophrenics, producing a 'pink spot' during urine analysis (Hollister and Friedhoff, 1966). In another study it was found that 92 % of schizophrenics excreted this compound, but also that 40 % of normal subjects produce it (Takesada, Kakimoto, Sano and Koneko, 1963). However, it should be noted that both these percentages were greater than those of Friedhoff, that different analysis procedures were used and that an unequivocal identification of 3,4-dimethoxyphenethyl-amines was not made. In the most comprehensive study of 'pink spot' (Bourdillon, Clarke, Ridges, Sheppard, Harper and Leslie, 1965) it was found that 60 % of schizophrenic patients excreted the abnormal product but only 0·3 % of normal control subjects and that more non-paranoid schizophrenics excreted the compound than paranoid. It is possible that the dimethoxy compound may be detected in schizophrenics merely because they are excreting more amines, including epinephrine as a result of stress and anxiety (Friedhoff and Van Winkle, 1965), but in the study of Bourdillon et al. (1965) various other sorts of patients under stress were tested and the 'pink spot' compound was not detected.

These studies suggest there may be a failure of the normal metabolism of some catecholamine so it is available for methylation. The dimethoxy compound is not psychoactive itself, but as we saw the trimethoxy compound mescaline is a potent hallucinogen. Studies have also shown that 3,4-dimethoxyphene-thylamine can be converted to another potent hallucinogen, N-acetyldi-methoxyphenethylamine, which is normally demethylated, but this process may be slowed up in schizophrenics although there is no direct evidence for this at present (Friedhoff, 1969). Another possibility is that the dimethoxy compounds may be just evidence of a general transmethylation disorder, and that the actual symptoms may be the result of transmethylation of another compound, such as an indoleamine.

Indoleamine hypothesis

In a study to test the transmethylation hypothesis, Pollin, Cardon and Kety (1961) administered large doses of amino acids, including *l*-methionine, together with the monoamine oxidase inhibitor, iproniazid, to chronic schizophrenics. The iproniazid was given to increase the tissue concentrations of catecholamines. Behaviour was observed continuously and the patients given tests daily. The major clinical effects of methionine were an increasing flood of associations, word salad, anxiety, hallucinations, disorientation and agitation. Pollin et al. point out that it was not clear whether the methionine and iproniazid exacerbated the schizophrenic symptoms, or whether toxic delirium was superimposed on chronic schizophrenia. Nevertheless, *l*-methionine is a precursor of S-adenosylmethionine, which donates its methyl group in transmethylation, and so the finding is consistent with the Harley-Mason hypothesis of transmethylation.

There was also evidence in this study that *l*-tryptophan also intensified the symptoms in some schizophrenics, and this amino acid is a precursor of serotonin (see Chapter 1). This research suggests that control of dietary amino acids might be one way of therapy. In a large study of the effect of amino acids in the diet on schizophrenic symptoms, Himwich and his group at the Thudichum Psychiatric Research Laboratory at Galesburg, Illinois, controlled the amounts of methionine and tryptophan in the diet to see if this would improve behaviour (Berlet, Matsumoto, Pscheidt, Spaide, Bull and Himwich, 1965). No clear pattern of improvement emerged in the nine patients on the limited diets. Of course, it was not possible to eliminate these amino acids completely from the diet without risking the breakdown of body tissues. It did emerge clearly that there was a correlation between disturbed episodes and increased urinary indoleamine excretion. These were tryptamine 5-hydroxyindole acetic acid, a metabolite of serotonin and 3-indoleacetic acid, a metabolite of tryptamine. They postulated that the increased amounts of tryptamine might form methylated indoleamines, including hallucinogens.

As a follow-up of this study the Himwich group have observed the formation of bufotenine in two out of four schizophrenic patients after they had been given *l*-cystenine and the monoamine oxidase inhibitor tranylcypromine, (Tanimukai, Ginther, Spaide, Bueno and Himwich, 1968). More recently (Himwich, Narasimhachari, Heller, Spaide, Haškovec, Fujimori and Tabushi, 1970) compared the excretion of tryptamine, of bufotenine, N-dimethyltryptamine and 5-methoxy-N-dimethyltryptamine in 24 hr urinary samples by six normal and two schizophrenics, a paranoid and a hebephrenic, after *l*-cystenine given together over a 3-week period. During the experimental period all the subjects lived in adjacent wards and were fed in the same ward, and their behaviour was rated by the same psychiatrists. It was found that only the schizophrenics showed any effects of the treatment on behaviour. Though both normals and schizophrenics showed rises in tryptamine during the combined treatment period, only the schizophrenics excreted the three

N-methyltryptamines, and that the appearance of large quantities was detected before and during periods when the symptoms were worst. It was interesting that N-methyltryptamines occurred sporadically in the schizophrenics prior to treatment. This suggests that normals cannot produce these hallucinogens in measurable quantities without the use of a monoamine oxidase inhibitor which blocked the normal breakdown of tryptamine to indole acetic acid and of serotonin to 5-hydroxyindole acetic acid. In schizophrenics the excessive quantities of these indoleamines are degraded by means of some N-methyl-transferase to the N-dimethyltryptamines (see Mandell and Spooner, 1968 and Kety, 1969 for further discussion of the precursor load strategy in psycho-chemical research).

The N-dimethyltryptamines are hallucinogenic compounds (Szara, 1957), and after an injection of N,N-dimethyltryptamine, there were visual illusions and hallucinations with euphoria. Auditory hallucination rerely occurred and anxiety was transient. Some depersonalization was experienced, and its action was similar to LSD but more intense and shorter lasting (Szara, 1957). Once again the hypothesized product of abnormal metabolism is a potent hallucinogen, but it mirrors the symptoms of schizophrenia rather poorly. Nevertheless, the studies of Himwich do suggest once again that excessive trans-methylation may be the biochemical disorder underlying the schizophrenias.

Accordingly there has been a 10-year search for an enzyme which would be able to N-methylate endogenous amines and produce hallucinogenic compounds. One enzyme was first identified in rabbit lung by Axelrod (1961), and he found that it would convert serotonin to bufotenine amd tryptamine to dimethyltryptamine, and dopamine to phenylethylamine derivatives. Mandell found a more specific enzyme in the brain of chickens and men which would only N-methylate indolamines, but not phenylethylamine compounds like norepinephrine (Morgan and Mandell, 1969; Mandell and Morgan, 1971). The presence of this enzyme in the human brain has been confirmed by Saavedra and Axelrod (1972), and they also identified tryptamine in the rat brain. If the tryptamine was injected in the ventricles of the rat brain, the latter experimenters were able to recover dimethyltryptamine. As we saw, this agent does not mimic the symptoms of schizophrenia very well, but it may be a related indole which interferes with the serotonin. It has been reported that some schizophrenics showed reduced slow wave sleep (Caldwell and Domino, 1967) which would be consistent with this. If it does interfere with the sero-toninergic raphé pathways, then it will be likely to disrupt the serotonin pathways in the lateral bulbar reticular formation which are believed to inhibit electrocortical activity (see last subsection of Section A this chapter). If this is true then we have a parsimonious explanation of hallucinations in terms of disturbances of electrocortical arousal, the mechanisms important in the suppression of irrelevant stimuli.

8
Biochemical Bases for Drug Dependence

For a number of years the World Health Organization Committee on Addiction-Producing Drugs have attempted unsuccessfully to define drug addiction for use in international control, and in 1964 they recommended that the term drug dependence be used instead (WHO, Committee, 1964). This was 'defined as a state arising from repeated administration of a drug on a periodic or continuous basis' and included both psychic and physical dependence. They felt that its characteristics varied with the drug involved so that there would be drug dependence of the morphine type, of the amphetamine type, of the cannabis type, of the barbiturate type etc. In this chapter some of the evidence concerning the biochemical basis for each type of drug use and abuse will be discussed. As a prelude to considering these drugs some discussion of tolerance and abstinence phenomena will be given.

A. TOLERANCE

Tolerance can be defined operationally in two different ways. First it can be defined as the decrease in biological response that results from repeated administration of the same dose of a drug. Second, it can refer to the fact that the dose of a drug has to be increased in order to produce the same biological response as the initial dose. Physical dependence refers to the changes in the organism underlying tolerance, such that withdrawal of the drug results in the manifestation of abstinence symptoms; the symptoms can be reduced or abolished by another dose of the drug. As an illustration of these phenomena some experiments dealing with the effects of repeated injections of anti-cholinesterase will be described.

These experiments were part of a series by Russell and Warburton, and were concerned with the effect of repeated injections of di-isopropylfluoro-phosphate on drinking behaviour (Russell, Warburton and Segal, 1969). Animals were deprived of water for 23 hr and then allowed access to a drinko-meter for 1 hr; after water intake had stabilized the drug was injected every third day. On the first day the water intake was depressed to two-thirds of the baseline level because the di-isopropylfluorophosphate had inhibited acetyl-cholinesterase activity to about 30 % of normal, and resulted in a depolarization block (see Chapter 1) of the cholinergic drinking system, described in Chapter 2. During the 2 days after the injections there was a recovery of water intake to baseline levels. The second injection depressed intake once again, but this time it decreased to only 80 % of normal. Similar effects were found in stimulus processing tasks (Warburton and Segal, 1971; Russell, Warburton, Vasquez, Overstreet and Dalglish, 1971), which are also thought to be mediated by cholinergic pathways in the brain (see Chapter 4).

A number of hypotheses could account for the tolerance effects observed in this study (Russell et al., 1971a). Firstly there could be feedback control of acetylcholine production so that less transmitter was synthesized and released preventing a depolarization block. Secondly, there could be decreased sensitivity of the postsynaptic receptors in the sense of increased threshold or loss of a proportion of receptors. Thirdly, there could be a mediation of drinking by a non-cholinergic system. Each of these mechanisms has been demonstrated in studies using cholinergic drugs. For example, end-product inhibition of acetylcholine synthesis by excess transmitters was found by Kaito and Goldberg (1969); decreased sensitivity of cholinergic receptors by excessive amounts of acetylcholine in the synaptic cleft was demonstrated by Meeter (1969) among others; and neural redundancy as a tolerance mechanism was found in the peripheral cholinergic system by Martin (1970). Tests of each of these explanations were made by Russell and coworkers.

The most likely explanation of the tolerance in this particular instance was end-product inhibition (Russell and Warburton, 1973), but assays of choline acetylase showed that there was no change in the activity of this synthesizing enzyme during the development of tolerance (Russell, Cotman, Carson, Overstreet, Doyle, Dalglish and Vasquez, unpublished). However, decreased sensitivity of the cholinergic receptors was demonstrated by tests with atropine (Russell, Vasquez, Overstreet and Dalglish, 1971a). In this study cholinolytics had a greater effect on the tolerant animals than on the normal rats, suggesting that there was a decreased sensitivity of the receptor sites. This finding was inconsistent with the third hypothesis where it would be predicted that if a second non-cholinergic system had taken over less effect than, or at least no difference from, the controls would be expected in the tolerant animals. In another study it was demonstrated that withdrawal of the drug resulted in a deviation from baseline once more, and that performance had not returned to normal within 60 days (Russell, Vasquez, Overstreet and Dalglish, 1971b). Thus 'normal' performance depended on the continued presence of the drug.

From these studies it can be seen that some of the basic systems of the brain are capable of adjusting biochemically to maintain behaviour within normal limits, but it must be pointed out that the rate of adjustment of particular behaviours depends on the extent of their dependence on the neurochemical systems modified. Thus food intake which depends on an adrenergic system (see Chapter 2) was not impaired substantially by changes in cholinergic neurochemical systems (Russell et al., 1969; Russell et al., 1971b; Russell and Warburton, 1973) while discrimination performance which does depend on the integrity of a cholinergic system was disrupted (Russell et al., 1971a, b, c; Warburton and Segal, 1971). Adaptation of the organism to the repeated presence of the drug the organism did not result in a 'normal' biochemical system, and this was shown by withdrawal producing disrupted performance.

The magnitude of abstinence symptoms depended to a certain extent on the dose, and the direction of the effects was opposite to those of tolerance. Therefore, it might be expected that in the case of human dependence if a drug produced euphoria and anxiety reduction, then withdrawal would result in dysphoria. The dysphoria would be eliminated by a further injection and so, escape from, or avoidance of, the abstinence symptoms could form the basis for drug dependence. This possibility will be discussed in the following sections.

B. OPIATE DEPENDENCE

Opium has been known for thousands of years and its effects have been described by authors ranging from Homer to de Quincey. Medically opiates have been used as analgesics and also they produce a sensation of euphoria and calmness in the patient. The principal active ingredient of opium is the alkaloid morphine, and the effects of morphine are much more potent than those of raw opium. Attempts to obtain an analgesic without the dependence-producing properties of morphine led to the synthesis of heroin and over 400 other derivatives, but the search has not revealed a compound superior to morphine in terms of analgesia and low-dependence properties.

Dependence on these drugs develops even though the relief of pain is not involved, and this process seems to fall into two stages, the initiation phase and the physical dependence stage. The first dose of an opiate gives the subject elevated mood with the amount of euphoria depending on the dose administered and the route of administration. The effects of injecting morphine into naïve normal subjects were described by Binz (1895) as follows: 'After a few minutes there is an undefined feeling of general comfort. The mental faculties are agreeably stimulated, the brain seems more active and without any sense of oppression. Fantastic lights and glimmerings appear before the eyes. There is a desire to remain undisturbed; the slightest attempt at movement is a trouble. Questions are answered only indefinitely; glimpses of indistinct, agreeable visions appear'. These hallucinations last only a short time and then the subject falls asleep. This sleep is normal in the sense that he can be awakened although there is a suppression of paradoxical sleep (Oswald, 1968). The intensity of this euphoria seems to depend on the person tested, but typically there are two effects. There is an intensely pleasurable thrill for a few minutes when the subject is injected intravenously; these have been described as the 'impact effects' by Lindesmith (1970), and are followed by a feeling of pleasurable gratification that persists for hours after the injection, the 'coasting effect' of Lindesmith. The effects with other opiates are similar to these with perhaps more intense euphoria with heroin and less after other opiates like codeine, meperidine and methadone.

The opiate novice finds that if the subsequent injections are closely spaced the doses must be augmented to achieve the same impact effect. This represents the development of tolerance to the effects of the drug. After a while the impact and coasting euphoria effects of the injection diminish as physical dependence develops, and repeated injections are then required by the addict to alleviate the unpleasant depression that begins to occur as the drug effects wear off. The drug injection has become a different event in the life of the person; it has changed from being a positive reinforcement to one of postponement of negative reinforcement. In between doses the addict claims that he feels normal, but this may be contrasted with the addict's powerful attachment to the drug and the frequently criminal life style which enables him to obtain it (Lindesmith, 1970). However, as Lindesmith points out, the regular opiate user is not markedly different from normals from the organic point of view, and he suggests that the 'hook' in the opiate dependence is the prevention of withdrawal and the maintenance of 'normality' rather than the euphoria, the escape from negative reinforcement rather than obtaining positive reinforcing effects. In the next two sections we will infer the possible biochemical basis of the human euphoric effects from animal studies.

Positive reinforcing properties

If a drug has positive reinforcing properties it would be expected that the animal would prefer solutions containing it to pure water. Unfortunately, most animals will not drink morphine solutions because of their bitter taste so that they do not normally ingest enough of the drug to produce any effects. However, if rats are deprived of any other water source they will drink morphine solution to relieve their thirst (Stolerman and Kumar, 1970). Over a period of a month, choice tests every third day showed the development of preference for morphine solutions until the intake was 60 % of the total fluid intake. This is rather slow learning, but this is probably the consequence of the delay between ingestion and the onset of the reinforcing effects. Quicker learning has been found in the self-administration studies which were designed to overcome an animal's aversion to the taste of morphine solutions.

In one study (Deneau, Yanagita and Seevers, 1969) monkeys were allowed to move freely around their cage and during this exploration they discovered the lever which injected a small quantity of opiate intravenously. It was found that seven of eleven monkeys spontaneously initiated self-injection of morphine. There was a gradual escalation of the dose over 6-week period, and then the dose level stabilized. The remaining four monkeys initiated self-administration after they had received regular injections of morphine every 4 hr. During the period of escalating dosage they were drowsy and apathetic after injection, but in the stable phase they appeared normal although less active. None of the monkeys stopped self-injecting voluntarily due to the establishment of the physical dependence. This period will be discussed later, but first additional evidence will be presented on the positive reinforcing properties of morphine's effects. In an interesting study, Beach (1957) injected rats with morphine and immediately forced them to run to the non-preferred arm of a Y-maze. Once in the goal box they were confined for 1 hr while the drug took effect. After twelve daily runs they were tested for their preference for the maze arms, and it was found that the rats preferred to run to the goal box where they had experienced the effects of morphine. In summary, the effects of morphine are reinforcing, and will induce consummatory behaviour, in the form of self-injection, and appetitive behaviour, in the form of maze running. Next we will discuss the biochemical basis of these reinforcing effects.

From the evidence discussed in Chapter 3 and considering the euphoric effects in man it seems likely that the reinforcing properties of morphine may be due to its action on the norepinephrine pathways of the median forebrain bundle. Collier (1969, 1972) has argued that all 'addictive' drugs including morphine enhance activity in these norepinephrine pathways. Indirect behavioural evidence comes from the finding of increased motor activity in mice injected with morphine (Rethy, Smith and Villarreal, 1971) and this effect can be blocked by reserpine (Hollinger, 1969). Biochemical studies have examined the effects of morphine on the synthesis and release of norepinephrine in the rat brain (Clouet and Ratner, 1970; Smith, Villarreal, Bednarczyk and Sheldon, 1970). It was found that morphine increased the incorporation of radioactive tyrosine into dopamine and norepinephrine, but that at 2 hr there was a decrease in norepinephrine in the diencephalon (Clouet and Ratner, 1970). This suggests very strongly that morphine increased both the synthesis and release of norepinephrine, and so produced short-term enhancement of activity in the brain norepinephrine pathways, including median forebrain bundle. However, the long term outcome of this activation is norepinephrine depletion, and histochemical fluorescence techniques discussed in Chapters 1 and 3 have demonstrated that morphine decreases norepinephrine in the ventromedial tegmental region containing the median forebrain bundle (Heinrich, Lichtensteiger and Largemann, 1971). On the catecholamine hypothesis of mood, depletion of the amine would result in a depressive state and this is one of the first symptoms of morphine withdrawal as we shall see in Section B.

Physical dependence

The physical dependence phase is characterized by a need for the effects of the drug to maintain normal function. Withdrawal of morphine results in the appearance of an abstinence syndrome which is similar, in some ways, to influenza. The symptoms include nausea, vomiting, diarrhoea, increased respiration, temperature and heartrate, tremor, muscle cramps, sweating, running eyes and nose, dehydration as well as anorexia, with depression, anxiety and restlessness. These effects start to appear within a few hours of the last dose, reaching a peak intensity at 12 hr, and subsiding spontaneously in about a week, the specific course varying with the dose and the opiate used (WHO, 1964). Although the symptoms are merely those of severe influenza they are sufficiently unpleasant to act as negative reinforcers and to increase the probability of behaviour which postpones them. We are obviously precluded on ethical grounds from experimenting on humans, but abstinence symptoms occur in animals, which provides us with the opportunity of using animal behaviour as a model for human dependence. In rats severe abstinence symptoms are characterized by agitation, restlessness, rapid respiration, hypersensitivity to touch, diarrhoea and weight loss, a group of symptoms which are similar to those described in man. This gives us a good model for studying symptom-escape and symptom-avoidance.

In an experiment on symptom-escape by Nichols (1963) rats were given 25 daily injections of morphine to establish dependence and were then placed in a cage and deprived of morphine and water for 24 hr. When offered morphine solution they drank it and continued to drink it whenever they were offered it, presumably to reduce the withdrawal distress. A yoked control rat was then injected with the same amount of morphine as the experimental rat had drunk, so that their morphine intake was identical. Both rats were given tests to determine their preference for water or morphine solution, and the active-intake animals preferred the morphine solution, while the passive-intake rats selected the pure water. The dependent rats were then withdrawn from morphine for 2 weeks and when retested they still showed a preference for the morphine solution even though the withdrawal symptoms had subsided. Moreover, some experimental animals 'relapsed' even after 7 weeks. In the above experiment it was the reduction of abstinence symptoms which reinforced the drinking and this was also demonstrated in an experiment in which Davis and Nichols (1962) injected morphine into the rats before each drinking session. The prior injection reduced the abstinence symptoms and so the animal drank less morphine in the preference tests. In these experiments it was noticed that some animals drink more than others (Nichols, 1963). This susceptibility was studied further (Nichols and Hsiao, 1967) and it was found possible to breed 'addiction prone' and 'addiction resistant' groups of rats.

We have seen that the reduction of abstinence symptoms will also sustain consummatory behaviour. In order to study appetitive behaviour Weeks (1962) devised the self-injection equipment discussed in the previous section which enabled direct injection into the blood stream. Rats were made dependent on morphine and then placed in a standard lever pressing box and soon they pressed the lever at regular 2-hourly intervals. Halving the concentration of the morphine solution resulted in an increased response rate, but not an exact doubling. If the reinforcement schedule was changed to a ratio of ten presses for an injection, the rat pressed ten times every 2 hr or so, and rested in between. He did not press more and more frequently as the time since the last injection increased, and Weeks (1964) has argued that rats do not anticipate their needs but wait until the onset of the abstinence symptoms, although these were not obvious to the experimenter. They do not appear to be responding for the 'pleasurable' effects because the interresponse times are much longer than those found in electrical self-stimulation studies. Thus, these experiments demonstrated clearly that escape from abstinence symptoms can be established in rats and, as we saw in the experiments of Deneau et al. (1969) in the last section, in monkeys. Self-administration has also been observed with other opiates and is not specific to morphine (Deneau, 1972).

Tolerance for opiates has been described in considerable detail by Seevers and Deneau (1963). The interested reader is referred to that paper for an extended discussion of the topic, but a summary of the major conclusions is included here. The first sign of tolerance is a shortening of the duration, and a diminution of the biological effects of the drug opiates, in particular the euphoria, lassitude, malaise, loss of appetite and analgesia. However, a stage is never reached with morphine where the drug is totally ineffective. The rate and degree of development of tolerance is dependent on a number of factors, including, the individual, the species, biological response tested, the type of opiate, the dose and its frequency of administration. Recovery from tolerance will similarly depend on the same sorts of factors.

In the remainder of this section the development of tolerance in the central nervous system pathways will be discussed. Tolerance of the central nervous system has been demonstrated by intraventricular injections of morphine into monkeys (Eidelberg and Barstow, 1971). Cannulae were permanently implanted and the effects of morphine tested on a fixed ratio schedule. The measure taken was the infused dose which was required to eliminate responding, and it was found that this dose had to be increased progressively over days. Withdrawal of morphine resulted in hyperactivity, hyperirritability, retching, vomiting, diarrhoea and behaviour suggesting abdominal pain. These findings suggest that many of the tolerance effects of opiates may be due to changes in the pathways adjacent to the ventricular system, e.g. hypothalamus and brain stem.

As we have seen already from Section A, single injections of opiates produce clear changes in the catecholamine systems in the brain, including the adrenergic pathways (Clouet and Ratner, 1970). The same authors also went on to examine the changes in catecholamines after tolerance had been established by injecting doses of morphine for 10 days with the dose increasing from 20 mg/kg to 60 mg/kg. The incorporation of radioactive tyrosine into catecholamines was increased in the hypothalamus, basal ganglia and midbrain compared with control rats or rats injected once. In all these areas the rate of turnover of norepinephrine was higher in the control animals, which suggested that the biosynthetic pathway for norepinephrine was operating more effectively in the tolerant animals. This increased synthesis prevented the tolerant rat from being depleted of brain amines, and so the rate of disappearance of radioactive norepinephrine from the brains of tolerant rats was the same as that of control animals. This finding means that both activity in rats and mood in humans will be 'normal' provided the injections are continued.

In addition, tolerance to morphine's effects seem to develop in the other three major transmitter systems, dopamine, serotonin and acetylcholine, and in time it should be possible to relate the behavioural changes observed in dependence and withdrawal to these systems in the brain.

Physical treatment of dependence

Early efforts to rehabilitate opiate addicts by psychotherapy had brought discouraging results, and by the middle of the 1960's pessimism and defeatism were common among both patients and therapists. In an evaluation of the effectiveness of psychotherapy in the treatment of addiction, thirty analysts pooled their data: there was only a 14 % remission rate which, in this sort of illness, approaches the spontaneous remission rate (Nyswander, Winick, Bernstein, Brill and Kaufer, 1958). In this climate of defeatism the husband and wife team of Dole and Nyswander (1965) induced a small group of addicts, for whom psychotherapy had not worked, to substitute the oral intake of methadone for intravenous opiates. Methadone seems to block the euphoric 'high' found after opiates, does not itself produce impact effects when taken orally (Dole, Nyswander and Kreek, 1966) but it blocks the abstinence symptoms. This procedure of methadone maintenance resulted in a dramatic decrease in the use of illicit drugs by this group of patients, and the procedure was extended to large groups of patients (Dole, Nyswander and Warner, 1968; Jaffe, Zacks and Washington, 1969; Gordon, 1972). The results were impressive as the patients took up regular jobs, stopped using heroin and so decreased the criminal activity needed to obtain drugs. Some of these patients seemed to function so well on maintenance doses that it was suggested that the legal status of methadone be redefined in the United States (Jaffe, 1970). In most states doses of methadone could only be prescribed under the most strictly controlled conditions. Although the Bureau of Narcotics and Dangerous Drugs frowned upon addiction maintenance programmes where there were no plans to withdraw the drug, considerable discussion took place in 1970 about the possibilities of easing the restrictions (Walsh, 1970).

In recent years the initial enthusiasm produced by apparent success of maintenance therapy has been tempered by a number of findings. These were summarized by Lennard, Epstein and Rosenthal (1972) in a paper entitled 'The Methadone Illusion' which attacked the three main successes of methadone maintenance, the absence of impact effect with methadone; the block of the euphoric effects of heroin; the 'normal' life of the patient on maintenance. They point out that methadone does not block the euphoric effects of opiates but merely increases the dose required to produce an effect. In addition, it is only the oral intake of methadone which does not produce impact effects; intravenous doses do. Many methadone patients do 'cheat' by taking heroin and other drugs and in a study (Chambers, Russell and Taylor, 1972) on patients who had been on methadone therapy for 6 months, it was found by urine analysis that 82·5 % of them had abused at least one of the detectable drugs. Specifically, 77·5 % of the patients had been taking heroin, 30 % had taken barbiturates and 25 % had been taking amphetamines. Fourteen months later the incidence of cheating had increased to 97·4 % with 92·3 % taking heroin. Interviews with the patients showed that it was futile to rely on their reports since on the 14-month check only a third of them admitted any cheating to the therapists.

Evidence for the 'normality' of patients on methadone maintenance is surprisingly poor, but it does not seem to be substantially different from the 'normality' of the stabilized opiate addict. Lennard et al. (1972) noted that these patients appear somnolent and their reflexes abnormal. They perspired more, were constipated and were often sexually impotent. From this it can be seen that methadone, at least in the usual outpatient situation, does not live up to the claims of its supporters. Accordingly, in 1972, the Food and Drug Administration in the United States tightened up the distribution of methadone to eliminate abuse (Holden, 1972). Unfortunately, the emphasis in most clinics is still on methadone as a *treatment* for dependence rather than as an adjunct to be used with other rehabilitative services to reduce dependence on opiates.

Another possible chemical approach to the control of dependence would be the use of opiate antagonists like nalorphine and cyclazocine. These compounds block the behavioural effects of opiates in animals and man (Weeks, 1964), and prevent the release of norepinephrine by morphine (Rethy et al., 1971). Unfortunately, nalorphine has too short a duration of action to be suitable, but clinical trials of cyclazocine were more successful and 40 % from a population of 450 addicts continued treatment with this drug. Almost all of these patients reported that although they had experimented with opiates since beginning treatment, they had not become dependent on them again (Fink, 1970) and urine tests have supported these reports. Wikler (1968) has argued that the antagonists prevented relapses by blocking the reinforcing effects of the opiates and allowing extinction of the opiate-seeking behaviour to occur. It is obvious that an important part of this reconditioning process is an extensive rehabilitation programme and this formed an integral part of all the studies discussed by Fink (1970). Group therapy in various forms has now become an important part of the therapeutic armoury. In the United States many of these programmes (Cancellaro, 1972) are based on the conceptual precedents of Synanon (Yablonski, 1965), Phoenix House (Biase, 1972) and Reality House (Kaufman, 1972). In these the community exerts group pressure against any member guilty of adverse behaviour, and the emphasis is on rejection of the underworld value system that the addict has adopted. Drug therapy is not sufficient by itself, but it may help considerably in making the patient more receptive to psychotherapy and 'self-help' schemes. Schemes like these must be considered as part of any attempt to return the drug-dependent individual to a normal life in the community.

C. AMPHETAMINE DEPENDENCE

The amphetamine is one of a group of drugs similar to epinephrine and used to control asthma. In 1937 articles were published warning students against using 'pep pills' containing amphetamine for examination studying and in the 1940's, while restrictions on the distribution of amphetamine pills were tightened up, amphetamine remained readily available from druggists in inhalers. Eventually abuse reached epidemic proportions in the United States, and in 1959 Federal legislation was passed making the inhalers available on prescription only (Jackson, 1971).

This particular example of abuse was a prelude to the large scale amphetamine epidemics that have flourished in the 30 years since World War II, in which amphetamines were used by the military on both sides to keep soldiers awake during prolonged missions. It has been estimated that British soldiers were given 72 million amphetamine pills and administration of stimulants on a major scale also occurred in the American, German and Japanese forces (see Jackson, 1971). After the war unused stocks of the pills found their way to the civilian population via the returning soldiers. In Japan amphetamine abuse spread first among the prostitutes and criminals, but since there was no restriction on their production and sale the abuse spread widely among the young. The peak of the epidemic was in 1954 (Masaki, 1956) when it was estimated that 2 % of the population were abusing the drug and in some areas up to 5 % of the 16–25 age group. Public opinion in Japan developed against the drug and strong legislation controlling the manufacture, sale and possession of amphetamines was passed. In 1954, 55,000 people were arrested for drug abuse, but by 1958 the figure had fallen to 271.

During the Japanese epidemic there was little concern in Britain and the Ministry of Health pronounced that amphetamines were non-addicting and non-toxic so that they could be prescribed without risk. In a revealing study by Kiloh and Brandon (1962) in Newcastle upon Tyne it was found that 10 % of the population were having amphetamines prescribed for various disorders ranging from depression to bedwetting. Kiloh and Brandon estimated that in the year 1960 each of these patients consumed about 80 tablets each month. About 20 % of the users were dependent on the drug, and the majority of these were housewives so that the disruption of family life when the women attempted to obtain more drugs can be imagined.

In 1964 a survey of young people sent to remand homes showed positive urine tests for amphetamine were obtained for 16 % of the girls and 18 % of the boys (Scott and Willcox, 1965). A more recent survey in 1969 at the same remand homes has shown that the amphetamine epidemic is on the decline, and that now only 5 % of the admissions had amphetamine in their urine on admission (Scott and Bucknell, 1971).

In the United States since World War II amphetamine use and abuse has been high, with an estimated ten million users, but there is very little exact information on the extent of its illicit use (Rockwell, 1968), although it has been estimated that over 50 % of the annual production finds its way into illicit channels (Walsh, 1964). In the United States the average amphetamine abuser is white, male, middle or upper class, and in the 20–25 age range (Kissin, 1972). Among university students, the undergraduates (10 %) are less likely than graduates (14 %) to use them (Millman and Anker, 1972), and about 35 % of medical students have taken amphetamines on more than one occasion (Smith and Blackley, 1966). The personality of abusers can be characterized as bored and depressed, with derives towards euphoria and sensation seeking (Kissin, 1972; Zuckerman, 1972).

Amphetamine users report that the drug gives them feelings of excitement and omnipotence. The following account of the subjective experience of amphetamine comes from a paper by Gioscia (1972).

'Hey man, dig it here's how it feels. . . . Do you like to drive fast in your car, see, with no windshield, see, and they say you can have NYC (New York City) all to yourself with all the other cars gone. So you go speeding around corners at 90 and open up to 200 miles an hour along Park Avenue, man, whizzing, and spinning around the whole city all to yourself. You can do anything as you want, an' you can go as fast as you want to go. Dig it man, imagine all that power just walking man, or screwing. Wow' (p. 167).

This intense euphoric experience is usually obtained by oral dosing in the first instance, but often users progress to intravenous injections to intensify the effect, i.e. to produce impact effects like heroin. In order to prolong the pleasant sensations the abuser has to give himself another dose after 3 to 4 hr· and often he will take the drug repeatedly for 2, 3 or more days. During this time he will appear manic (see Chapter 3) talking and moving continuously. Repeated intravenous injections result in the occurrence of psychotic behaviour.

Typical behavioural manifestations of amphetamine psychosis are stereo-typed searching and examining movements (Ellenwood, 1967). The amphetamine user engages in repetitive examining, dismantling and rearranging of objects, such as watches and television sets which seems to be accompanied by an intense pleasurable sense of curiosity. In some users this activity takes the form of picking at the skin which is sometimes the result of a delusion that minute parasites have hidden under the skin and in others the looking takes the form of furtive glances at the people around which the individual may describe as attempts to watch the other person without being noticed. Episodes of suspiciousness evolve from this enhanced looking behaviour and this can develop into severe paranoia, with the individual continually looking for signs of persecution. Thus the behavioural patterns induced by the drug interacting within the context of a particular environment can lead to the evolution of paranoid delusional beliefs (Ellenwood, 1972).

Amphetamine psychosis was thought to be rare in Britain, and prior to 1956 only six cases were reported, but in 1958, Connell (1958) published a monograph in which he described 42 cases which he found to have symptoms very much like paranoid schizophrenia, but induced by amphetamine. McConnell (1963) in Northern Ireland found amphetamine effects in 2 % of all psychiatric referals, and distinguished three categories of these. The first type showed symptoms that were indistinguishable from paranoid schophrenia. The second group were dependent on amphetamines and showed some paranoia. The third group showed various side effects of amphetamine dosing but no paranoia. This latter group without paranoia formed 16 % of the group as a whole and McConnell concluded that whatever the initial effects may be, amphetamine does produce paranoid reaction with the intensity depending on the dose and probably the personality of the individual. Intravenous administration was more likely to produce the effects because this route of administration enhances the drug action on the nervous system.

A comparison by Bell (1965) of amphetamine psychosis with those of paranoid schizophrenia revealed a strong resemblance between the two disorders. Delusions of persecution and ideas of reference were present in every case, and delusions of influence were present in both categories. Neither group had any insight into their delusions, although the amphetamine psychotics regained their insight within 10 days of withdrawal. Auditory and visual hallucinations were present in both groups, with the visual type being more common among amphetamine psychotics than among schizophrenics. Disturbances of affect was frequently encountered with depression being more common than euphoria in both groups. The distinguishing characteristic of the schizophrenics was their thought disorder, in the sense of splitting and loosening associations, concrete and bizarre meanings in abstract thought and impairment of goal directed thought. No amphetamine psychotic showed this sort of thought disorder. The amphetamine psychosis can be identified by eight characteristics present in the first 2 days (Jonsson and Sjöstrom, 1970). They are lack of concentration, hallucinatory behaviour, disorganization of thought, increased motor activity, anxiety and fear, suspiciousness, delusions of persecution and lack of insight. Criminal behaviour is likely to occur during this phase of drug taking which often leads to the arrest of the amphetamine absuers. Some of this criminal behaviour is probably a direct consequence of the abuser's overestimation of his own abilities, and an underestimation of the consequences of his actions.

When the amphetamine user stops taking the drug he begins to 'come down' or 'crash' as the drug wears off. Fatigue and depression occur, sometimes to the point of suicide, and after long periods of intoxication the user may lapse into a coma for a day. The only way he can achieve the sensations he seeks is by taking the drug repeatedly, and if the interval between doses is short then the dose will have to be increased. Interestingly, although tolerance develops, the only abstinence symptoms are the extreme fatigue, so that dependence seems to be based on the positive reinforcing effects of the drug rather than on escape from the aversive consequences of withdrawal. This is an important difference between the effects of opiates and those of amphetamine and a similar drug, cocaine.

Positive reinforcing effects

The findings with the self-administration of amphetamine in animals are similar to those observed with opiates; animals given an opportunity to self-inject amphetamine will do so (Deneau et al., 1969). One interesting observation was that the monkey injection pattern was exactly like that of human abusers; they injected continually for days at a time while eating and drinking very little. Then they would 'crash', sleep for a long period, eat and then start the whole cycle again.

In the last section it was argued that heroin's positive reinforcing effects were the result of releasing norepinephrine at synapses in the median forebrain bundle. In Chapters 1 and 3 the effects of amphetamine on the norepinephrine system were discussed in some detail and so only an outline of these will be given here. It is a well established fact that amphetamine enhances adrenergic function by increasing release of norepinephrine from the neuron, and by preventing inactivation of the transmitter by re-uptake (Glowinski and Axelrod, 1966; Glowinski, Axelrod and Iversen, 1966). The result of enhanced activity in the adrenergic neurons ascending in the median forebrain bundle is an increased tendency to respond as a consequence of a reduction in the reinforcement thresholds (Stein, 1964b), discussed already in Chapter 3. The immediate consequences of the increased activation in the hypothalamic reward system in animals was repeated self-injection, just as Deneau et al. (1969) found. In the long term the increased release and reduced re-uptake produced by amphetamine results in a depletion of norepinephrine since it is inactivated by catechol-0-methyltransferase (Smith, 1965). As depletion occurs, mood will change and the 'crash' into depression will occur, and this state will continue until the depleted stores reach functional levels once more.

Chronic effects of amphetamines

There is some evidence (Mandell, 1970) that the effects of chronic administration may be more complex than depletion of norepinephrine. In a number of studies with Morgan he investigated the biochemical consequences of injecting amphetamine into chicks every 12 hr for 5 days. This procedure resulted in increased activity of tyrosine hydroxylase, the synthesizing enzyme for norepinephrine, and choline acetylase, the synthesizing enzyme for acetylcholine in the brain stem of the chick. Thus, the euphoria seen in some patients after withdrawal from amphetamines may be due to the increased capacity of the nervous system for synthesizing norepinephrine as a result of the induction of tyrosine hydroxylase.

The production of hallucinations in amphetamine psychosis has not been fully explained. However, we saw in Chapter 7 that there was some evidence for the production of 3,4-dimethoxyphenylethylamine in some schizophrenics (Friedhoff and Van Winkle, 1962). This compound was thought to be produced by the 0-methylation of dopamine which is in turn converted to N-acetyldimethoxyphenylethylamine by N-methylation. Normally the N-acetyldimethoxyphenylethylamine is quickly broken down, but it is conceivable that if dopamine is being produced in large quantities due to the induction of tyrosine hydroxylase, then it could build up and induce hallucinations. Once again, this argument is highly speculative, but it indicates a possible line of future research.

Physical treatment

As we have seen, the major reason for amphetamine dependence is the positive reinforcing effects which result from the enhanced activity in the median forebrain bundle produced by the increased release and blocked re-uptake of norepinephrine. One method of treatment, particularly when the individual has paranoia, is the use of phenothiazines which block adrenergic postsynaptic receptors; phenothiazines like chlorpromazine ameliorate the effects of hallucinogens, including d-lysergic acid and 3,4-dimethoxyphenyl-ethylamine that we mentioned in the last sub-section. In addition, repeated use of phenothiazines will reduce the euphoric effects of amphetamines by prevent-ing the increased norepinephrine at the adrenergic synapses reaching the receptors, with the disadvantage of inducing Parkinsonian symptoms (see Chapter 5).

Another approach, working on the same principle of reducing the euphoria, has used a blocker of norepinephrine synthesis. This is based on a study (Jönsson, Änggård and Gunne, 1971) in which amphetamine abusers were given alpha-methyltyrosine which inhibits tyrosine hydroxylase. The self-rated euphoric effect of intravenous amphetamine in the amphetamine abusers was reduced at the peak effect time of 15 min with the amount of reduction depend-ing on the dose; in other words the impact effect could be reduced and in some cases abolished. Clinical trials suggest that doses of alpha-methyltyrosine, suitably spaced to prevent tolerance, are effective in treating amphetamine abuse. These results are theoretically important because, if amphetamine dependence can be broken by preventing the euphoria, then other types of dependence may result from the euphoric effects as well as the escape from abstinence symptoms.

D. ABUSE OF CANNABIS DERIVATIVES

The major versions of Indian Hemp (Cannabis sativa) are marijuana and hashish, hashish being the stronger compound because it is derived from the resin of the female plant. Both types are usually smoked, but marijuana is rolled into cigarettes ('reefers') while hashish is always smoked in a pipe. The active ingredient which gives hashish and marijuana their particular properties is tetrahydrocannabinol. In the United States Cannabis derivatives are the most frequently used illicit drugs. A Gallup poll interview of 15,000 adults of all ages from all walks of life disclosed that 6 % of the men and 2 % of the women admitted that they had taken marijuana at least once. In the 21–29 age group 12 % said they had taken the drug; among undergraduates the figure was even higher at 27 %, and among graduate students it was even higher still at 42% (Anker and Milman, 1972). It appears that the use of marijuana is prevalent among the younger members of the upper socio-economic groups especially among those from the East and West coasts (Mizner, Barter and Wermer, 1970).

One consequence of the drug's widespread use is that it is difficult to define the personality of the user clearly. Among women its use was significantly higher in high-sensation seekers, but among men the use was so high among low-sensation seekers that the difference was not significant (Zuckerman, 1972). Sensation seeking was measured in terms of factors like thrill and adventure seeking, experience seeking, disinhibition and boredom susceptibility. A similar pattern was found in another study (Kissin, 1972) which showed that heavy marijuana users tended to suffer from boredom, depression, anxiety and repressed hostility and to have drives towards euphoria, grandiosity and oblivion. However, the cannabis epidemics cannot be simply explained in terms of the personality of the individual user, and part of the explanation must lie in the positive properties of cannabis itself.

The subjective effects of cannabis are rapid in onset with smoking, and the magnitude of the effects depends on the tetrahydrocannabinol content of the mixture smoked. At low doses the hallucinogenic properties are not manifest, instead the user has feelings rather like alcohol intoxication and he experiences heightened mood and increased self-confidence. One characteristic of the person smoking marijuana is that he is given to unmotivated laughter and bouts of giggling. There is a general feeling of relaxation, loss of tension and inhibitions and a desire to remain undisturbed in contemplation. After smoking about 12 mg of tetrahydrocannabinol there is a floating sensation and loss of concentration in 3–4 min. Between 20–30 min there is marked euphoria, mental impairment and loss of time sense. From 30 to 60 min there is sleepiness and then recovery by 90 min (Hollister, 1971). The precise symptoms seem to depend on the degree of experience with the drug and Becker (1967) has shown that the user must 'learn to experience' the euphoric effects because of the low doses of tetrahydrocannabinol in marijuana, e.g. less than 1 % (Jones, 1971). A subjective evaluation of marijuana and a placebo at 30 min after smoking by experienced users showed that the users could estimate the potency of the marijuana quite accurately, but the placebo scores, although lower, had a wide range, showing the importance of expectations (Jones, 1971). This was confirmed by the finding that the higher potency rating was made by frequent users, but the frequent users also tended to underestimate the absolute potency of the active marijuana, suggesting that they had developed tolerance for the euphoric effects. Jones (1971) concludes that the short-term effects of marijuana depend on the attitude, set, expectations and experience of the user as well as the potency.

The effects of a tetrahydrocannabinol dose are a decrease in the judgment of time, and distance as well as psychomotor errors (Weil, Zinberg and Nelson, 1968; Crancer, Dille, Delay, Wallace and Haykin, 1969; Rafaelson, Bech, Christiansen, Christrup, Nyboe and Rafaelson, 1973). In a set of tests of mental functionings, it was found that there was impairment on the digit span and goal directed serial attenuation (Melges, Tinklenberg, Hollister and Gillespie, 1970). Digit span depends on recent memory, while goal-directed serial attenuation depends on recent memory and serially indexing these with reference to the goal. The deficit in recent memory demonstrated here may account for the disorganization of speech patterns that occurs in the intoxicated individual, whose speech is characterized by loose associations and failure to organize words and phrases hierarchically. In the most sophisticated study of simulated driving tests by Rafaelson et al. (1973) different doses of tetra-hydrocannabinol were administered orally to the subjects 105 min before testing in a driving simulator for 10 min. Behavioural measurements included brake time, start time and number of gear changes. The tetrahydrocannabinol increased brake time and start time with the amount of increase depending on the dose. The effect would have been even greater on brake time because one subject had to be excluded because he failed to 'stop' at eight out of ten red lights. This subject was also excluded from start time scores because there were only two out of ten start scores! The number of gear changes were reduced, but not significantly. These findings can be compared with those of Crancer et al. (1969) who had found speedometer monitoring errors but no braking, signalling or steering errors in a *passive* simulator using much smaller doses. The changes in judgment observed in the simulated driving seem to extend to judgment of the consequences of his actions, but these effects are not as large as those found with the amphetamine users and the consequences to society are accordingly less severe in terms of criminal activity. Both Weil et al. (1968) and Crancer et al. (1969) observed that the experienced user was less susceptible to the effects of a given dose than inexperienced users, and this was probably the result of tolerance to tetrahydrocannabinol.

Chronic use does lead to the development of tolerance so that more drug has to be taken to obtain effects. Dependence seems to develop rapidly, and it seems difficult to break the habit even though physical dependence does not appear to occur, and in this sense it is similar to amphetamines. Part of this difficulty in discontinuing marijuana use may be due to the changed inter-pretation of reality which occurs in the heavy user; it seems to him that the only meaningful experiences are those induced by the drug. One consequence of this change is that the quality of remembered experiences deteriorates, and the user may remember a happy childhood as a nightmare (Bejerot, 1972). Dependent individuals lose interest in the everyday world becoming more and more passive and prone to daydreaming, even when not under the direct influence of the drug. Tetrahydrocannabinol is probably not the cause of this change, rather it is the result of an interaction of the personality of the user and the drug effects.

This conclusion is almost certainly true of the psychotic episodes that seem to be attributable to the effects of the drug and these episodes resulted in 22·5 % of the psychiatric admissions in Brazil, India, Morocco and Nigeria between 1957 and 1964 (Wilson, 1968). These episodes can last for months, and in some cases the personality changes seem to be irreversible. Part of the disorder is the appearance of mini-delusions (Browne-Mayers, Seelye, Brown and Fleetwood, 1972); these are plausible but false beliefs with normally associated affect that do not occupy the whole spectrum of the patient's mental life. Browne-Mayers et al. (1972) describe a patient who complained that whenever she smoked marijuana she believed that there was a man in the bath beside her. She never looked to see if he was there, but she was certain that he was. This mini-delusion was accompanied by feelings of excitement and anxiety. It is clear from the surveys of the literature (Hollister, 1971) that the subtle personality changes mentioned earlier are more common and potentially more disturbing for society than the rarer psychotic episodes.

Biochemical correlates

It is only recently that the tetrahydrocannabinols have become available for research purposes (see Mechoulam, 1970). Analysis of the brain chemistry has shown that 45 min after low doses of the drug there was a depletion of norepinephrine in the mouse brain (Holtzman, Lovell, Jaffe and Freedman, 1969). Higher doses showed much less spontaneous activity. Injections of comparable doses in monkeys produced increased responding in continuous avoidance situations with the low doses, like amphetamine, but decreased responding in the later phases with higher doses (Scheckel, Boff, Dahlen and Smart, 1968). It seems reasonable to suggest that the depletion of norepine-phrine reflected the release of this transmitter by the tetrahydrocannabinol, and it is this release, occurring in the median forebrain bundle, that produces the euphoric effects.

Obviously the behavioural effects of tetrahydrocannabinol are not the same as those of amphetamine. Besides the hallucinogenic effects there is the onset of sleepiness which is reflected in the decreased spontaneous activity at higher doses in mice. These behavioural effects are probably the direct consequence of a rise in brain levels of serotonin produced by tetrahydrocannabinol (Holtzman et al., 1969). In addition to this rise in serotonin there is also an elevation in the major metabolite of serotonin, 5-hydroxyindoleacetic acid (Holtzman et al.; 1969), showing that the serotonin was being released. As we saw in Chapter 6, hypnotic agents produce an increase in brain serotonin and slow-wave sleep until the levels return to normal (Anderson and Bonnycastle, 1960). It is clear that marijuana is not increasing release substantially otherwise sleep would be observed; instead relatively small amounts are released in the raphé neurons inducing sedative effects.

E. BARBITURATE ABUSE

The extent of barbiturate abuse is much larger than the incidence of opiate abuse, but it has not been subjected to the same intensive study. The wide extent of abuse is due to overprescription by physicians, and very often the disorder is not detected except when severe intoxication occurs. Acute intoxi-cation from barbiturates results in 25 % of all deaths from acute poisoning admitted to hospitals in the United States (Ban, 1969). Some of these were the result of suicide attempts and some were the result of accidental overdoses. Barbiturates tend to be used by the older generation so that in the United States it is estimated that 11·5 % of all adults aged 30–49 use barbiturates (Parry, 1968) and the figure is probably much higher in urban areas and among women. This is certainly true in Canada where 19·7 % of women and 9·9 % of men use barbiturates (Smart and Fejer, 1972). Their personality characteristics seem to be those of anxiety with repressed hostility and a drive towards oblivion rather like marijuana users.

The initial effects of barbiturates were described by Smith and Beecher, (1960a) in a study on the effects of barbiturates on athletes. The major effect, seen at 30 min after an oral dose of secobarbital, was one of 'intoxication', shown in the subjective reports of feeling dopey, lightheaded, cheerful, elated and happy. After some months of use, tolerance develops (Seevers and Deneau, 1963) and the dose must be increased. The effect of doses like 600–800 mg/kg is characterized by prolonged hypomania and irritability. In addition to the emotional lability, the 'intoxicated' subject will show loss of judgment, confusion and ataxia. Abrupt withdrawal of the drug demonstrates the marked physical dependence because within 36 hr of the last dose a set of abstinence symptoms occur including anxiety, insomnia, nausea, vomiting, intention tremor, involuntary muscle twitching, and sometimes convulsions of the grand mal type with hallucinations, resembling delirium tremens. The rate of appearance and intensity of the symptoms is greater for the shorter acting agents. Without treatment, recovery is complete within 10 days, except for some sleep disturbance which consists of an early onset to paradoxical sleep and excessive paradoxical sleep, a pattern which can last for 2 months (Oswald and Thacore, 1963). The abstinence symptoms can be blocked by a dose of any barbiturate in its early stages, but not when they are fully developed (Jaffe, 1970).

Dependence in animals

As we observed with the other dependence-producing drugs, barbiturates will reinforce responding and in a study by Deneau (1972) some, but not all, monkeys voluntarily initiated self-injection of pentobarbital. As the effects of one dose wore off the monkey staggered back to the lever to inject another dose, so that the animals maintained themselves in a state of extreme intoxication. If the drug was withdrawn, the animals developed abstinence symptoms which were physically identical to those observed in man, including convulsions and bizarre behaviour suggesting hallucinations. These symptoms could be abolished by another dose, and the monkeys would always reinject rather than have withdrawal symptoms. Thus we have the same pattern of dependence as in morphine addiction except that initiation of self-injection was more difficult.

Mechanism of dependence

As we observed in the discussion of human barbiturate abuse, some tolerance develops for the behavioural effects of the barbiturates, and this is also true for the behavioural effects in animals. In a study of sleeping time produced by pentobarbital, Aston (1965) injected a series of hypnotic doses in rats. He recorded the time from the loss of the righting reflex to the time when the animals righted themselves and made spontaneous crawling efforts. A second group were preinjected with a large dose of pentobarbital 24 hr before the test. It was found that this group had consistently shorter sleeping times, showing that a single dose was sufficient to produce tolerance. In an extension of this study Warburton (1968) confirmed this effect using smaller doses and testing in a continuous avoidance situation. The mechanism underlying barbiturate tolerance is not certain, but Seevers and Deneau (1963) have shown that there is markedly increased activity of the hepatic enzymes after chronic injections, suggesting that barbiturates are metabolized quicker in tolerant animals. In Warburton (1968) it was found that the recovery rate of tolerant and non-tolerant rats was not significantly different, suggesting that increased enzyme activity did not play a part in tolerance effect obtained after single doses, and that the effects were probably due to changes in the sensitivity of the central nervous system. The nature of this change is unclear at the present time.

F. ALCOHOLISM

In 1970, the Vice President of the United States, Spiro Agnew, observed that 'Alcohol has been known for thousands of years and it has won the approval of people and governments'. Nevertheless, the abuse of alcohol poses an appalling social problem, seen in its most acute form in the case of the alcoholic. The World Health Organisation Alcohol Subcommittee (1952) has defined an alcoholic as an excessive drinker whose dependence on alcohol has attained such a degree that he shows a noticeable mental disturbance or an interference with his mental and bodily health, his interpersonal relations and his social and economic functioning. The consequences of alcohol abuse are similar to other forms of drug dependence excpet that alcoholism is more socially acceptable than any other form, because of the attitudes embodied in Agnew's observation. A consequence of the complacency of society is that alcoholism is not a notifiable disease and so it is difficult to estimate the incidence of alcohol abuse. A critical review of indirect methods of estimation is given in Popham (1970), and he argues that the Jellinek method which is based on the incidence of cirrhosis of the liver is the best method. From this method it has been estimated by Popham (1970) that the incidence in England and Wales is $1·1 \%$ of the adult population. In the United States it ranges from $2·3 \%$ (Kansas) to $5·2 \%$ (Illinois), and in France it is $5·2 \%$, although other data suggest it may be as high as 7%. An interesting statistic, not often considered, is that of the percentage of users who become addicted. For alcohol the figure is $5-7 \%$ compared with $5-7 \%$ of marijuana users, $5-10 \%$ of barbiturate users and $80-90 \%$ of heroin users (Kissin, 1972). Alcohol dependence seems to develop more slowly compared with these other drugs, and it may take from 3 to 10 years to become alcoholic.

Attempts have been made to explain the development of alcohol dependence in terms of self-medication for emotional instability. This is difficult because of widespread incidence of the disorder among different social groups and obviously it cannot be concluded that a high incidence of alcoholism in France indicates a high incidence of emotional disturbance in this society. However, it may be argued that different societies have different modes of self-therapy. A number of studies have shown that many alcoholics are depressed and anxious, and use alcohol as a form of self-medication (Kessel and Walton, 1969; Chafetz, 1970; Kissin, 1972) to produce euphoria, to reduce social tensions and inhibitions, and to promote self-expression and so lessen frustrations. Most individuals report that alcohol makes them 'high together with feelings of calmness, haziness, relaxation and absentmindedness' (Goldberg, 1970). The drinker will attempt to maximize these sensations without upsetting his stomach or getting a hangover. Repeated imbibing results in the development of tolerance, so that larger quantities have to be taken to reach the intoxication maximum.

Kessel and Walton (1969) chart the progressive development of alcohol dependence from this point. At first the potential alcoholic starts to take the first two or three drinks rapidly to achieve the effects as quickly as possible; he grows away from friends who are critical of his drinking habits, and so the process of alienation begins. His drinking starts to interfere with his work and he feels guilty about his over-consumption. Soon he finds that he requires alcohol to function effectively even in routine activities. The dependence phase is characterized by an inability to stop drinking, and the alcoholic may be intoxicated for days. Chronic alcoholic intoxication can result in the appearance of acute auditory and visual hallucinations, paranoid episodes, confusional states and episodes of psychotic excitement. Most of these are short-lived, but in about 10 % of the cases they persist into the non-intoxicated state (Wieser, 1970).

After a drinking episode there is often partial or complete amnesia for certain events occurring during the intoxicated period, and recent research discussed in Chapter 9 suggests that these may be state-dependency effects. Abstinence also results in the appearance of withdrawal symptoms consisting of tremor, hallucinations, disorientation, restlessness and anxiety in about 80 % of those dependent (Isbell, 1970; Mendelson and Mello, 1970; Mello and Mendelson, 1971). These symptoms could be abolished by further ingestion of alcohol, and it seems likely that the continued ingestion of alcohol by an alcoholic represents an attempt to reduce the withdrawal symptoms as well as to maintain the intoxication; thus alcohol dependence is similar in this respect to opiate and barbiturate dependence.

Positive reinforcing effects

It will come as no surprise to the reader to discover that alcohol has reinforcing properties. Using the self-injection technique described in the previous sections, Deneau (1972) demonstrated that most monkeys would initiate self-administration of alcohol. As in the case of barbiturates the monkeys maintained themselves in a state of extreme intoxication by repeatedly pressing the injection lever. One difference was that the 'alcoholic' monkeys would voluntarily abstain even though they showed severe abstinence symptoms including convulsions and bizarre behaviour like the monkeys withdrawn from barbiturates.

Biochemical bases of alcohol's acute effects

The reinforcing effect of the other dependence-producing drugs had its origins in increased activity in ascending adrenergic neurons from the reticular formation. Alcohol does not appear to differ from the others. Gursey and his coworkers have claimed that alcohol increased the release of norepinephrine in the brain stem (Gursey, Vester and Olsen, 1959; Gursey and Olsen, 1960) which would result in increased activity in the median forebrain bundle in the short term. Later there was depletion of the norepinephrine (Gursey and Olsen, 1960) which would explain the depression. It is only fair to point out that Gursey's results have not been supported by a number of investigators (Corrodi, Fuxe and Hökfelt, 1966; Duritz and Truitt, 1966; Haggendahl and Lindquist, 1961). Duritz and Truitt (1966) have demonstrated that it may be the acetaldehyde, the first step in the metabolism of alcohol, which is responsible for the release and depletion of brain stem norepinephrine. The effects of depletion would be accentuated if the synthesis of norepinephrine was inhibited (Corrodi et al., 1966; Duritz and Truitt, 1966). The mechanism of increased release of norepinephrine is unknown, although Gursey and Olsen (1960) postulated that it might be blockade of storage rather like reserpine. In Chapter 3 it was shown that reserpine does not produce adrenergic activation unless monoamine oxidase is inhibited. It seems that acetaldehyde inhibited this enzyme in the brain, while alcohol had little effect (Towne, 1964). Thus the euphoric sensations produced by alcohol can be attributed to the increased activity in the median forebrain bundle produced by acetaldehyde, the metabolite of alcohol, which releases norepinephrine.

As one might expect from a chemical that has reserpine-like effects and inhibits monoamine oxidase, acetaldehyde releases serotonin. Gursey and Olsen (1960) found that alcohol produced a depletion of serotonin from the brain stem, and Duritz and Truitt (1966) showed that this effect was mediated by acetaldehyde. Increased release of serotonin in the brain stem would be expected to produce drowsiness, which is one of the symptoms of alcohol intoxication. However this explanation must remain tentative until the failures to substantiate Gursey's findings with serotonin (Bonnycastle, Bonnycastle and Anderson, 1962; Corrodi et al., 1966; Haggendahl and Lindquist, 1961) have been explained.

By a similar argument it may be predicted that a compound with reserpine-like and monoamine oxidase inhibiting properties would modify dopamine function. Increased dopamine activity in the basal ganglia (see Chapter 5) could produce the disturbances in motor control which have been observed during intoxication. Similarly the resulting depletion of this amine would result in tremor, which is usually observed after a drinking episode. Unfortunately, this hypothesis has not been examined biochemically so far.

The effect of alcohol on the other amine, acetylcholine, has not been studied in detail. *In vitro* studies have revealed that alcohol reduces the spontaneous release of acetylcholine in the cerebral cortex by up to 30 % (Kalant and Grose, 1967; Kalant, Israel and Mahon, 1967). From the evidence discussed in Chapter 4 on the behavioural properties of the cholinergic pathways from the reticular formation, one would predict that reduced functional acetylcholine would induce lapses in attention, and might even produce hallucinations. As we have seen already, both these effects have been reported with lapses in attention being the most common. It is interesting to speculate that some people may smoke more during alcohol consumption because they are using nicotine to prevent the blockade of cortical acetylcholine release, and so avoid the lapses of attention without reducing the euphoric effects.

These biochemical studies give some explanation of the acute effects of alcohol, but do not explain tolerance or physical dependence. It seems likely that the mechanisms of tolerance and physical dependence will turn out to be very similar to those found after chronic barbiturate use. Certainly, cross-tolerance is found between the two drugs in animals and man (Seevers and Deneau, 1963). As we said in the last section on barbiturates, increased metabolism by the liver enzymes seems to be a negligible factor in this tolerance, and this seems to be true of alcohol too. Thus, we are left with changes in the central nervous system to explain tolerance. These changes are clearly much less than those observed after chronic opiate administration; however, there do not seem to be any plausible hypotheses for the role of the central nervous system amines in the development of tolerance and physical dependence, and the area must still be regarded as one for extensive investigation (Kalant, Leblanc and Gibbins, 1971).

G. CONCLUSIONS

In the introductory section the hypothesis was proposed that drug dependence was a consequence of the avoidance of abstinence symptoms rather than of the transient euphoria obtained by the chronic user (Lindesmith, 1970). This hypothesis was supported most clearly in Section B on opiate dependence where unpleasant abstinence symptoms were shown to act as negative reinforcers, and to increase the probability of behaviour patterns which postpones them. One of the interesting examples of this were the self-injection studies where rats and monkeys would administer opiates to themselves each time the abstinence symptoms start to occur (Weeks, 1962; Deneau et al., 1969). Abstinence symptoms also occur after withdrawal from barbiturates and alcohol and dependence on these drugs could also be explained in terms of abstinence symptom-escape. However, this explanation would not explain the initiation of drug use, nor the relapses that frequently occur after treatment. If the abstinence symptoms were the only motivation, relapses would not be expected. In addition, drug dependence is also found with groups of drugs like amphetamine, cocaine and marijuana that do not produce physical dependence. Studies have shown that animals will self-inject amphetamine and cocaine in the same way as opiates. Obviously, some other principle is involved.

The one factor common to all drugs of dependence mentioned here, except barbiturates, is that they produce euphoria in man. Studies of the personality characteristics of drug users (Kissin, 1972) suggest that the euphoriant drugs are taken to fulfil a drive towards euphoria and away from anxiety and depression, and in some cases as a form of self-medication. The choice of the drug depends on cultural factors to some extent. For example, Kissin (1972) reports that in the Puerto Rican area of Brooklyn the drug of dependence is heroin, while in the Irish–Italian part barbiturates and amphetamine are used predominantly. Alcohol abuse occurs mainly among older males from Irish, Scandinavian and Negro groups. In some Muslim countries where alcohol is forbidden to the faithful, marijuana is the drug of choice. Drugs which give the greatest impact euphoria, like opiates, are more likely to produce a dependence and only a minority become dependent on alcohol and marijuana.

The euphoria is the result of these drugs, acting on the adrenergic pathways ascending from the reticular formation via the median forebrain bundle causing the release of norepinephrine. The median forebrain bundle is the pathway mediating 'reward', so that increased activity will act as a reinforcer increasing the types of behaviour, in this case drug administration, which result in the reinforcement. This process can be seen most clearly in amphetaminists where withdrawal symptoms do not occur, but the dependence is still very powerful. However, we saw the dependence can be broken by using drugs which reduce the production of norepinephrine and block the euphoric effects (see Section C). Thus the major factor in dependence is the euphoric effects, although avoidance of abstinence symptoms almost certainly contributes to the motivation of intake.

9

Biochemical Bases of Learning and Memory

The study of the biochemical mechanisms underlying the storage and retrieval of information is one of the most popular fields of research in the neuro-chemistry of behaviour, and the preliminary results of research in this area have been quick to hit the headlines. This popularization has led the general public to believe that causal relations between learning and changes in cell chemistry have been established unequivocally. Unfortunately, this is untrue, although there is an established body of knowledge that suggests some neuro-chemical systems that may be involved. Many of the advances in this area of research have been solutions of methodological problems that beset the re-searcher in trying to demonstrate a relationship between biochemical changes and learning. One of the central problems has been the definitional one of distinguishing between 'performance', 'learning' and 'memory'. McGaugh and Petrinovich (1965) point out that although 'learning' is inferred from a change in performance, that is brought about by practice, or more broadly experience, there are clearly some performance changes that result from experience that are not learning such as physical growth, effects of fatigue and changes in motivation. Conversely, there are cases, such as latent learning, where learning is not always manifested in performance during acquisition and so performance is not an unequivocal index of learning but it is the only one that we have. In the analysis of learning, important distinctions must be drawn between the drug's effect on performance and those on learning.

Memory is also an inferred process, it is inferred from the persistence of the learning from the training experience until manifested in performance at the time of test. It implies that information has been registered, retained and retrieved (Jarvik, 1964) and loss of memory could be due to impairment of these stages. Chemical manipulations have played an important role in dis-tinguishing the types of neurochemical process that may underlie each of the three stages of registration, retention and retrieval.

A. REGISTRATION

The process of registration in this context is taken to mean the input of information, storage of the information in a transient form and then the transfer of some of this information from a transient to a permanent form.

This latter process is known as consolidation and its rate may depend on the sort of information, as well as the species, age and sex of organism, tested. This schema was outlined by Hebb (1949) in his trace memory theory although this need not imply a labile trace and a permanent trace but rather a short term trace, and a long term trace (McGaugh, 1968).

Prevention of information loss prior to consolidation

In classical psychology, learning was believed to be the result of assiduously delivering reinforcement to organisms appropriately deprived and research on reinforcers was devoted to the demonstration that drive-reduction was a prerequisite for effective reinforcement, where the efficacy was judged from the performance on some retest. However, later research in the 1950's demonstrated that reinforcing outcomes need not be drive-reducing or even related to drive-reduction; for example, the studies on light-onset reinforcement or self-stimulation of the limbic-midbrain systems (Kimble, 1961). In conjuction with earlier studies of latent learning, these sorts of experiments led to many psychologists discarding the law of effect. Nevertheless it is undeniable that behaviour is strengthened and weakened by its outcomes, and that the amount of strengthening or weaking depends on the interval between the response and its outcome. In simple operant conditioning experiments, the optimal interval seems to be between 0·5—1·0 sec (Landauer, 1969) which fits in very well with the optimal CS–UCS interval in Pavlovian conditioning of around 0·5 sec (Kimble, 1961). It may be more than coincidence that a minimum period of cortical activation lasting from 0·5–1·0 sec is necessary for conscious experience of a weak stimulus (Libet, 1965), and Libet (1966) has suggested that an activation period could be involved in the elaboration or fixation of a memory trace. This is related to the idea that information must produce some minimum reverberation in neural networks for consolidation to occur (Gerard, 1955) and from this idea it is only a short leap to the hypothesis that the importance of an outcome is its prolongation of the after-discharge of stimulus input, and so any input to the organism which is contingent upon a response and produces prolongation of the sensory after-discharge will act as a reinforcing stimulus. This hypothesis has been proposed in slightly differing forms by Miller (1963) and Landauer (1969). Landauer suggests that 'while the reverberatory trace remains active an ordinarily ineffective input may reinstate part or all of the neural activity originally initiated by the events of the learning trial' (p. 83). As a consequence a reinforcing stimulus by enhancing the reverberation will promote the formation of the structural alterations which store the information permanently.

Our hypothesis differs from this mainly by emphasizing that the outcome is 'ineffective' only in the sense of not inducing the precise neural activity produced by the sensory input. It certainly is very effective in changing the neural activity in the brain (Marczynski, 1969). In a series of studies Marczynski observed that, in the presence of light, positive reinforcement of a hungry cat with milk produced cortical synchronization, a burst of 180–200 V activity of 5–9 Hz over the primary and secondary visual projections. This change was named the postreinforcement synchronization, and it was always associated with an abrupt and transient (3–5 sec) surface positive steady potential shift of 200–400 µV in the same region. This second shift depended on some sort of visual input and on the taste and appropriateness of the reward and it was therefore termed the reward contingent positive variation (RCPV). Recently it has been suggested that both shifts result when reward produces a transient decrease in the tonus of the brain stem arousal system (Marczynski, Hackett, Sherry and Allen, 1971).

In another study Marczynski (1971) demonstrated that scopolamine and atropine blocked the post reinforcement synchronization and simultaneously slowed lever-pressing performance. However, there was normal lever pressing in cats when there was physostigmine blockade of both shifts with concomitant increases in the background synchronization during non-rewarded lever pressing. These data are consistent with two functionally opposed cholinergic systems at the cortex. The first produces the recurrent phasing of the PRS–RCPV shifts by recurrent hyperpolarization, and the second normally blocks this change by desynchronization of the electrocortigram, and the surface negative steady potential shift. It is this second system which is transiently decreased by reinforcement, and is probably the system producing electro-cortical arousal. The transient inhibition of the second system results in an enhancement of any evoked potentials present at this moment, regardless of whether they are from 'relevant' stimuli or not (Marczynski and Hackett, 1969). Stimuli presented during non-reinforced lever pressing produced poorly developed evoked potentials. In other words, we have a mechanism which enhances and prolongs evoked potentials that occur in temporal conjuction with the reinforcing stimulus. The suggestion of involvement of the electro-cortical arousal system is significant because of the independent evidence for stimulus modulation by this discussed extensively in Chapter 4. The anatomical pathway involved with cortical desynchronization is the cholinergic reticular pathway ascending from the ventral tegmental area (Shute and Lewis, 1967) with perhaps some involvement of the hippocampal formation. The evidence given above suggests that reinforcement inhibits this pathway and initiates the consolidation process.

Consolidation process

Consolidation refers to the changes in the nature of the memory trace which make it relatively insensitive to external interference. The evidence for a period of memory fixation is strong; for example, it is commonly observed clinically in cases of retrograde amnesia that can occur after a blow on the head, and these may last up to 30 min or more after severe brain injury. The type of information loss depends on the severity of the blow, but in general the more complex material is most susceptible to disruption. It is important for the theory of consolidation that patients examined just after the injury may give information which is not available at a later time (Russell and Newcombe, 1966). Many laboratory studies of retrograde amnesia in animals have used electroconvulsive shock, and it has been found that there is a temporal gradient of susceptibility whose slope seems to depend on the shock intensity, complexity of the task, strain, sex and even time of test. (See Discussion, pp. 264–265 in Kimble, 1961). Other treatments which have retroactive effects are hypothermia and hypoxia. These studies clearly support the hypothesis that there is a time-dependent change in the memory trace.

Another approach which has supported this hypothesis is the use of drugs to impair or facilitate the process of consolidation. In these studies the drugs are administered after the training and so obviously the drug must be rapidly absorbed and be metabolized before the test session in order to be able to draw conclusions about its action on consolidation. In a typical study on impairment (Pearlman, Sharpless, and Jarvik, 1961) rats were trained to lever-press for food until performance was stable. On the test day the rat was given an electric shock whenever he pressed the lever and after the session several drugs, including ether and pentobarbital, were administered. When their rate of lever-pressing was measured 24 hr later, it was found that ether abolished memory for the shock and so prevented depression of the rate when it was administered within 5 min, while the maximum effective period for pentobarbital was 10 min. In each case there was greater susceptibility of the consolidation process to disruption with the shorter intervals. This study is consistent with other experiments with neural depressants, showing that memory consolidation can be disrupted by drugs during the first hour after acquisition (McGaugh and Petrinovich, 1965; McGaugh, 1966). These impairment studies give little evidence for the neural mechanisms involved.

It is obvious that studies of retrograde memory enhancement excite the imagination more than studies of impairment, in view of their implications for the therapy of the memory disorders. Studies of strychnine will be used to indicate the ways in which work with an effective agent proceeds. In the first of these (McGaugh, Thomson, Westbrook and Hudspeth, 1962) the effect of variations in the training-injection interval on the degree of facilitation was studied in addition to sex and strain differences. A four-unit, eight-blind-alley Lashley III maze was used, and a 1·0 mg/kg dose of strychnine was injected either 1 min, 15 min or 90 min after each trial. Maze learning was enhanced by injections given as long as 15 min after training, but the degree of facilitation varied with sex and strain of the rats. It seemed from the study that Tryon S strain 'Maze-Bright' were stable in the consolidation process compared with the 'Maze-Dull', S_3 strain.

The most thorough investigation on the effects of strychnine on memory consolidation is that of McGaugh and Krivanek (1970). This study examined the effects of dose and time of injection on black-white discrimination acquisition in mice. It was clear from the results that most doses of strychnine, injected immediately after three trials in the maze, facilitated consolidation as measured by the number of errors to a criterion during training on the following day.

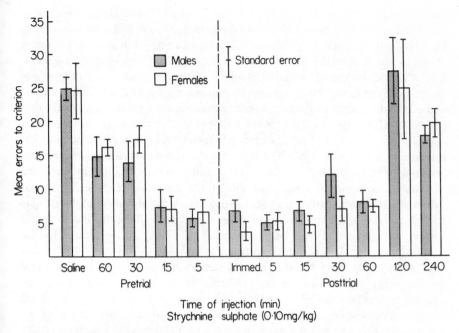

Figure 26. Mean errors to a criterion as a function of time of injection of 0·10 mg/kg of strychnine sulphate before or after daily training on a black–whilte discrimination. From McGaugh and Krivanek (1970). *Physiol. Behav.,* **5**, 1437–1442. Reproduced by permission of Brain Research Publications Inc.

It was interesting that an intermediate dose (0·40 mg/kg) was without effect, suggesting that there were two different mechanisms of facilitation—one occurring at low doses (0·025–0·20 mg/kg) and one at high doses (0·80–1·25) mg/kg). In a second part of this study, doses intermediate in the low and high dose ranges were injected at 0, 5, 10, 15, 30, 60, 120 and 240 min after training. It can be seen in Figure 26 that the errors to a criterion increased with increases in the training-injection intervals up to 1 hr indicating that the consolidation mechanisms affected by the strychnine could only be activated in the first hour after acquisition, with the greatest effect at 5 min. Studies with other drugs have supported a time-dependent effect (McGaugh, 1966), although the gradient varies with the drug used (McGaugh and Dawson, 1971). It is unclear whether this represents drugs acting on different facilitatory mechanisms or differential sensitivity of the same system in those studies using the same training situation.

In an attempt to locate the anatomical pathways involved in consolidation, Alpern (1968) injected strychnine into the mesencephalic reticular formations of rats just dorsal to the ventral tegmental region, and obtained clear facilitation of learning. Facilitation of memory storage has been obtained by post-trial electrical stimulation of the mesencephalic reticular formation close to the ventral tegmental region (Bloch, Denti and Schmaltz, 1966; Bloch and Deweer,

1968; Deweer, Hennevin and Bloch, 1969; Alpern, 1968; Denti, McGaugh, Landfield and Schinkman, 1970) and hippocampus (Stein and Chorover, 1968; Erickson and Patel, 1969). It is possible that the improved memory is the result of the stimulation of the cholinergic pathways of the limbic-midbrain system. The converse finding of impaired memory storage has been obtained by disruptions of the limbic-midbrain system and changes in both the hippocampus and midbrain reticular system produce retrograde amnesia in a similar fashion to lesions (Isaacson, 1972), spreading depression in the hippocampus (Avis and Carlton, 1968), and seizures from hippocampal puncture (Bohdanecka, Bohdancký and Jarvik, 1967). For this reason it is not surprising that Deutsch, Hamburg and Dahl (1966) produced retrograde amnesia by bilateral puncture of the hippocampus and the injection of the anticholinesterase, di-isopropylfluorophosphate, 30 min after acquisition. The administration of this drug is known to enhance the epileptiform electrical activity of the brain, and so would ensure that amnesia would be produced. This would also explain why the cholinolytic, scopolamine, failed to produce amnesia in this study and others (e.g. Wiener and Deutsch, 1968). Consistent with this picture in the limbic system, electroconvulsive shock produces brain seizures and retrograde amnesia is counteracted by doses of scopolamine, but its effects are potentiated by the anticholinesterase, physostigmine (Davis, Thomas and Adams, 1971).

From the point of view of our story these results are crucial because recent experiments on the facilitation of learning have been focused on the relations between cerebral electrical activity and memory consolidation. In a study of the electrical effects of strychnine, Luttges (1968) found that facilitating doses of both drugs activate the posterior hypothalamus, while the hippocampus and reticular formation showed slower sustained slow activity. On the basis of his data, Luttges suggested that the facilitation was the result of high internal activation, with a reduction of retroactive interference. The reduction of the retroactive interference would depend on inhibition of a hippocampal-tegmental–cortical system which would result in hypersynchronization of cortical activity seen in the form of theta rhythms (4–9 Hz). Recordings of this activity have been made after a training and, in some cases, training and electroconvulsive shock (Landfield, McGaugh and Tusa, 1972). It was found that the amount of post-training theta activity was correlated with subsequent retention scores, and that the electroconvulsive shock appeared to lower the probability of theta activity during the 30 min period after training and the retention scores. Landfield also stimulated the ventral hippocampal formation electrically and found that theta rhythms were produced, and that post-trial stimulation facilitated the acquisition of a black–white maze discrimination (Landfield, McGaugh and Tusa, 1972). They noted that many of the drugs, which facilitated learning when administered in the post-trial period, increased the amount of theta rhythm, and the cortical frequency and/or theta activity. It remains to be seen what are the transmitter pathways mediating these changes, and thus aiding consolidation, although Landfield et al. (1972) argued that cholinergic and adrenergic pathways are involved. If it was inhibition of the cholinergic pathway from the tegmentum to the cortex, then consolidation would be linked with reinforcement, since both seem to invoke 4–9 Hz activity in the cortex. Thus, facilitating drugs would induce or prolong post-reinforcement synchronization increasing the probability of initiating the chemical processes involved in the permanent storage of information.

B. RETENTION

The process of retention refers to the maintenance of the information in a form that makes it resistant to the normal changes occurring in the central nervous system. It starts with the onset of the consolidation process, and ends with the death of the neural tissue involved in the storage and overlaps with the registration stage to some extent. The possible nature of the neurochemical change that makes memory permanent will be discussed in this section, and it is important to bear in mind some of the criteria for 'memory molecules' during the discussion. They must be molecules that can be modified by neural activity; they must be of sufficient complexity to store information; they must be stable and remain undistorted for a lifetime, or alternatively be able to reproduce themselves precisely; and they must modify the pattern of later firing in neurons, mediating retrieval with appropriate inputs.

The obvious candidates for memory molecules are the macromolecules, especially those involved in protein synthesis, and the hypothesis that information storage might be accomplished by protein synthesis was proposed by a number of scientists in the late 1940's. Biologists had long stressed the similarity between memory and heredity (see Russell and Warburton, 1973) because they recognized the immense coding possibilities of the protein macromolecules. More recently, as a result of the advances in molecular genetics, the focus has shifted to the deoxyribonucleic acid (DNA) and ribonucleic acid (RNA) molecules in which the genetic information is encoded in the sequences of nucleotide bases, and Crick (1959) has calculated that there are 10^{200} possible combinations of nucleotides in one DNA molecule and so there would be no problem of coding the information of the organism's lifetime. Important to its role in storing genetic information is the ability of such molecules to reproduce themselves at each cell division, which would be expected if genetic information within a species is to be transmitted over countless generations. The DNA molecules are also very stable and resistant to changes in their chemical environments.

During the read-out of information in protein synthesis DNA imposes some part of its pattern upon a second nucleic acid, ribonucleic acid, molecule synthesized in the nucleus of the cell with the synthesis catalysed by RNA polymerase. Each of these messenger RNA is a copy of genetic information held in the nucleus coded in the sequence of nucleotide bases of which there are four, adenosine, cytosine, guanine and uracil. The nucleotide base sequence on the messenger RNA is the complement of the bases on the DNA and it is now known that three nucleotides, a triplet, are required to specify one amino acid. The messenger RNA leaves the cell nucleus and attaches itself to a ribosome in the cytoplasm where protein synthesis occurs. In addition to messenger RNA, a second type of RNA participates and this is called either transfer RNA or soluble RNA. Each transfer RNA molecule has an amino acid attached to it so there are 20 types of transfer RNA corresponding to the 20 amino acids. The transfer RNA passes over its amino acid as specified by the triplet, and it is added on to the chain of amino acids already in the protein. The messenger RNA moves along the ribosome and the next triplet takes up amino acid from the appropriate transfer RNA in the cytoplasm to add to the protein chain molecule. Thus a protein which is composed of 100 amino acids linked in sequence will be specified by a messenger RNA with 300 base nucleotides.

In the next subsections we will discuss attempts to test the hypothesis that protein synthesis may be involved in memory storage. Most research has concentrated on the early part of the retention interval, during or just after consolidation. These experiments will be considered in three subsections; firstly, studies which have tried to manipulate protein synthesis with drugs and impair retention; secondly, studies which have tried to measure the specific changes in protein synthesis produced by learning; and thirdly, studies that have attempted to transfer stored information to an untrained organism by means of brain extracts from trained donors.

Manipulation of protein synthesis

In the mid 1950's biochemical pharmacologists developed a group of drugs, as antibiotics and anticancer agents, whose biochemical mode of action involved inhibition of protein synthesis and some of these have been used as tools for the experimental manipulation of protein synthesis in memory studies. In the earliest of the experiments drugs were used to produce fraudulent nucleotides; one of these, 8-azaguanine, acts as an analogue of the base, guanine, producing non-replicating messenger RNA. Dingman and Sporn (1961) examined the effects of $136 \mu g$ of 8-azaguanine, injected into the brain on the acquisition of a water maze. Although the control and drug groups were matched on their performance on two other mazes, the 8-azaguanine injected animals made a significantly greater number of errors on the test maze. An examination of the results showed that the greatest differences between groups in mean errors were found during the first five trials, and that the most marked difference was on the first trial. However, this trial cannot validly be considered as demonstrating learning, and after the first few trials the learning curves were essentially parallel for both groups. When the scores on each trial were considered as percentages of the first trial, the Vincent curves for the two groups overlapped to a great extent and the drugged animals, when equated for 'exploratory' activity, showed no impairment in learning. A possible explanation of these results is that the experimental treatment affected exploratory activity, and this is supported to some extent by the results of an experiment reported in the same paper (Dingman and Sporn, 1961) where rats were tested for recall after 8-azaguanine, and it was found that the mean number of 'errors' made by the drug group was double the control, although the variance was so large that it affected the tests of significance. Thus, there was no direct evidence that 8-azaguanine affected retention because of the evidence of non-specific effects of the drug influencing performance.

One approach to this problem of non-specific effects is to use an experimental design which enables separate assessment of the general effects of the drug's effects on performance from those on learning, as well as controlling for the irreversability of the learning experience, and Warburton and Russell (1968) studied the behavioural effects of 8-azaguanine using a design of this type which was developed for situations where a subject can be used as its own control, but the treatment order cannot be reversed. As a result it was possible to show that 8-azaguanine injected into the ventricles of the brain delayed the acquisition of a new temporal discrimination response by rats and the length of this delay was the same as the duration of protein synthesis inhibition by 8-azaguanine reported by Dingman and Sporn (1961). The site of inhibition was determined by Warburton and Russell (1968) using a staining technique and they found that protein synthesis in the hippocampus was inhibited at the same time as the impairment in learning occurred.

The importance of the hippocampus in the long-term storage of information was also emphasized in a series of studies using puromycin, another inhibitor of protein synthesis; the puromycin molecule is an analogue of the transfer RNA–amino acid complex and, thus, interferes with the final stage of protein synthesis. Puromycin was used by Flexner and his colleagues and in their early studies (Flexner, Flexner and Stellar, 1963; Flexner, Flexner, Stellar, Roberts and de la Haba, 1964); they found that, although there was 33 % inhibition of protein synthesis after a subcutaneous injection of puromycin, acquisition of a shuttle-box avoidance response or a discriminated avoidance response in a Y-maze was not impaired, showing that registration did not depend on protein synthesis. However, injections directly into the hippocampus and into the caudal cortex 1–3 days after learning impaired retention. The most convincing demonstration of this phenomenon used a reversal procedure in which animals were trained to run into one side of a Y-maze, and then 21 days after the original learning the position habit was reversed. The day following the reversal training the animals were given bilateral injections of puromycin into the hippocampus and adjacent cortex. When retested the following day, the reversal was abolished and the animals performed consistently in accordance with the habit learned 3 weeks previously. These results supported the view that the functional integrity of the hippocampus is important either for the storage of information or for retrieval during the retention test.

Another protein synthesis inhibitor, acetoxycycloheximide, exerts its action by impairing the transfer of amino acids from the transfer RNA to the protein preventing the formation of peptide bonds. In a study with acetoxycycloheximide, avoidance training was carried out with a criterion of five out of six correct responses in a T-maze and animals injected with drugs into the

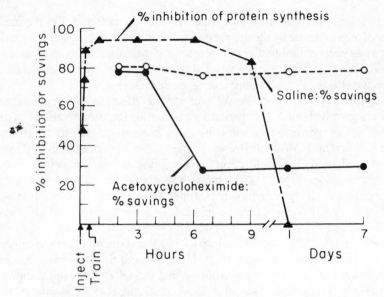

Figure 27. Effect of acetoxycycloheximide on protein synthesis in the cerebrum and on memory. Different groups were tested for percentage saving at each of the indicated states. From S. H. Barondes (1970). In F. O. Schmitt (Ed.) *The Neurosciences: Second Study Program.* Reproduced by permission of the author and Rockefeller University Press

ventricles of the brain showed normal performance when tested at 3 hours, despite 95 % inhibition of protein synthesis, but retrograde amnesia at 6 hours and thereafter (see Figure 27) (Barondes and Cohen, 1968a). This study supported the hypothesis that during the first 3 hours after training while consolidation was occurring information was not stored by changes in protein synthesis, but in some transient form. In a later study, Barondes and Cohen (1968b) trained the animals and gave immediate subcutaneous injections which produced rapid short lasting impairment of protein synthesis. They were then tested after recovery from inhibition 3 hours later, and it was found that they remembered, but there was amnesia at 6 hours while injections of acetoxycycloheximide at 30 min after training produced no amnesia. One fascinating finding in the study was that if the animals were injected with acetoxycycloheximide immediately, and then given amphetamine 3 hours later, then they did not show amnesia suggesting that the amphetamine reinitiated the consolidation process leading to the long-term storage of information (see Figure 28). Added evidence for the involvement of protein synthesis came from the abolition of the effect by an injection of acetoxycycloheximide at the same time as the amphetamine. One explanation for amphetamine reinitiating storage may lie in the drug-releasing cyclic AMP since the norepinephrine has been found to release cyclic AMP in the brain (see p. 173).

The complementary approach to the impairment studies would be experi-

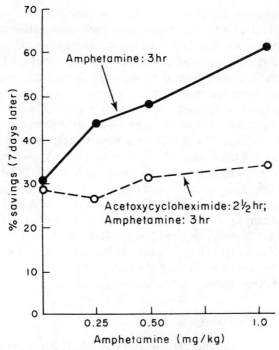

Figure 28. Effect of acetoxycycloheximide on memory and its antagonism by amphetamine (Upper curve). However, reinstatement of protein synthesis inhibition by acetoxycycloheximide blocked the effect of amphetamine. From S. H. Barondes (1970). In F. O. Schmitt (Ed.) *The Neurosciences: Second Study Program.* Reproduced by permission of the author and Rockefeller University Press

ments using injections of protein synthesis enhancers. In the section on consolidation we have already discussed the use of strychnine, which has been shown to increase the concentration of brain RNA (Carlini and Carlini, 1965). Another compound, magnesium pemoline, believed to stimulate RNA polymerase and thus increase RNA synthesis (Glasky and Simon, 1966) has been tested for its effects on acquisition. Plotnikoff (1966) measured the response latencies on the first three trials of a discriminated avoidance and rats with the longest latencies, i.e. those who only escaped the shock, were assigned to one of four groups, who received either saline, magnesium pemoline, or one of the stimulants methamphetamine or methylphenidate. The three groups were compared on the number of trials required to reach the group criterion of a mean latency of 15 sec. Pemoline animals reached the criterion significantly earlier than the other four groups. A retention test 24 hr later showed that pemoline animals had shorter latencies than the other groups. Plotinkoff interpreted these data as showing that magnesium pemoline enhanced both learning and memory, and that these effects were not due to general motor stimulation.

Later studies have failed to support this conclusion and Frey and Polidora (1966) showed that Plotnikoff's experimental group consisted of rats which tended to 'freeze' when shocked, so that the facilitation of responding by pemoline was directly related to amount of freezing behaviour. Control and pemoline groups who were trained to the same latency criterion showed no differences in performance in the later retention test so that pemoline given prior to acquisition had no effect on retention if animals were trained to the same level. In our laboratory, animals who were lever-pressing at stable rates in a continuous avoidance schedule were used to test magnesium pemoline and it was found that the drug markedly increased the response rate suggesting a general stimulation effect (Segal, Cox, Stern and Maickel, 1967). In addition, potentiation of this effect by a monoamine oxidase inhibitor, pargyline, strongly suggests that magnesium pemoline's effects were due to the release of norepinephrine rather than to a direct effect on protein synthesis. Later studies have been unable to confirm pemoline's enhancement of protein synthesis (Stein and Yellin, 1967). These studies on pemoline have been included to emphasize the importance of ruling out performance effects in investigations of the effects of drug on learning and memory.

Measurement of protein synthesis

This subsection will consider experiments which have manipulated learning as the independent variable and examined changes in RNA and protein synthesis as the dependent variables. In one group of studies Hydén (1959, 1961) examined the glial and nerve cell RNA and the proportions of the four nucleotide bases in them, i.e. adenine, cytosine, guanine and uracil, and then he examined the base ratios in the nerve cells of rats that had learned (Hyden and Egyházi, 1962, 1963). Rats were trained to walk up a steel wire inclined at 45° to reach food, then the RNA from Deiter's nucleus, the neurons involved in balance, and the surrounding glial cells was analysed and compared with control rats whose Dieter's cells were merely stimulated by rotation. It was found that the adenine–uracil ratio of the Deiter's and the glial cell RNA increased significantly, and there was also an increased amount of RNA per nerve cell. In a later experiment (Hydén and Lange, 1964) rats were trained to use their non-preferred paw to reach for food in a narrow tube, e.g. right-handed rats were trained to use their left paw. Analysis was made of the base ratios of RNA extracted from the fifth and sixth cortical layers of the right somatosensory cortex; lesions in these areas prevent transfer of handedness in a right-handed rat. A significant RNA increase and a change in the base ratios was found in this area, using the comparable area in the left cortex as a control. These experiments by Hyden and his coworkers indicate that there is a relation between brain cell RNA and learning. Hydén was then interested to see if there were any differentiated changes occurring in the neurons with respect to RNA during the early and late stages of learning. It was found that, during the first 5 days of training on the wire climbing, the cortical neurons synthesized

RNA with high adenine and uracil values, which in this respect has DNA-like composition (Hydén and Lange, 1965). Later, at 9 days, a more ribosomal type of RNA characterized by high guanine and cytosine was formed. It was thought that these two stages might represent the labile, short-term memory phase and the fixation of long-term memory. In a complimentary set of studies by Glassman and his coworkers in North Carolina, the rates of incorporation of radioactive amino acids into the liver, kidney and brain RNA were measured using a double isotope method with uridine-C^{14} and uridine-H^{3}. One mouse of a pair was injected with carbon-14 uridine, and the other with tritiated uridine. One was trained in a box where he had to jump on to a shelf when a light came on, but before a shock 3 sec later, while the other, the yoked control, received the shocks, but could not avoid or escape. Then the brains of both animals were homogenized together, the RNA extracted together and the ratio of tritiated to carbon-14 uridine in the RNA of the brains measured to give the amount of incorporation in the trained subject compared with the yoked control. There was clearly greater incorporation in the brains of the trained rats (Zemp, Wilson, Schlesinger, Boggan and Glassman, 1966) and comparison of the yoked controls with unshocked controls showed no difference, suggesting that shelf-jump learning was the important factor. The changes in RNA occurred in both nuclear and ribosomal RNA, with the greater increase in the latter after 15 min training, and analysis of the incorporation in different parts of the brain (Zemp, Wilson and Glassman, 1967) suggested that most of the incorporation occurred in the diencephalon and limbic system, including the hippocampus, rather than the cortex.

In view of this evidence, and that in the last two sections implicating the hippocampus in memory storage, Hydén and Lange (1968) studied protein synthesis of the hippocampus in relation to the transfer of handedness. Subjects were injected with radioactive leucine intraventricularly, and an hour later trained on the transfer of handedness. Immediately after training the animals were killed, and 300 pyramidal cells dissected out of the hippocampus. The specific activity of the protein sections of trained animals was higher than that of control subjects and there was also a trend towards greater synthesis to occur in the side of the hippocampus contralateral to the used paw, which was surprising in view of the interaction between both sides of the hippocampus. Thus, there seemed to be a correlation between increased protein synthesis in the hippocampus and the neural activity occurring during learning.

The protein fractions studied by electrophoresis in the latter study were immediately behind S100 protein which is a unique type of soluble protein found in the nervous system, but no other organ of the body (Moore and McGregor, 1965). In an extension of the previous studies the formation of this protein was studied during training of handedness (Hydén and Lange, 1970). After training, pyramidal cells from the hippocampus were removed from the decapitated rat, weighed and the sample separated into its component proteins by electrophoresis. It was found that in trained rats there was a double protein band at the anodal front of the protein column compared with only a single band in the untrained rats. These front anodal bands were analysed, and it was discovered that the extra band in the trained rats was exclusively S100 protein (Hydén and Lange, 1970).

The next question was whether the newly synthesized S100 protein was specifically related to the learning occurring during training, rather than some other effect of training such as the stress of switching paws. In order to examine this possibility, an antiserum for S100 protein was injected intraventricularly after 4 days of training. The rats were not impaired in their sensory or motor functions, but the injected animals showed no further improvement in performance. Some control animals were injected with the S100 antiserum that had been placed with S100 extract, and it was found that the antiserum absorbed the S100 protein and the injected animals showed the same performance as non-injected controls. This control gave strong evidence for the idea that the serum was having a specific effect against S100 protein. Specific immunofluorescence tests showed that the S100 antiserum was localized in the hippocampal nerve cells, whereas other antisera had no effect (Hydén and Lange, 1970). These sets of studies point to the involvement of protein synthesis in the hippocampus in the processes involved in the long-term storage of information.

Theoretical models of information storage

It seems clear from the preceding sections on retention that selected sensory information is fixated into specifically structured macromolecules which control protein synthesis. There are two possible ways in which this might occur; *instruction* whereby qualitative changes occur in the nucleotide sequences of the DNA, or more likely RNA, to produce new protein, or *selection* whereby specific portions of the DNA molecules were activated to increase selected species of protein (Schmitt, 1962). An instructional theory was proposed by Hydén (1959) based on the fact that DNA controls the production of cytoplasmic proteins through the mediation of RNA. Neuronal RNA could change as the result of the pattern of neural impulses affecting the ionic equilibrium of the cytoplasm, or change the stability of one or more bases of the RNA, and which results in an exchange with other bases in the intraneuronal pool. The specification of the new RNA would be completed if the new bases were stable under the influence of the new pattern of neural impulses. The specified RNA with its novel sequence of bases would control the production of a novel protein. Subsequent stimulation activates the previously specified protein which releases a transmitter substance and transmits an impulse to the next neuron. This model would explain the changes in base ratios reported by Hydén and his co-workers, and the disruptions observed with protein synthesis inhibitors. At the present time there is no very good evidence to support this type of instruction of RNA and arguments against the specification of novel protein have been put forward by Briggs and Kitto (1962), who have pointed out that changes in the genetically determined RNA would disrupt normal cellular metabolism, and that the novel protein synthesized would elicit a foreign protein reaction. Morrell (1964) also noted that an instructional theory could only apply if there was an atypical DNA–RNA specification process in nerve cells which allowed uncoupling as the result of neural activity, and that the theory did not explain how neural activity could induce a molecular rearrangement which was thereafter immune to further electric charges.

As an alternative, Morrell (1962) proposed that all possible RNA nucleotide sequences might be available already in the nerve cell nucleus, and that neural activity selected some of these at the expense of others. This selectional theory has been extended as the gene expression theory of Flexner and his colleagues (Flexner, Flexner and Roberts, 1967) which proposes that protein molecules are the final storage molecules and their synthesis is controlled by a species of messenger RNA by particular genes, i.e. from DNA. Thus, the essential biochemical change in memory storage is not modification of molecules but increased production of already available species of messenger RNA. The molecule synthesized is the outcome of a change in the pattern of gene expression produced by neural activity during training and the newly synthesized proteins would modify the characteristic of synapses to facilitate future neural transmission. In addition, the proteins or their products act as inducers of their specific messenger RNA, maintaining the concentration of the inducer proteins above the critical level for further gene induction.

In this model a number of steps are unclear at the present time but are the subject of active research. We have already mentioned the part that reinforcement or a similar significant event may play in initiating storage by modifying electrocortical arousal and enhancing the sensory-evoked potentials occurring in temporal conjuction with the biologically significant stimulus. The next question is how increased neural activity can initiate selective transcription from the cell DNA. It is now known that in certain neurons such as sympathetic ganglia, presynaptic stimulation increases the production of adenosine $3'$, $5'$-monophosphate (cyclic AMP) in the ganglia (McAfee, Schonderet and Greengard, 1971). It is also known that cyclic AMP increases the capacity for RNA synthesis at least in rat liver, and this effect seems to depend directly on gene transcription and not on RNA polymerase (Dokas and Kleinsmith, 1971). As a result, a new species of protein, or more of an old species, will be produced in the cell and these changes would explain the shift in RNA base ratios found by Hydén and Egyházi (1962, 1963), and the increased production of RNA and specific proteins (Hydén and Lange, 1965, 1968, 1970). These changes would be localized in some of the neurons active during the acquisition session; obviously there must be some neurons which cannot be modified. Changes in protein could be in the nature of increases in transmitter-synthesizing enzymes which would increase the presynaptic stores of transmitter and would make more available for release at each synapse. This, in turn, would increase the probability of the postsynaptic neuron reaching threshold and transmitting nerve impulses. An alternative sort of neural change produced by increased protein synthesis charge might be either the development of new synaptic connexions in the form of more terminals and more receptors, or changes in presynaptic releasing mechanisms, or increased sensitivity of the postsynaptic receptors (Barondes, 1965). The outcome of these changes would be functionally equivalent to increased transmitter synthesis in terms of enhanced transmission.

It must be emphasized here that the selectional model does not require that the molecular changes occurring in memory store the specific information, rather it is the increased probability of firing of a set of neurons which is important. These neurons would not necessarily be localized in one particular part of the brain; it would depend on the pattern of sensory stimulation at the time of acquisition. These conclusions on a selectional mechanism are at variance with the experiments on memory transfer that are to be discussed next. The underlying assumption of these studies is that there is an 'instructed' molecule that can transfer information to a naïve organism.

Memory transfer

These neurochemical studies represent the science fiction end of experiments which were designed to examine the macromolecular storage hypothesis. The idea of extracting 'trained' molecules from one animal and passing on the information to a naïve animal certainly seems far fetched, but the idea is soundly based in molecular genetics. In 1928 it was found that a culture of pneumococci could be heated and killed, but if the dead bacteria were added to living non-pathogenic bacteria, then the non-pathogenic bacteria would become pathogenic. It is now known that the DNA carrying the pathogenicity entered the chromosome of the host and by some sort of genetic recombination, transferred specific information so that this new DNA directed protein synthesis producing pathogenic bacteria. Therefore, it was argued that the same sort of incorporation might occur in more complex living organisms.

The first memory transfer study with a more complex organism was performed by McConnell, Jacobsen and Kimble (1959) on planaria. They gave these animals training on a classical aversive conditioning, using a light and a shock until they reached over 90 % responding, and then each planarian was cut transversely. The most interesting capacity of these animals is regeneration, with the tail section growing a new head and the head section regenerating a new tail, and it was found that both regenerated sections showed significant retention of the conditioning. This was particularly surprising because the tail section grows a new brain and replaces most of its neural tissues during regeneration, and it suggested that the usual neurally oriented theories of memory might have to be extended. These findings were as great a surprise to the investigators as the rest of the scientific community, and led to a spate of follow-up studies and a planarian journal, the *Wormrunner's Digest*.

Unfortunately, other investigators Bennett and Calvin (1964) concluded after 2 years investigation that reliable and reproducible training methods have not yet been developed, to demonstrate learning in planaria. Some investigators (Jensen, 1965; Brown, 1964) have found that planaria can be sensitized by light and shock so that the acquired responses reported in the literature cannot validly be attributed to conditioning in the precise sense of this word. It must therefore be concluded that despite their intriguing possibilities planaria are not the species of choice for 'memory' experiments where demonstrably consistent learning is a prerequisite.

In a transfer experiment on rats (Jacobsen, Babich, Bubash and Jacobson, 1965) two groups of rats were trained to approach a food cup when a discriminative stimulus, a light or click, was presented. When this training was completed the trained animals were sacrificed, a portion of the brain removed and a sample of RNA was extracted from the tissue and 8 hours later injected intraperitoneally into a food-deprived test rat. A research design was used whereby the experimenters were unaware of which extract, i.e. from animals trained to click or light, the test rat had received. Responses of the rats to magazine clicks were counted and it was found that, on the average, rats injected with 'click' RNA tended to approach the food cup significantly more often after the magazine click than rats receiving injections of 'light' RNA. These differential response tendencies were attributed to differences in the RNA extract injected and were considered as evidence for memory transfer.

Simultaneously in Denmark, Fjerdingstad, Nissen and Røigaard-Petersen (1965) trained rats in a light–dark discrimination in a maze to a criterion of 95 % correct choices, over 3 successive days. Extracts of either 'trained' or 'untrained' RNA were made and injected into the ventricles of the brain and the two groups compared with untreated control subjects in subsequent maze learning. It was found that acquisition of the maze was facilitated by extracts from the brains of trained rats. In a more specific experiment, the same authors (Nissen, Røigaard-Petersen and Fjerdingstad, 1965) trained one group to run to the lighted arm and one group to run to the dark arm. Extracts from both sets of rats were made and tested on two different groups and it was found that the 'dark' RNA increased the performance under reinforcement of light preference, while the. 'light trained' RNA gave better performance under reinforcement of dark preference as the test situation. This was the opposite of what was expected from the Jacobsen studies.

These experiments were followed up by a massive research effort attempting to replicate and extend the findings. These attempts resulted in a few successes and a total of 18 experiments by 23 investigators (Byrne and 22 others, 1966) which obtained no clear evidence of transfer of any kind of training from donors to recipients. An alternative explanation of the positive transfer experiment was offered by Halas, Bradfield, Sandlie, Theye and Beardsley (1966), who were unable to transfer a differential response tendency, but found a change in level of general activity. They suggested that this change in behaviour may well have been dependent upon the presence of some stimulating contaminate in the RNA extract, and it has been shown that, in the normal animals, radioactivily labelled RNA does not pass the blood-brain barrier (Luttges, Johnson, Buck, Holland and McGaugh, 1966), indicating that positive results obtained in some experiments were not the result of unmetabolized RNA carrying encoded information into the brain.

In a significant examination of the 'trained' extracts Rosenblatt and his coworkers (Rosenblatt, Farrow and Herblin, 1966; Rosenblatt, Farrow and Rhine, 1966; Rosenblatt and Miller, 1966) found that their subjects showed weak transfer of training, even with extracts which had been incubated with ribonuclease which had destroyed all the RNA in the extract. Accordingly, Rosenblatt (1970) argued that there was little evidence to support the conjecture that the active ingredient of the trained extract was RNA. Their evidence showed that the information-bearing molecule had a molecular weight of between 1,000 and 5,000, which would be consistent with it being a polypeptide, and peptides of this size would pass through the blood-brain barrier. If it was a polypeptide that transferred the information this would explain the lack of success of the specific RNA extracts used by the investigators listed above.

One of the recent sets of experiments that have been performed (Ungar, Galvan and Clark, 1968) have been based on the preference of rats for a dark box rather than a lighted area. When the rat ran into the dark box he was given a shock which reversed the preference and then extracts were prepared from the brains of trained and untrained donors by a procedure designed to extract RNA. This extract was injected intraperitoneally into rats who had been tested for their dark box preference, and it was found that the time the recipients spent in the dark box was reduced. The amount of reduction depended on the dose of extract injected and the 'strength' of the extract, in the sense of the amount of training given to the donors. Biochemical analysis suggested that the active substance in the extract was a peptide that had formed a complex with RNA. Ungar (1970) identified this substance as a pentadecapeptide which he named scotophobin, and he claimed that it is specific to dark avoidance. It has now proved possible to synthesize scotophobin and two independent groups (Malin and Guttman, 1972) have tested its properties using equipment that was a replica of that used by Ungar et al. (1968); they found that the synthetic scotophobin reversed the dark preference of rats and so was not specific for dark avoidance.

At the moment there are a number of positive results but the psychological and chemical variables are still poorly understood. The biochemical and psychopharmacological experiments on scotophobin have been criticized by Stewart (1972) and Goldstein (1973). Stewart (1972) has argued that the biochemical studies are so poorly described that they cannot be replicated, and that the work will not be accepted until the work is repeated with more documentation and with more attention to important detail. Goldstein (1973) makes comments in the same vein and points out that the chemical is probably produced as the result of extreme stress. He suggests that the experimental conditions should be recorded in detail to check the reproducibility, and that all experiments should include a stressed control group. This is crucial because the studies of de Weid and others (see Chapter 11) have demonstrated that pituitary peptides, like adrenocorticotrophic hormone have an effect on avoidance performance. In view of the dispute over the relationship between the transfer effects and memory, it is probably better to refer to the phenomenon as 'behavioural induction' rather than 'memory transfer'.

C. RETRIEVAL

Retrieval means that the stored information influences behaviour at some time after registration, and it presupposes that retention has occurred. It is paradoxical that it is the least studied and least understood memory mechanism, although most experiments in psychopharmacology could be the result of drug effects on retrieval. In many studies training is given without a drug, and then various doses of drug are tested on stable performance which involves use of information from the training situation as well as the test situation, and translation of both sets of information into observable behaviour. Some separation of drug effects on retrieval from the rest could be achieved by concomitant testing, using unlearned behaviour patterns, but this has rarely been done and, in normal circumstances, it is probably impossible to separate the effects of drugs on information from the test situation from retrieved information, since environmental information plays a part in initiating re-trieval. Thus, if environmental information activates a subset of the 'trained' neurons, then this would result in some probability of an output to the effectors controlling the response. Obviously the closer the subset to the training set the more probable the response will be, which is only a restatement of the psychological phenomenon of generalization (Kimble, 1961). It follows from this notion that in a retest situation the response probability will be unity if the pattern of activity in the nervous system is identical with that occurring during training. Retrieval will be modified if drugs either decrease the similarity between the patterns of activity or reinstate the activity present during training. A number of studies demonstrating this have been mentioned in Chapter 4; they are the experiments on state-dependent learning.

State-dependent learning

State-dependent or dissociated learning refers 'to a body of data which indicates that retrieval of information stored in the brain while a special "state" of the organism was maintained is made more difficult or even im-possible if this critical state is altered' (John, 1967, p. 67). State-dependency can be demonstrated unequivocally only by using a factorial design where half the subjects are trained after a drug and half after a placebo injection, and then half of these two groups are tested after either a drug dose or the placebo. Other experimental designs which have demonstrated the phenomenon include partial versions of the factorial design where drug trained subjects show better performance tested with the drug than with the placebo, and designs where incompatible responses have been trained in the same subject under drug and no-drug conditions.

The latter design was used by Overton (1964) in studies of pentobarbital, in which he alternated training under drug and placebo in a T-maze and the amount of response conflict induced by transfer from one state to another was evaluated by comparing performance with a comparable control group. No significant evidence of conflict due to transfer was found in either of the two states when drug doses were high, but graded conflict was found with decreasing drug doses. Then Overton (1964) examined whether the differences in performance could be due to the drug changing the stimulus input by manipulating exteroceptive and interoceptive stimuli, and found that the dissociation effect was much smaller with these manipulations than with pentobarbital, and so it seems to be the *total* pattern of activity in the nervous system which is crucial in retrieval. Pentobarbital acts on a number of biochemical systems including the serotonergic pathways (see Chapter 6), and the cholinergic pathways by blocking acetylcholine synthesis at the cortex (McLennan and Elliott, 1951). Since the latter action would reduce the functional acetylcholine it might be expected that it might be producing state-dependency because of its action on the ascending cholinergic pathways, and thus have the same neurochemical basis as atropine and scopolamine. However, Overton (1966) found that he could train rats to run to one side of the T-maze with pentobarbital, and to the other with a huge dose of atropine (150 mg/kg) and so the atropine and pentobarbital do not mimic each other, probably because they produce their respective dissociated states by different neurochemical mechanisms. In the case of atropine this is as a result of changes in the activity in the ascending cholinergic pathways resulting in decreased electrocortical arousal. In the case of pentobarbital and other barbiturates the dissociation is probably a function of changes in the serotonin systems as well as in the cholinergic pathways. One test of this suggestion would be that alcohol which also blocks cholinergic transmission and releases serotonin should have the same effects as pentobarbital in producing state dependency. Alcohol has been shown to produce state-dependent learning in man (Goodwin, Powell, Bremer, Hoine & Stern, 1969) and rats (Overton, 1966). In the latter study Overton tested rats trained to run to one side with pentobarbital and to the other with saline. Tests with alcohol showed in every case the rats ran to the drug side of the maze, showing that the two compounds are essentially equivalent as far as their effect on retrieval was concerned, which means that the two drugs were having comparable efforts on the patterns of brain activity, i.e. cholinergic blocking and serotonin activation.

The state dependency produced by alcohol has important implications for the therapy of alcoholism, so the evidence for its occurrence will be considered in some detail. The most extensive study is that of Goodwin et al. (1969) in which retrieval in four situations was tested, including a verbal rote learning task to measure recall, a word association test to measure recall of self-generated learning, and a picture task to measure recognition. The clearest effects were found in the two tasks measuring recall, while there were no significant effects on the recognition task (memory for pictures of people). This study demonstrated that information stored by a subject while he is intoxicated is more easily retrieved when he is drunk than when he is sober. Conversely, but to a lesser extent, information acquired when sober is less easily retrieved when he is drunk. Goodwin et al. relate this to clinical reports that alcoholics often reported hiding bottles while drunk, but had no recall when they were sober. This state dependency would also give some explanation of the Dr. Jekyll-Mr. Hyde behaviour of alcoholics if the alcoholic learns different sorts of behaviour for the two states due to the different social reinforcement contingencies applicable in the drinking and non-drinking situations (Storm and Smart, 1965) then loss of control of drinking after a few drinks typical of alcoholics would represent the transition from the restrained normal state to the unrestrained alcohol state. Since therapy for alcoholism is given when the patient is sober and this information will be less available after a few drinks, consideration should be given to therapy while the patient is under the influence of alcohol (Storm and Smart, 1965).

This argument can be extended to all psychotherapeutic drugs because most psychotherapy is combined with drug therapy, and it is a common finding that withdrawal of the drug results in the recurrence of the symptoms. Part of this effect may be due to state dissociation between the drug state and the undrugged state, with small transfer of the therapeutic experience between the two. One solution to this may be the progressive reduction of the drug dose while psychotherapy is continued until the complete transfer to the non-drug state is assured.

Truth drugs

Truth drugs and truth sera are one of the common weapons of the thriller novel villain in his attempts to break the hero, and the possibility of facilitating retrieval against the subject's wishes is something that has excited the imagination of policemen and psychiatrists as well as the novelist and his readers. Two drugs often quoted in these stories are scopolamine and pentothal. At the turn of the century, scopolamine was used by obstetricians to induce a sleep-like state during childbirth. It was noticed that women were able to answer questions in this state, and appeared to be indiscreet (House, 1931). House suggested that scopolamine might be used in interrogation, and this was tried out on two suspects in Dallas, Texas; both denied the charges, and House claimed that under the influence of scopolamine the suspect cannot create a lie, and they were duly acquitted. This experiment and its conclusion excited the public imagination and led to the use of scopolamine in police interrogation in the United States (Larson, 1932). However, any success it may have had was overshadowed by its disturbing hallucinogenic properties mentioned in Chapters 4 and 7, and so the use of scopolamine was superseded by the barbiturates pentothal (thiopental) and amytal (amobarbital).

There are few published studies on the efficacy of these drugs in facilitating retrieval and extracting information. In one interesting study (Redlich, Ravitz and Dession, 1951) nine subjects were asked to describe a true shameful incident from their life. They were then asked to invent a cover story to be told to an interrogator, after an amytal injection. Six of the subjects stuck to their cover story while two revealed their true story and one gave a mixed version, and although all cover stories contained elements of the true story, the interrogator was only able to infer the true story in the case of those who confessed. Significantly, it was the subjects with more 'normal' personalities who lied best. In a more extensive study (Clark and Beecher, 1957) subjects were required to withhold information during an 8-hr period while they were interrogated with a number of drugs, including pentothal, amytal, scopolamine and atropine. In spite of the subjects becoming semicomatose, delirious, panicky, verbose and euphoric, none of the subjects revealed the 'military secret' that they had been entrusted with. Two subjects did reveal their mother's maiden name, an item of personal information that the subjects had been requested to conceal, and two subjects betrayed their 'shameful secret' in the course of telling their cover story, as in the experiment of Redlich et al. (1951) and, in this study too, it was the most neurotic subjects who revealed the truth. In summary, the published evidence on interrogation shows that there is no truth serum, but the last two studies suggest that the drugs may be of some use in psychiatric interviews.

The technique of narcoanalysis and narcodiagnosis in psychiatry was developed in the 1930's by Lindemann (1932) and Horsley (1936). In typical narcodiagnosis the drug, usually pentothal, amytal or another barbiturate is injected intravenously into the recumbent patient. When the patient is relaxed, some innocuous questions are asked first and then more pertinent questions. Lindeman (1932) reported that the drugs increased the communications by the patient and that they readily revealed psychotic material. The success rate depended to a great extent on the skill of the examiner, even though the patient is usually not aware of the significance of the material released (Hoch and Polatin, 1952).

It is popularly believed that pentobarbital acts as a truth serum and that the injected person invariably reveals the truth about his actions and motives, but this is not the case (Hoch and Polatin, 1952). Suggestible individuals have agreed to questions improperly formulated or phrased; deluded patients have confessed to crimes that they had fantasized but had not committed; and many individuals, including psychopaths, have been able to lie and fabricate convincingly even when they are almost unconscious. Thus drug induced confessions are useless from the legal standpoint. Only those individuals who have conscious or unconscious reasons for co-operating are liable to confess, and it is unlikely that during police interrogation the suspect will be motivated towards co-operation.

For the psychiatrist it is the uncovering of psychological reality as opposed to factual reality which is important, for it is the psychological reality with its mixture of truth and fantasy which is the raw material of the therapist. The patient's revelations of his fantasies, delusions and hallucinations are as useful as his recollections of the past. It seems to work particularly well with paranoid and psychoneurotic schizophrenics where the normally unavailable delusions and rationalizations underlying their behaviour can often be easily elicited under the drug (Hoch, 1946). It is uncommon for the release of this material to precipitate a psychotic state, because when the patient wakes up there is amnesia for the session (Hoch and Polatin, 1952).

The perceptive reader will have noted that the 'truth drugs' in this section are the same as those that produce retrieval deficits in the state-dependency studies. This apparent contradiction can be resolved when it is realized that 'truth drugs' are producing retrieval in inappropriate situations, and not facilitating normal retrieval. Thus, these drugs are interfering with control by the situational stimuli on the responses, as in the state-dependent situations. As a result the probability of verbal recall increases in situations where the subject would not normally retrieve the information. Thus, strictly speaking, 'truth' drugs do not facilitate normal retrieval processes but impair them by reducing their situational specificity.

10
Development of Intelligence

Intelligence is a term used to refer to certain features of human behaviour; the features frequently included under this heading are learning, memory, problem solving, form discrimination and language ability. Various intelligence tests have been devised to assess an individual's overall performance tasks involving these aspects of behaviour. In this chapter we will consider the ways in which intellectual functioning can be explained in terms of the maturation of the nervous system and the various influences on maturation. Thus the chapter will consider the influence of genetic endowment and the environment on the development of problem-solving proficiency, and studies presented which emphasize that gene structure does not impose an unmodifiable developmental pattern on the organism which sets a limit to the ability which may be reached in an ideal environment. Instead, it will be seen that an individual's intelligence score depends on the interaction of the unique set of characteristics coded by the genes with the environment, where genes set the limits but these limits vary according to the environment. It will become clear that biochemical maturation does not provide a neutral base on which experience is superimposed but rather there is a reciprocal relation between maturation and experience. This becomes clear when it is realized that the neonate's brain has only a tenth of the solid matter of the adult brain, and its organization proceeds under the influence of factors in both the internal and external environment of the organism (McIlwain, 1955). The first section will discuss the maturation of the nervous system and the ways in which this may explain the concomitant changes in intellectual ability.

A. BIOCHEMICAL DEVELOPMENT AND INTELLIGENCE

In the last chapter the following hypothesis of learning and memory was summarized. Sensory information entered the nervous system and if a reinforcing event occurred, cortical desynchronization was blocked and any cortical evoked potentials present at the time of reinforcement were enhanced. The increased neural activity initiated a process of selection from the available genes and this led to protein synthesis. The new proteins synthesized might be transmitter enzymes, terminals, or postsynaptic terminals, but the outcome of the increased synthesis was improved transmission along specific pathways in the central nervous system. It follows from this scheme that registration, retention and retrieval will not occur until there is effective protein synthesis and the capacity for producing enzymes and synaptic structures. It should be noted that registration, retention and retrieval may proceed at different rates, and Campbell and Spear (1972) have listed a number of studies which demonstrate that, when the degree of original learning was matched, younger subjects showed poorer performance on a retest.

Most of this work was performed on rats, so that it is convenient for present purposes to consider the maturation of the rat brain. McIlwain (1955) divided up the development into four stages: the foetal stage, when 97 % of the brain cells are synthesized but the brain has only reached 15 % of its adult weight; the first 10 neonatal days, when the cell processes differentiate and the rat's cortex nearly reaches adult size and weight; days 10 to 20, when the enzyme activities rise rapidly to adult levels and axonal myelination begins; and days 20 to puberty when myelination is completed and the brain finally reaches adult size and weight. Let us now examine in more detail the processes listed above.

As we said, the vast majority of the adult neurones are present at birth, but the DNA and RNA content of the brain continue to increase reaching their maxima in the middle of the third stage, i.e. 14 days (Mandell, Rein, Harth-Edel, Mardell, 1964). This is long before long-term storage of information is possible in the rat showing that it is not simply the total DNA content which is important but its function in protein synthesis. Thus, the amount of protein in the brain starts to increase at the beginning of the second stage and reaches a maximum at about 30 days (Agrawal and Himwich, 1970), when the neural processes have proliferated. This proliferation consists largely of growth of the dendritic tree of the neurons and the synaptic contacts on them. In the cortex Caley and Maxwell (1971) have described the development of the first synaptic contacts during the first 10 days, during the third stage they observed synapses which were indistinguishable from those in the adult, and by the fourth stage they could find no immature synaptic contacts. It is important that it is during this period of 20 days that the rat shows the greatest increase in long-term memory (Campbell and Spear, 1972) suggesting that a mature cell structure is a necessary condition for effective memory consolidation. Campbell and Spear (1972) discuss two possible ways in which the number of synaptic connections may be related to the poorer memory of the infant rat. Firstly, memory may depend solely on the number of functioning synaptic connections. Secondly, the development of synaptic contacts may in some way interfere with retrieval of information. As we saw in the last chapter, retrieval depends on environmental stimuli activating a subset of the originally 'trained' neurons and the closer the test subset to the 'trained' set the more probable retrieval will be. Obviously, the rapid development of synaptic contents would change the pattern of neural activity and decrease the probability of retrieval.

As well as these qualitative changes in the neurons there are also dramatic changes in the transmitter substances as a result of increased protein synthesis and consequent production of enzymes. Thus at birth norepinephrine was barely detectable in the whole brain, 0·12 mg/brain (Karki, Kuntzman and

Brodie, 1962), was clearly present in the median eminence at 5 days (Hyppä, 1969), had reached 0·353 mg/brain at 12 days and 0·367 mg/brain at 24 days (Schoemaker & Wurtman, 1971) and the adult concentration at 6 weeks (Karki et al., 1962). Serotonin seems to appear at about the same age as norepinephrine and reaches adult concentrations at about 21 days (Karki et al., 1962). Although there seems to be no direct evidence on the neonatal development of acetylcholine levels in the brain, the cholinesterase activity augments markedly, tripling in the first 21 days and increasing five-fold by 7 weeks (Im, Barnes and Levitsky, 1971). From this one would predict changes in the reinforcement mechanisms described in Chapter 3, the attentional mechanisms described in Chapter 4 and slow wave sleep control described in Chapter 6. As we have seen, all have been said to be involved in memory, and since their development parallels memory development they must also be considered as sufficient conditions for memory storage.

As well as an involvement in memory the ascending cholinergic pathways have been implicated in discrimination performance, and problem solving, another aspect of intelligence. Although we have no direct evidence on the changes in brain acetylcholine, some evidence can be obtained from the development of electrocortical arousal. A careful study of the autogenesis of electro-encephalographic activity shows that it differentiates between 15 and 30 days with the faster (20–30 Hz) desynchronized waves appearing after 20 days (Deza and Eidelberg, 1967), and this related very nicely to the finding that scopolamines had no effect on activity until rats were 20 days or more (Campbell, Lytle and Fibiger, 1969). This result contrasted with the finding that amphetamine markedly enhanced activity at 10 days after birth, showing that the measurement techniques were sensitive and supporting the idea that the cholinergic system develops more slowly than the adrenergic system.

The sensory processing involved in problem solving and discrimination performance also depends on rapid and accurate transmission of information, and this is achieved by myelination of axons which also increases tremendously in the first 35 days, doubling from the age of 14 days (Balasz, Cocks, Eayrs and Kovacs, 1971). Similar changes are found in the kitten and this results in a marked change in the rate of conduction (Purpura, Shofer, Housepian and Nobach, 1964). It is interesting that this appears to be a relatively late event compared with neuronal and synaptic development (Purpura et al., 1964). Long ago Flechsig (1896) noticed that myelination takes place according to systems of fibres rather than brain regions, and that these systems tended to have different functional significance. This finding supports the ideas outlined in the Chapter 1 that the brain is organized in neuronal networks rather than neural centres. In the brain the first pathways to be myelinated are the electro-cortical arousal pathways from the brain stem, and this will have similar consequences as the development of synapses on retrieval and discrimination performance. In addition, as Campbell and Spear (1972) point out, the neurally immature organism will process less information than when he is more mature, because of the slower conduction time. Accordingly, during the retrieval test at a later age the pattern of neural activity will be different and generalization decrements will be observed.

In summary, the rapid changes in information processing and storage due to maturation will result in both an increased ability to register, retain and retrieve new information, but will concomitantly interfere with the ability to retrieve previously stored information. These factors account for the marked changes in 'intelligence' as exemplified in memory, problem solving and discrimination tasks.

So far in this discussion we have only discussed the maturational aspect of development but an equally important phase of development is old age, which is inevitably accomanied by deteriorating function. Ageing has been viewed by Gordon (1971) as 'a self-enhancing deteriorative intracellular, as well as intercellular process', whose major feature is loss of control by the genetic material over cellular synthesis. This loss of control would be manifested in the transcription of information from the cell DNA to the messenger RNA, and so there would be decreases in the quantity and quality of protein synthesis. Gordon (1971) has determined that ribosomal changes occur in the ageing rat brain which suggest a reduction in the efficiency of protein synthesis. This deterioration in cell function will disrupt transmitter enzyme production, and so will impair transmission of information throughout the neural networks. As well as changes of information transmission the evidence presented in Chapter 9 would suggest that transcription errors within the neuron will impair information storage because the selective activation of the DNA molecules will not increase the appropriate species of protein.

Recently Gordon (1971) has developed compounds which appear to ameliorate the disordered function. A group of these, the inosinealkylamino alcohol complexes enhance protein synthesis in ageing animals by acting on polyribosomes; the effects are complex but it appears that these compounds reverse the changes in the polyribosomes accompanying ageing. Concomitantly with this enhancement there is an improvement in avoidance acquisition and memory consolidation in rats over 2 years old. As yet no evidence has been presented on acquisition of appetitive tasks but clearly this is the next step in the development of a drug which is potentially of great importance.

B. INHERITANCE OF THE BIOCHEMICAL CORRELATES OF PROBLEM SOLVING

One of the oldest attempts to select for a specific behavioural characteristic was the breeding experiments of Tryon (1942) for maze learning ability. The parental population were unselected heterogeneous population of rats which were tested in a T-maze, and then bidirectional shifts in the first filial (F_1) generation were produced by means of assortative breeding of the 'low error' rats with other 'low error' rats and the 'high error' rats with other high scorers. The two groups separated until by the eighth generation the scores of the two groups scarcely overlapped and further breeding produced negligible effects.

The strains have remained stable with respect to the traits for which they were bred up to the present day. There is some doubt about what particular ability was selected in the Tryon studies and later experiments have shown that the two strains are not simply different in maze-learning ability and so the names 'maze-bright' and 'maze-dull' will not be used in this account. Krechevsky (1933) found that the low-error scorers tended to show preferences for spatial cues while the high-error scorers tended to respond to visual cues. In the Tryon, 17 blind T-maze, a solution depended on the animal responding to cues of location and thus spatial animals would be more adaptive in this equipment. As well as discrimination differences the Tryon strain differed in memory consolidation.

In a careful series of studies, McGaugh examined the effects drugs and electroconvulsive shock on the acquisition performance of S_1, S_3 and the F_1 hybrid of the Tryon strains. He found that large, small and intermediate doses of picrotoxin injected pretrial facilitated acquisition of a 14-unit T-maze in terms of a decrease in blind-alley entries in the S_3, descendents of Tryon's high-error scorers but significant effects were obtained with the S_3 strain using the larger dose (Breen and McGaugh, 1961). These results suggest that the drug may be improving discrimination or consolidation. In an experiment using post-trial injections of 1757 IS (1-5-diphenyl-3-7-diazadamantan-9-01) in a Lashley III maze, it was found that the greatest facilitation was found with S_3, the least with S_1, with the F_1 hybrid intermediate (McGaugh, Westbrook and Burt, 1961). Broadhurst (1964) analysed these data biometrically and found that the performance of the strains under the placebo demonstrated a strong heritable component with no significant dominance, i.e. F_1 not closer to one strain than the other and that the drug eliminated the inherited component by bringing the S_3 and F_1 strains' performance up to that of the S_1 strain. This latter study showed that the two strains did differ in memory consolidation. This was confirmed by a comparison of the two strains on performance after electroconvulsive shock administered at different intervals after learning (Thomson, McGaugh, Smith, Hudspeth, and Westbrook, 1961). The study demonstrated that consolidation was faster in the S_1 strain and so there was greater resistance to the shock. In the last chapter one neurochemical system which seemed to be intimately involved in both discrimination and consolidation was the cortical cholinergic system, and research has focused on differences in this system as the basis for the Tryon strain differences.

In 1953 a group of workers led by Krech (formerly Krechevsky), Rosenzweig and Bennett started a programme at U.C., Berkeley, in which they tested descendents of the Tryon strains in various spatially cued mazes and correlated their scores with the activity level of the enzyme cholinesterase in different regions of the brain. The results were consistent with the hypothesis that there was a positive correlation between efficiency in adaptive behaviour and the overall levels of cholinesterase at the cortex (Rosenzweig, Krech and Bennett, 1960). Unfortunately the differences in cholinesterase activity were only of the order of about 5 % and were close to the probable limits of sampling error.

Nevertheless, the values were obtained in a large number of studies and the differences were in the same direction and of the same order of magnitude, giving greater confidence to the validity of the result.

In addition to this set of studies the Berkeley team developed several strains, the K, M_2 and S_3 strains, by crossing members of the Tryon strains and allowing them to breed randomly. It was hoped that this would lead to a wider range of variation than the pure strains provided and increase the likelihood of obtaining significant correlation if one existed. If there was no relation between cholinesterase and maze learning then the correlation for the K-strain would be close to zero, since the genetic determinants of each trait would be reassorted randomly. The results turned out exactly opposite to those found previously; there was a significant positive correlation between cholinesterase activity and errors (McGaugh, 1959; Rosenzweig et al., 1960). The original hypothesis was revised proposing that maze-learning ability was related directly to the levels of both acetylcholine and cholinesterase, such that, within limits, the greater the amounts of acetylcholine functioning at the synapse, the greater the efficiency of transmission and, consequently, the greater the learning ability (Rosenzweig et al., 1960). This hypothesis was examined independently by testing the S_1 and S_3 strains for acetylcholine and the results confirmed the hypothesis showing that the S_1 strain had a significantly greater concentration of the transmitter (Rosenzweig et al., 1960).

In an experiment that was designed to complement these studies, Roderick (1960) developed a set of strains with high and low cholinesterase by using a selection programme in which he began with two heterogeneous strains and bred them selecting the progeny of high and low cholinesterase parents for further interbreeding. After six generations, the four strains (Roderick-Dempster High, Roderick-Dempster Low, Roderick Castle High and Roderick Castle Low) were tested behaviourally on the Hebb-Williams, Dashiell and Lashley III Mazes. In general the low cholinesterase activity strains had significantly lower error scores than the high cholinesterase activity strains. So once again the simple hypothesis was contradicted by new data. Biochemical analyses showed that the concentration of acetylcholine in the Roderick-Dempster High strain was about 9 % greater than the Roderick-Dempster Low rats, but there was no difference between the other two strains and this shows that acetylcholine alone was not related to maze-learning ability. However, if we examine the ACh/ChE ratios it can be seen that, in general, the strain with the greater ratio is the one that performs better in the mazes (Rosenzweig et al., 1960). In Chapters 4 and 9 we gave evidence for the importance of the cortical cholinergic system in sensory processing and in consolidation and these studies give supporting evidence for this hypothesis.

C. DIFFERENTIAL EXPERIENCE AND BIOCHEMICAL DEVELOPMENT OF THE CORTEX

In a previous section we have discussed half of the research work by the

Berkeley team of scientists, and in the studies to be discussed here the inverse type of experiment was performed using the amount of experience as the independent variable. As in the other studies, most of the biochemical measures were made on the cholinergic system, and it was hypothesized that increased cerebral activity produced by certain environmental experiences would lead to a greater cortical liberation of acetylcholine, and the resulting increased amount of the substrate would induce the synthesis of the inactivating enzyme acetylcholinesterase.

The independent variable of experience was manipulated by forming two groups of 10 to 12 littermates after weaning at 25 days. One group, the environmentally enriched group, were kept in a large cage with ladders, turrets, wheels and bars, and each day they were allowed to explore open-field environments and run in mazes, while their less fortunate littermates, the environmentally impoverished group, were kept socially isolated in bare individual cages in sound-attenuated rooms and were handled rarely. At 105 days the rats were sacrificed and their brains coded for biochemical analysis. Samples were taken from the visual and somaesthetic cortices, 'ventral cortex' (two-thirds of which was hippocampus and one-third was amygdala and corpus callosum), and the remainder of the subcortex (Rosenzweig, Krech, Bennett and Diamond, 1962) and analysed by the Ellman method for acetylcholinesterase activity.

The initial results were discouraging because the enriched-experience group had significantly lower acetylcholinesterase activity per unit weight in the cortical samples but significantly higher activity in the remaining subcortical samples. However, the experimenters then noticed that the cortices of the enriched group were 6 % thicker and heavier compared with the impoverished groups. It turned out that this increased cortical weight could be attributed to 20 % larger neurons, 14 % more glial cells and more large blood vessels (Diamond, Krech and Rosenzweig, 1964). Although there was a slight decrease in the amount of acetylcholinesterase activity per unit weight, there had obviously been an increase in the total acetylcholinesterase concomitantly with increase in cortical cells showing that there was probably an increase in cholinergic neurons as well. Unfortunately, the assay for acetylcholine did not reveal any differences in the two groups (Rosenzweig, 1966). However, the assay for acetylcholine was not as sensitive as the acetylcholinesterase determination so that differences of a few percent could not be detected reliably. Studies of other brain amines in rats given enriched and impoverished experience have indicated other sorts of changes in brain chemistry. For example, norepinephrine concentration was elevated in the cortex of the isolated group by 30 %, the amount in the colliculus was doubled, while the concentration in the caudate increased five-fold (Geller, Yuwiler and Zolman, 1966). These are the largest effects of differential experience ever reported, but they have not been replicated.

In summary, the effects of environment were manifested in different changes in the cortical mass, so that the ratio of cortical to subcortical weight (a measure of corticalization) was higher in the rat who developed in an enriched environment. Concomitantly there was an increase in the total cortical acetylcholinesterase, but a slight decrease in the cortical/subcortical ratio of acetylcholines-terase per unit weight. Since there were more glial cells there was more cortical, non-specific cholinesterase, and so a large ChE/AChE ratio in the enriched subject compared with his impoverished brother.

It is obviously of interest in view of these differences to know if the effects can be reduced by decreasing the amount of experience after weaning. The results have been compiled by Rosenzweig, Krech, Bennett and Diamond (1968) from their own data and work in the laboratory by Zolman and Morimoto (1962, 1965). The percentage value by which the enriched group mean exceeded the comparable impoverished mean was greater for the 80-day group than the 3-day group on the measures of cortical weight and total acetylcholinesterase activity, and the values for the 60, 30, 14 and 7-day groups were intermediate. Preliminary data cited in this paper suggested that 30 days of differential experience was sufficient to produce significant changes in cerebral acetylcholinesterase and cholinesterase activity.

In a behavioural study of rats given 30 days differential experience, it was found that the two groups of rats did equally well on the first simple light–dark discrimination problem, but the impoverished group made more errors with each successive reversal of the problem. An interesting unexpected finding of this study was that the brain levels of acetylcholinesterase for the two groups, sacrificed at the end of this study, did not differ very much. This suggests that training on a complex task at 55 days of age has reversed the effects of isolation on the brain chemistry. This brings us to the question of whether there is a critical period for producing the changes.

The most convincing study on this aspect of the problem was performed by exposing adult rats to 80 days of the enriched or impoverished environments (Rosenzweig, Krech and Bennett, 1964) and it was found that the cortical weights of the enriched rats were greater than those of their impoverished littermates, and the total acetylcholinesterase activity was greater for the enriched group as well. Thus there were similar changes in the brains of adult rats, and there was no evidence that there is a critical period after weaning for the effects of experience on cortical chemistry using these measures. In contrast, a second sort of treatment did have effects depending on the age that treatment was initiated. Tapp and Markowitz (1963) handled one group of rats from 2 days to 10 days after birth, while the other set were left untouched. At weaning the rats were sacrificed and their brains analysed in the usual manner, and the results indicated that handling in the pre-weaning period increased both cortical and subcortical weights with a proportionally greater increase in the subcortical sample. Total AChE increased in the sensory cortex and subcortex but not in the remaining cortical areas. Handling in the post-weaning period had no effects on either the cortical weight or the AChE activity (Rosenzweig et al., 1968).

In Section B it has been shown that there were differences in maze learning abilities of the Tryon S_1 and S_3 strains, the four Roderick strains bred for high and low cholinesterase, as well as four strains selectively bred to differ in brain acetylcholine concentration. A comparison of the strains' anatomical and chemical characteristics shows that the brains from the good problem solving strains have a high cortical to subcortical weight ratio, low AChE to weight ratio, high ChE and a high ChE/AChE cortex to subcortex ratio. This pattern is exactly the same as that observed in the environmentally enriched group (Krech, 1968) and the question can now be raised whether the environment can be used to counteract the 'inferior' genetic endowment or whether an impoverished environment can prevent the development of the 'superior' brain. In the first study (Krech, 1968) the two Tryon strains were raised in either of the two environments and the results showed that the differences between the strains were maintained within each of the environments. However, when we compare the brains of the 'inferior', S_3, strain kept in an enriched environment with the brains of the 'superior', S_1, strain raised in an impoverished environment, it can be seen that the environmental manipulation erased the strain differences. Conversely, the 'inferior' strain, S_1, brought up in the impoverished environment is even more disadvantaged, by a factor of two, compared with animals from the superior strain, S , kept in the enriched environment. Unfortunately, Krech (1968) did not present any behavioural data from which we could judge the effects of the interaction of the genetic endowment and the environment in the determination of the behavioural characteristics, and so we are left rather unsatisfied by the programme of research because it is so obviously incomplete.

D. PHENYLKETONURIA

In 1908 Garrod published the classic book *Inborn Errors of Metabolism* in which he discussed four rather obscure disorders, albinism, alkaptonuria, cystinuria and pentosuria. The first disorder to be studied was alkaptonuria (see Figure 29) with its characteristic changes in urine colour; it turns black. This disorder was detected in early infancy which suggested to Garrod that this was a congenital abnormality in metabolism, and after reviewing the case histories he discovered that although their parents were normal the parents of eleven of the children were first cousins which suggested to Garrod that the disorder was determined by a rare recessive Mendelian factor. He went on to suggest that this factor was a deficiency of a specific enzyme, which we now know to be homogentisic acid oxidase, and in the absence of this enzyme, unmetabolized homogentisic acid is excreted in the urine and turns black when exposed to the air. Nowadays about thirty of these inbred errors of metabolism have been identified and there are probably many more to be discovered. In the remainder of this section we will consider only one of these disorders, phenylketonuria, which results from an inability to oxidase phenylalanine to tyrosine in the liver, and autopsy has shown that this is due to the absence of the enzyme, phenylalanine hydroxylase (Mitoma, Auld, and Udenfriend, 1957). As a result phenylalanine and its more unusual metabolites, e.g. phenylpyruvic acid (see Figure 29) are enormously increased in the blood and urine.

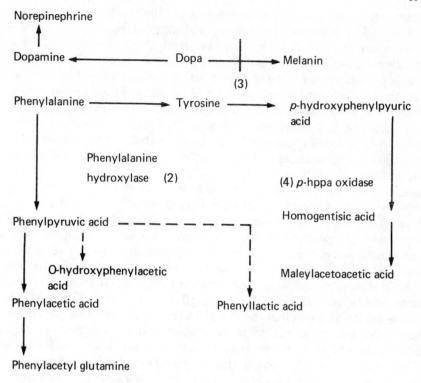

(1) Alcaptonuria
(2) Phenylketonuria
(3) Albinism
(4) Tyrosinosis

Figure 29. Diagram showing the enzymes that are absent in four types of inborn errors of metabolism, alcaptonuria, phenylketonuria, albinism and tyrosinosis

The inheritance of phenylketonuria is thought to be due to a Mendelian recessive trait which is the result of mutation of a specific gene producing an abnormal allele, giving the enzyme deficiency. The disorder will only be exhibited if the individual is homozygous, i.e. receives a recessive trait from both partners. If the parents are clinically normal they must be heterozygous unless the mutation occurred within them, so their children will be one-half heterozygous (Aa) and one-half homozygous split equally with dominant (AA) and recessive (aa) genes, i.e. one-quarter will show the disorder and from this we can estimate the number of heterozygotes in the normal population. The incidence of phenylketonuria is about 1 in 5,000 births in Britain, and as the frequency of phenylketonurics is one-quarter of the children of carriers, so the frequency of marriages between two carriers must be about $1/3 \cdot 770$. The frequency of carriers in the general population is the square root of $1/3 \cdot 770$, which is approximately 1 in 60. These carriers are mentally normal, although the single mutant allele does result in a plasma phenylalanine level about one-and-a-half times the average, showing that there is some enzyme deficiency (Hsia, 1967), but this is a small increase compared with the twenty-fold increase in the homozygote. When we consider that there are about thirty of these 'inborn errors of metabolism' known, the incidence of phenylketonuria carriers suggests that there must be a considerable number of the 'normal' population that are heterozygotes for one or another of these metabolic abnormalities. Minor biochemical abnormalities in the heterozygotes must play a significant role in the biochemical substrate underlying individual differences, and Garrod (1908) predicted that the idiosyncratic responses to drugs would probably turn out to be the result of individual differences in metabolism.

In all developed countries routine urine tests are made on infants to check for these compounds which are only present in small quantities in normal individuals, and if the tests are positive the phenylketonuric is placed on a diet low in phenylalanine. As all proteins normally contain phenylalanine, their food must be specially treated to remove most of this amino acid, although some must be left in for normal protein synthesis in the body. It has been found by several investigators that children given the special diet from early infancy had higher intelligence test scores than siblings whose treatment was begun after the age of 2 years, but lower scores than unaffected siblings (e.g. Berman, Waisman and Graham, 1966). In this study the average developmental rate of the early treated group was only 90 % as rapid as the rate of a normal population while their unaffected siblings were slightly faster, although there was considerable overlap between the groups and some early treated phenylketonurics surpassed their unaffected siblings in intelligence. The lower average score suggests that the special diet does not offer complete protection even though the blood phenylalanine levels are controlled, and it may be that the toxic metabolites cause damage to the brain before birth and neonatally prior to diagnosis.

Evidence supporting this hypothesis has become available recently as a consequence of a survey of children born to women who had phenylketonuria but had been on a restricted diet since birth; of the 60 cases examined only three of the children were of normal intelligence (Waisman, 1970), so that part of the critical period for impairment occurs prenatally. Nevertheless, it appears that the effects of postnatal exposure of the brain to phenylalanine and its metabolites has a more deleterious effect on the brain than prenatal exposure (Waisman, 1970). After about 2 years the disorder is permanent although further deterioration does occur up to the age of 8 years. The sorts of changes that have been observed in the central nervous system are similar to the demyelinating diseases, where there are areas of altered myelination in the central and peripheral nervous system (Poser and Bogaest, 1959; Alvord, Stevenson, and Vogel, 1950). Autopsy results demonstrate depletion of cerebral lipids which could be due to a lag in development as well as loss of lipids with advancing age (Gerstl, Malamud, Eng and Hayman, 1967). In this study there seemed to be a failure to synthesize lipid with long-chain fatty acids and/or a differential loss of these fatty acids. The reason for this failure of myelination is not clear although one can speculate; for example, phenylàlanine metabolites inhibit enzyme systems like the decarboxylase group of enzymes and a decarboxylase enzyme is essential for the process of lipogenesis from decarbohydrates (White, Handler, Smith and Stetten, 1959) and myelin is composed of lipids.

In the last 15 years there have been numerous attempts to produce a rat model of phenylketonuria (see review of the early literature by Karrer and Cahilly, 1965). The favourite technique has been the administration of large doses of *l*-phenylalanine by itself, or with other amino acids, in order to simulate some of the biochemical criteria of PKU. such as high urinary phenylketones and high plasma phenylalanine. Although biochemically sound, there has been considerable disagreement about whether the treated animals show the typical symptoms of the untreated phenylketonuric (Karrer and Cahilly, 1965). For example, a rat's phenylalanine hydroxylase activity is at least ten times that of men and rhesus monkeys (Woods and McCormick, 1964), and this may preclude the severe irreversible damage that is produced in human infants by phenylalanine. This may explain why some studies have found reversible deficits with prenatal and postnatal loading with phenylalanine (Polidora, Cunningham and Waisman, 1966). The biochemical basis for the reversible deficits has not been studied in detail, but part of the impairment may be due to a disruption of serotonin synthesis.

As stated earlier part of the deficit in humans was attributed to inhibition of decarboxylase enzymes. Another decarboxylase enzyme, 5-hydroxytrypto-phan catalyses the synthesis of serotonin from its precursor, 5-hydroxytrypto-phan decarboxylase and evidence for the inhibition of this enzyme in human phenylketonurics is seen in the low excretion of 5-hydroxyindolacetic acid, the final breakdown produce of serotonin, suggesting that serotonin synthesis has been inhibited (Baldrigde, Borofsky, Baird, Reichle, and Bullock, 1959), and thus impairing the serotonin pathways in the brain. The first direct evidence for this idea came from a study by Woolley and Van der Hoeven (1964) in which they treated newborn mice with dl-phenylalanine and *l*-tyrosine until maturity, when they were tested in shock avoidance in the Flexner Y-maze described in Chapter 9. Normal mice received about 3·7 shocks before they learned to avoid, while the phenylalanine-treated mice gave scores of 4·7, but when they were treated with 5-hydroxytryptophan, from birth to maturity, the average score was 3·6. Thus the precursor appeared to protect the mice from the ill effects of excessive amounts of phenylalanine but, unfortunately, the authors present no data on the brain levels of serotonin or on the blood levels of phenylalanine, so that they are open to the criticism that 5-hydroxy-tryptophan had induced the production of phenylalanine hydroxylase which had removed the excess phenylalanine and protected the brain. A more con-vincing study used diets of either leucine, phenylalanine, tryptophan, or phenylalanine and tryptophan (McKean, Schanberg and Giarman, 1962).

They demonstrated that cerebral serotonin was lowered below control in rats given the phenylalanine or leucine diets but significantly higher in the rats fed tryptophan. Correlated with the fall in brain serotonin in the two groups was an impairment in water maze performance, while the tryptophan group with elevated serotonin had fewer errors. The combination of phenylalanine and tryptophan produced rats who were slightly better than control rats in terms of maze errors and had slightly higher than normal levels of brain serotonin, so tryptophan was effective in blocking the effects of phenylalanine. There was no evidence that the abnormal pattern of amino acids in the brains of rats treated with phenylalanine was altered by the tryptophan. In addition it was found that phenylalanine elevated brain dopamine levels while norepinephrine levels remained constant and that tryptophan had no effect on either of these transmitters. Clearly the deficits correlated best with changes in brain serotonin levels.

These studies show that expression of the recessive gene for the absence of phenylalanine hydroxylase depends on the presence of phenylalanine in the diet. The consequence of the excessive unmetabolized amino acid during neural development is on interference with the process of myelination, which seems to occur during a critical period. Myelin is important in the rapid processing of information in the nervous system. In Section A we saw that myelination was correlated with the development of memory and problem solving, two important aspects of intelligence, and failure to develop myelin leads to an irreversible intellectual deficit. As well as this deficit the untreated phenylketonuric has chronic phenylalanine poisoning which would result in disruption of decarboxylase enzyme activity including serotonin synthesis. In rats this poisoning results in problem-solving deficits, but these appear to be reversible. In humans untreated from birth the introduction of a low phenylalanine diet produces only minor improvements showing that it is early infant exposure which is critical for intellectual improvement.

E. MALNUTRITION AND INTELLECTUAL DEVELOPMENT

It is a well-established fact that inadequate diet during infancy results in retarded physical growth and no amount of supplementation will correct the results of deprivation during the critical growth period of the neonate, and this effect has been seen clearly in mice, monkeys and men (Scrimshaw and Gordon, 1968). The same effects have been observed when measures of brain weight have been made after malnutrition, suggesting that there may be impaired neurochemical development. Very often the stage of development corresponds to that of younger, but normally nourished, organisms of the same height and weight, but they never reach the height and weight of normals at maturity. It is considered that the brain is particularly vulnerable at the time when it is developing most rapidly. The time of this critical period of growth varies from species to species so that the growth spurt in rats is from 5–20 days, compared with 6 weeks before birth to about 5 weeks afterwards in pigs, while in humans the last few weeks of foetal life and the first few postnatal months are critical (Dobbing, 1968). When attempting to extrapolate from animal studies to man it is important to consider the time of malnutrition relative to the growth spurt. An advantage of the rat in these studies is that the growth spurt for this species occurs postnatally, and so is accessible to direct manipulation.

One method of manipulating nourishment in rats has been increasing the litter size, and using this technique Winick and Noble (1966) found that there was a reduction in the number of brain cells at maturity, even when the rats were given a normal diet after weaning. This irreversibility was supported by a study (Culley and Lineberger, 1968) which showed a permanent reduction in brain weights and brain cholesterol levels of rats suckled in litters of fifteen compared with rats in litters of three. There is some evidence that 'catch up' is possible to some extent with normal cholesterol concentration per unit weight (Dobbing, 1964; Culley and Lineberger, 1968), although there were still lower total amounts of cholesterol and DNA than normal (Culley and Lineberger, 1968).

The latter authors were unable to enhance these deficits by postweaning undernourishment and this was confirmed in the next study of Dobbing (1968). Newborn rats were subdivided into three main groups—Group I (3 weeks) were placed in either litters of 3 or 15 to 20 until weaning and then fed normally; Group II were placed in large litters for 3 weeks and then fed a restricted diet for 8 weeks. Group III (3–11 weeks) were kept in small litters for the first 3 weeks and then were given the restricted diet for 8 weeks. All three groups were then given ad lib feeding 'rehabilitation' until 28 weeks then they were sacrificed and various growth measures made. These results are shown in Table 1, with the values for the undernourished rats expressed as percentage deficiency compared with the ad lib fed rats for each group. It can be seen that the 3 weeks mild undernourishment prior to weaning at 3 weeks produced a reduction in the number of brain cells, reflected by the amount of DNA phosphorous, and the amount of myelination of the neurons, shown by brain

TABLE 1

The percentage deficiency in various factors as a result of under-nutrition during three phases in the development of a rat. From Dobbing (1968)

	Body weight	Brain weight	Brain DNA-P	Brain cholesterol concentration
Group I (3 weeks)	23 %	7 %	12 %	7·6 %
Group II (11 weeks)	33 %	13 %	22 %	7·4 %
Group III (3–11 weeks)	23 %	12 %	10 %	0·0

cholesterol concentration. Increases in the period of undernourishment up to 11 weeks (Group II) produced no further deficits in myelination, suggesting that the pre-weaning period is critical, and this was confirmed by the post-weaning deprivation animals (Group III) where there was no deficiency in cholesterol. Although the number of brain cells was further reduced in Group II, the greatest proportional loss was during the first 3 weeks, so these results supported the hypothesis that the effects of undernutrition on the brain, in both the short-term and the long-term, depend on the timing of deprivation in relation to the period of fastest brain growth. In rats the brain size reaches 80 % of adult size by weaning although body weight is only 20 % of adult value when human infants reach 80 % of their adult brain weight the brain is only 20 % adult body weight at the age of 3 years. The effects of undernutrition on the development of the neurochemical systems in the brain were studied by Schoemaker and Wurtman (1971) using a method of perinatal undernourishment starting at midgestation with 24 % and 8 % protein diets. After birth the pups were distributed eight to a mother so that no mother suckled her own young, and also 50 % of the pups were cross-fostered by a different diet mother. There were thus four groups for which comparisons were made of the total brain content of norepinephrine (NE), dopamine, dopa, tyrosine and the activity of the catecholamine-synthesizing enzyme tyrosine hydroxylase and these are shown in Table 2.

TABLE 2

Amounts of catecholamines 24 days after birth following normal gestation and normal postnatal feeding (CC), normal gestation and deprivation postnatally (CD), maternal undernourishment during gestation and normal postnatal feeding (DC) and doubly deprived (DD). From Schoemaker and Wurtman (1971)

	Body weight	Brain weight	Tyrosine hydroxylase activity	Dopamine (mg/brain)	Norepinephrine (mg/brain)
C-C	79·5	1·63	$8·8 \pm 0·8$	$0·51 \pm 0·008$	$0·397 \pm 0·023$
C-D	30·8	1·36	—	$0·410 \pm 0·005*$	$0·342 \pm 0·018$
D-C	77·3	1·61	—	$0·500 \pm 0·005$	$0·368 \pm 0·01$
D-D	17·3	1·12	$11·8 \pm 0·9*$	—	$0·287 \pm 0·012*$

*Differs from C-C group, $p < 0·02$

These results showed that the accumulation of dopamine and norepinephrine was impaired in these rats. It is not clear from the experiment whether there are a decreased number of neurons or decrease in the norepinephrine content per neuron, or both. Whatever the loss one would expect impairment of all types of behaviour involved with the norepinephrine systems, e.g. behavioural arousal and stress (see Chapter 11).

A number of studies have attempted to assess the types of behavioural impairment especially in problem solving. In a multigeneration study of malnutrition Cowley and Griesel (1959, 1963, 1964) have indicated some of the complexities of the problem of undernourishment. Initially a group of Wistar female rats were fed on a low protein diet from weaning and then mated at 90 days with males of the same strain. The first filial generation was reared with the mother until 35 days and then fed on a reduced protein diet until testing. From our point of view the first relevant observation is the mother–infant interaction; it was observed that the deprived mothers spent more time away from the nest than the normally fed mothers, and as a result there was a reduction in the overall mother–infant interaction (Cowley and Griesel, 1964). However, there were no other differences in maternal behaviour, such as grooming, retrieving and feeding, and in fact the low-protein mothers built better nests for the young, perhaps reflecting their own heightened thermal sensitivity due to less adipose tissue.

The first filial generation was tested on open field exploration, emotionality, hoarding, problem solving, and later on maternal behaviour, and it was observed that the deprived progeny showed longer latencies for cage emergence, significantly less ambulation and less exploration of the centre of an open field, although the mean number of faecal boluses was greater in the low protein group. These observations are consistent with the idea that the low protein rats were more emotional and timid than comparable control rats, and this would fit it in with the idea proposed in the next chapter that norepinephrine is involved in controlling stress steroids and anxiety. The surviving females were then mated with normal males in order to test their interaction with their own pups; it was observed that the first filial generation suckled their litter more frequently than the control group and more frequently than their own mothers had done. This suggested that undernourishment may have cumulatively altered either the quantity, quality, or both, of the mother's milk (Cowley and Griesel, 1963). The inferior problem-solving performance in a Hebb-Williams maze of pups from normal mothers, but fostered by first filial generation mothers, suggests that the quality of the milk was reduced (Cowley and Griesel, 1959). The second-generation offspring, both males and females, also showed impaired problem-solving performance with the difference being significant only in the case of the males (Cowley and Griesel, 1964). Unfortunately, the males and females were tested at different ages so we cannot conclude that there were sex differences. In summary, these studies show that pre-weaning malnourishment impaired problem solving probably as a result of myelination deficits. They also showed that the nutritional history of the parents plays a part in producing behavioural changes as well as the present environmental conditions.

The observation of Cowley and Griesel on the changes in emotional responsiveness after malnutrition were extended by the experiments of Levitsky and Barnes (1970) with aversive stimulation. They tested postnatally deprived animals for their startle response, in passive avoidance and continuous avoidance. They observed a greater response to noise in terms of defecation and urination in the deprived animals and they also made longer latency passive-avoidance responses, but responded faster in the Sidman avoidance. The authors suggest that one effect of malnutrition during early development is to produce a lowering of the threshold of the stress-response system and this is supported by the finding of enlarged adrenals in the undernourished rat.

So far we have discussed the effects of malnutrition on non-human mammals, but similar effects have been observed in humans as the result of famine or neglect. Clinically the acute effects are known as marasmus and kwashiorkor. In marasmus the diet is insufficient in calories and protein so that the child's body responds to the lack of dietary carbohydrate by mobilizing amino acids, usually from the muscle, for gluconeogenesis in the liver, and so there are adequate amounts of glucose and amino acids in the blood stream. In kwashiorkor the diet has insufficient protein, but is only moderately poor in calories so that the blood glucose levels do not fall sufficiently to mobilize amino acids, although the protein deficiency results in inadequate protein synthesis. Studies of Chilean children who had died from marasmus before the end of the growth spurt, about 1 year, revealed that their brain weights were lower than a comparable age group who had died in accidents (Winick and Rosso, 1969). The ratio of DNA, RNA and protein per unit weight was normal showing that there was a reduction in the number of brain cells, but no cholesterol determinations were made, so that it is not known if there was reduced myelinization. The data also showed that the younger the child when malnutrition occurred the more marked the effect, suggesting that the neonatal brain was more sensitive to undernourishment, and it seems likely that severe early malnutrition retards cell division in human brain development irreversibly.

Survivors of clinical marasmus were examined by Mönckeberg (1968) after rehabilitation in hospital. The age of admission of the children ranged from 3 to 11 months with weight ranges upwards from 2·7 kg for a 5-month old child, which was 0·5 kg below its birth weight. A comparison of these children with normal Chilean children showed that their height was more than two standard deviations below normal, and the values for head circumference, a measure of brain size, were clearly below normal. Measures of I.Q. ranged from 45 to 76 with a mean of 62, which was significantly less ($p < 0.001$) than the average of Chilean pre-school children of comparable socioeconomic status with the biggest deficits in language.

The results of kwashiorkor are just as debilitating in terms of apathy, list-lessness and irritability, but the malnurition occurs most commonly in the second and third year of life when the rate of brain growth is decreasing, so that the effect on intelligence is smaller than marasmus. An interesting pattern of urinary amino acids has been observed in children suffering from kwashiorkor, which consists of abnormally high ratios of phenylalanine to tyrosine similar to phenylketonuria, and perhaps the behavioural symptoms in the two conditions may have their origins in myelination deficits. In a study of children recovering from kwashiorkor, Cravioto and Robles (1965) found that their performance on the Gesell tests of motor behaviour, adaptive behaviour, language, personality and social development was much lower than that of children of the same chronological age and ethnic group who had not been affected by protein–calorie malnutrition. During a 6-month recovery period the difference was found to decrease except for children admitted before the age of 6 months, again suggesting that this deficit might be irreversible in the early deprived children.

11
Anxiety and Stress

This chapter is concerned with two related areas in the neurochemistry of behaviour, and they are discussed because of their importance and not because they form a neatly worked out neurochemically based theory of behaviour. One consequence of the lack of a well-established theory is that some sort of coherence has been imposed by taking one hypothesis of emotion and selecting experiments which fit in with that viewpoint. Thus, following Pribram (1967), emotion will be viewed as a passive state representing awareness of a response to certain internal and external stimuli where the magnitude of the emotional experience will depend on the amount of arousal measured, in information theory terms, as the amount of uncertainty in the environment. Amount of arousal is thus viewed as the size of the match and mismatch between expectancy and stimulus input (Pribram, 1967).

In Chapter 4 we saw that electrocortical arousal was controlled by the cholinergic pathways ascending from the reticular formation and, in this chapter, the emphasis will be on the contribution of the hippocampal branch of these pathways. Warburton (1972) has characterized the hippocampal circuit as a feedback loop providing recurrent regulation of the internal and external inputs to the ascending cholinergic arousal systems, so that redundant, i.e. 'expected', stimuli from the external and internal milieu are prevented from arousing the cortex. This implies that the system is more than a simple negative feedback loop because there must be some sort of storage of information about stimulus regularities in the internal and external environments. This characterization was based on the hypothesis that the limbic midbrain maintains stability by diminishing the response to recurrent events in the internal and external milieux (Pribram, 1969). It will be part of the argument in this chapter that the hippocampus 'recognizes' a state of mismatch, and produces specific changes in the output to the hypothalamus and the tegmental region. One of the major consequences of the change in output to the hypothalamus seems to be the release of stress steroids from the adrenal glands. It is argued that it is the action of these hormones on the brain chemistry, combined with the cognitive interpretation of the situation that constitutes the response that is experienced as the emotion 'anxiety'. As a preliminary to a discussion of this hypothesis it is useful to outline the mechanism of release of corticosteroids.

A. CONTROL OF CORTICOSTEROID SECRETION

The adrenal gland, located just below the stomach in man, is controlled from the hypothalamus by means of hormones released into the blood stream from the pituitary. Hypothalamic regions close to the median eminence translate the neural stress information into chemical information in the form of corticotrophin-releasing factor secreted into a specialized blood vessel system, the hypophyseal portal, by cells in the hypothalamus. The blood flow carries the corticotrophin-releasing factor to the adenohypophysis portion of the pituitary where it triggers the release of adrenocorticotrophic hormone (ACTH) into the general blood circulation, and it is carried to the adrenal cortex, its target organ. The adrenal cortex secretes about thirty steroids which are categorized into three main groups. These are the androgens, involved in sexual behaviour and secreted by the zona reticularis; the mineralocorticoids, secreted by the zona glomerulosa and important in regulating the sodium and potassium balance of the body; and the glucocorticoids, like cortisol and corticosterone, which are secreted by the zona fasciculata and stimulate gluconeogenesis, synthesis of carbohydrate from protein. In the rat, cortisol is not present and only corticosterone is released by ACTH.

Neural Control. The levels of corticosterone are kept stable by a negative feedback system to receptors in the hypothalamus, the so-called 'hormonostat' (Yates and Urquhart, 1962). There is inhibitory control of corticotrophin-releasing factor by feedback of corticosterone and administration of corticosteroids into the hypothalamus (Chowers, Feldman and Davidson, 1963), and other regions like the mesencephalic reticular formation decreased the secretion of corticotrophin-releasing factor, causing lower levels of ACTH and eventually atrophy of the adrenal cortex. The region implicated in these studies and others like them include nuclei in the basomedial hypothalamus and cells in these regions were investigated by Ruf and Steiner (1967), using a potent synthetic corticosteroid called dexamethasone, introduced close to the cells by microelectrophoresis. The firing rate of these cells was recorded by means of another barrel of the pipette and a group of cells were found in the periventricular gray close to the third ventricle which were depressed instantaneously by the dexamethasone. These cells were either receptor cells monitoring the circulating corticosteroid levels and controlling the effect, or cells synthesizing the corticotrophin-releasing factor for release to the pituitary.

The studies of Ruf and Steiner (1967) found that the hippocampal cells that they tested were not responsive to dexamethasone, yet McEwen and his coworkers (e.g. Gerlach and McEwen, 1972) found that radioactive corticosterone was taken up and bound in the hippocampus of adrenalectomized rats, suggesting that adrenalectomy eliminated competition for binding sites in the brain by endogenous corticosterone that would have been released by the stress of injection. The greatest incorporation of the tritiated corticosterone was in the pyramidal cells, particularly in region CA1 and CA2, the alveus region next to the ventricles, whose fibres project to the mamillary bodies, among other structures (Raisman, Cowan and Powell, 1966). This study relates very neatly with an electrical recording experiment from the same laboratory (Pfaff, Silva and Weiss, 1971) which found that corticosterone produced decreases in hippocampal unit activity, starting 10 to 40 min after injection and lasting for at least 2 hr. This prolonged effect would be characteristic of a substance binding intracellularly in the hippocampal formation and inhibiting activity to the pyramidal cells in the CA1 and CA2 region. Direct data on hormonal control by the hippocampal formation was obtained in studies in which the hippocampus formation was stimulated and steroid release was inhibited. For example, Rubin, Mandell and Crandall (1966) found that stimulation of the hippocampus in human patients induced a fall in plasma corticosteroids within 5 to 30 min, while different patients stimulated in the amygdala showed increase plasma corticosteroids.

Cholinergic control of ACTH secretion

The neuronal pathways regulating anterior pituitary function are thought to converge on the hypothalamus, and in a search for the possible neurochemical systems involved in corticosteroid control it would seem logical to evaluate the effects of injecting various presumed transmitter substances into hypothalamic regions and the parts of the limbic system implicated by lesion and stimulation studies. This sort of study was performed by Krieger and Krieger (1970), and they found that carbachol, a cholinomimetic, implanted in the median eminence region, mid-mamillary body area and amygdala area induced an abrupt rise in corticosteroid levels reaching a peak within 45 min of injection, but little response was obtained from hippocampal and septal implants. The effects of carbachol were blocked by means of atropine, but not by dexamethasone, suggesting that corticosteroid feedback inhibition is not anatomically associated with this cholinergic steroid release system.

In an earlier investigation of cholinergic involvement in ACTH release, Hedge and Smelik (1968) had implanted cannulae in the septum, anterior hypothalamus and posterior hypothalamus. The stress of implantation produced a prolonged release of ACTH lasting over an hour, which was reflected in enhanced plasma corticosterone levels. Crystalline atropine placed in the anterior hypothalamic cannula reduced plasma corticosterone levels but atropine stimulation at the other two loci was without effect, which showed the specificity of the effect to the particular neural locus, in this instance the median eminence region, found to be sensitive to carbachol by Krieger and Krieger.

Adrenergic inhibition of ACTH secretion

Some of the first studies that gave indirect evidence for adrenergic inhibition came from the laboratory of Brodie and Maickel in which they examined the effect of reserpine on the stress response and found that reserpine induced a pattern of change similar to stress, including a release of plasma corticosteroids (Maickel, Westermann and Brodie, 1961). As we have seen earlier, reserpine impairs the binding of dopamine, norepinephrine and serotonin in the brain and so Westermann, Maickel and Brodie (1962) made concomitant measures of norepinephrine, serotonin and plasma corticosteroids which found similar dose-response and time-response effects for monoamine depletion and increased corticosteroid release. Pharmacologically this enhanced release of ACTH could also be produced by chlorpromazine which blocks norepinephrine and serotonin synapses (De Wied, 1967a), and in an attempt to separate the function of the two monoaminergic systems Westermann et al. (1962) injected alpha-methylmetatyrosine, which inhibits dopamine and norepinephrine synthesis, and observed that plasma corticosterone increased, although the increase was not as large as that produced by reserpine. In a more recent study alpha-methylparatyrosine, another inhibitor, produced an increase in plasma corticosterone in most animals, and in those animals in which cortico-sterone was not increased alpha-methylparatyrosine did not deplete hypo-thalamic catecholamines (Scapagnini, Van Loon, Moberg and Ganong, 1970). In order to distinguish between the two catecholamines intraventricular injections of l-dopa, dopamine, l-norepinephrine and l-isoproterenol, an adrenomimetic, were investigated for their inhibitory properties (Van Loon, Scapagnini, Cohen and Ganong, 1971); all compounds produced inhibition, and since the first two are precursors of norepinephrine but the last two do not form dopamine, it looks as if norepinephrine is the inhibitory transmitter for ACTH release. This was confirmed by means of an inhibitor of dopamine beta-oxidase, FLA-63, which prevented the synthesis of norepinephrine from dopamine, and so reduced hypothalamic norepinephrine and increased plasma corticosterone. At this point some contradictory evidence should be mentioned; an extensive study by Carr and Moore (1968) found that combined injections of reserpine and alpha-methyl-paratyrosine markedly reduced the brain levels of dopamine and norepinephrine, but did not change the baseline levels corticosteroids. However, the release of steroids induced by stress was slightly, but not significantly, increased over controls during the depletion.

Indirectly evidence for an adrenergic inhibitory system comes from studies of the effects of stress steroids on brain chemistry. Brain levels of norepinephrine have been found to decrease after exposure to cold (Maynert and Levi, 1964), electric shock (Levi and Maynert, 1964) and immobilization stress (Corrodi, Fuxe, Lidbrink and Olson, 1971). This suggests that feedback of stress steroids to the brain activates the adrenergic inhibitory neurons in the median eminence to reduce the production of corticotrophic releasing factor. This activation could be by increased release or reduced re-uptake and some evidence for the latter mechanism has been provided by Iversen and Salt (1970).

Serotonin control of ACTH secretion

The evidence for the involvement of serotonin in the control of stress steroids is consistent with the hypothesis that this transmitter is involved in the in-hibition of corticotrophic releasing factor which initiates ACTH secretion by the pituitary. Indirect evidence consistent with this hypothesis has come from assays of the brain after acute stress, such as immobilization stress for 3 hr (Fuxe, Corrodi, Hokfelt and Jonsson, 1970; Curzon and Green, 1968) depleted serotonin when tryptophan hydroxylase, the rate-limiting synthesizing enzyme, was inhibited. These results in themselves are not convincing, but cortisol (Curzon and Green, 1968) and corticosterone (Scapagnini, Moberg, Van Loon, De Groot and Ganong, 1971) also produce depletion showing that it is the feedback that releases the serotonin. This increased rate of release would be expected eventually to reduce the functional serotonin if there was not a compensatory increase in synthesis. Some evidence for corticosteroid control of synthesis has come from studies where synthesis inhibitors were not injected and the serotonin levels remained constant after stress (Levi and Maynert, 1964; Maynert and Levi, 1964) and injections of cortisone (McKennee, Timiras and Quay, 1968). Increased synthesis appears to be due to increased tryptophan hydroxylase activity produced by corticosteroids. In bilateral adrenalectomized rats with high levels of plasma ACTH, but low levels of plasma corticosterone, the activity of midbrain tryptophan hydroxylase was reduced by 75 % compared with sham operated rats (Azmitia and McEwen, 1969) and as a result there was reduced conversion of tritiated tryptophan to tritiated serotonin (Azmitia, Algheri and Costa, 1970). Injections of corticosteroids restored the activity of tryptophan hydroxylase towards normal in these adrenalectomized animals (Azmitia and McEwen, 1969) and returned serotonin turnover to normal (Fuxe, Corrodi, Hokfelt and Jonsson, 1970), suggesting that it was the low level of corticosterone that was responsible for the reduced serotonin synthesis. Cold and electric shock stress also stimulated tryptophan hydroxylase synthesis, but this did not happen in adrenalectomized rats (McEwen, Zigmond, Azmitia and Weiss, 1970) confirming that the feedback of corticosteroids enhances serotonin synthesis. At the moment the balance of evidence favours the hypothesis that there is adrenergic inhibition of ACTH release in parallel with a cholinergic releasing system. It seems that part of the adrenergic inhibition is mediated via the hippocampus. Another likely source of adrenergic inhibitory input is the median forebrain bundle mediating reinforcing effects and euphoria (see Chapter 3), and this would make sense because euphoria and stress would seem to be incompatible. The cholinergic control of secretion is probably mediated via the pathways ascending from the reticular formation involved in electrocortical arousal and attention (see Chapter 4).

B. ANXIETY, UNCERTAINTY AND CORTICOSTEROID RELEASE

In the introduction to this chapter it was stated that uncertainty, i.e. mismatch between the organism's expectancy and the stimulus input, induced the release of corticosteroids. In this section some experimental support for this statement will be given, using anxiety as an example.

Human anxiety

Anxiety can be characterized as an unpleasant emotional state which is directed towards the future, and this anticipatory element distinguishes it from fear which is the result of a response to an actual threat. All normal people are subject to anxiety prior to critical and uncertain events, like examinations, public appearances and so on. Pure pathological anxiety reactions are rarely seen in clinical practice (Mayer-Gross, Slater and Roth, 1954); more often, anxiety is found as an accompanying sympton of depression and occasionally schizophrenia where the patient is preoccupied with impending disaster, with either no idea of the source of the threat or, if the threat can be recognized, it is trivial compared with the reaction.

There have been a number of studies correlating cortisol production and psychiatric state, but few of these have focused on anxiety. In a summary of his own work Sachar (1970) found that patients who had little anxiety and emotional arousal also showed little change in cortisol production during recovery, while patients with symptoms of anxiety had a moderate 25 % elevation in their cortisol production compared with their recovery levels. In cases of psychotic disorganization in depression and schizophrenia there was disintegrative anxiety and very large, 50 %, increases in cortisol levels.

Experimental models of anxiety

Attempts to study 'anxiety' in animals have concentrated on behaviour performed while anticipating an aversive event. One experimental situation was developed by Estes and Skinner (1941) in which animals were trained to press a lever for positive reinforcement, and after this training the animals were placed in a box and a warning stimulus presented prior to foot shock. Thereafter, presentation of the warning signal while the animal was pressing resulted in suppression of responding in anticipation of shock. Measures of the corticosteroids during sessions, when the warning signal but no shock was given, showed that the adrenocortical secretion was increased progressively on successive sessions (Mason, Brady and Sidman, 1957). In the same study measures were made in non-discriminated avoidance situations (Sidman, 1953) where animals could reset a clock by pressing a lever, but if the clock timed out they received a shock. Here the animal has no signal but must press in anticipation of shock, and once again the plasma corticosteroids were elevated. However, in a later experiment when the animal was given a warning signal for the avoidance response, i.e. discriminated avoidance, the steroid secretion was decreased. Elimination of the warning signal resulted in an elevation of the steroids again, although they did not return to their original level.

These results can again be interpreted in terms of 'uncertainty' and 'control'. In the first situation a signal is presented, which had previously been paired with shock, but no shock is given, i.e. the animal's expectancy is contradicted, while in the second situation the animal has no exteroceptive cue which signals when an avoidance response should be made, so it is not predictable and greater amounts of steroids are secreted, but introduction of the warning signal makes the situation more predictable and the steroid levels fall. Another way of talking about this expectancy and control is proposed by Mandler (1967), when he argues that by 'anticipation', or expectancy, we mean that the subject has some available behaviour which can be performed when the proper stimuli occur. A situation is 'ambiguous' or unpredictable when no single set of behaviours is directly or dominantly relevant to the situation, and anxiety occurs when no situationally relevant behaviour is available to the organism.

In discussing the previous experiments it was argued that corticosteroid release was the result of the uncertainty in the situation. However, all of these experiments used shock as motivation which releases steroids, and unequivocal support for the uncertainty argument can only come from studies using appetitive reinforcement. In the first study uncertainty was introduced by changing the rat's schedule of reinforcement after training; rats were trained to lever-press on a continuous reinforcement schedule, and then reinforcement was withheld. Measures of the stress steroids demonstrated a rise in plasma corticosterone during the extinction phase (Coover, Goldman and Levine, 1971). That the phenomenon was not due to removal of reinforcement *per se* was shown by Levine in another study (Levine, Goldman and Coover, 1972), in which animals were trained on either a variable interval schedule or a fixed interval schedule with the same density of reinforcement, and then switched to the other schedule. Animals switched from unpredictable variable interval schedule to the predictable fixed interval schedule showed no change in steroid levels, but rats switched from the predictable fixed interval to the unpredictable variable interval showed an elevation of plasma corticosterone concentration.

A third experiment by Levine and his group (Coover, Ursin and Levine, 1972) confirmed that this phenomenon was not merely due to stock. During training on discriminated avoidance in a shuttlebox blood samples were taken from the rats during acquisition, stable performance and extinction when the response was prevented. During acquisition there was a high level of corticosterone which one might expect from the high shock density, but the levels fell when stable performance was achieved and the number of shocks decreased. However, shock was not the only reason for elevated steroid levels, because during extinction with the lever removed the levels were again elevated. Levine et al. (1972) suggested that during acquisition the situation becomes more predictable for the animal and the corticosterone levels fall, but during extinction 'when the predictability and control that the rats had acquired no longer exist, an increase in pituitary–adrenal function occurs'.

It is unclear from the human and animal studies whether the stress steroids cause of the anxiety or whether they are an epiphenomenon of anxiety. The next section will examine this problem.

C. BEHAVIOURAL SIGNIFICANCE OF THE STRESS RESPONSE

The physiological consequences of increased ACTH secretion, and the resulting enhanced corticosteroid release, are well documented, but until recently there have been very few behavioural studies because of the difficulties of obtaining pure extracts of corticosterone and ACTH. The effects of these two compounds are considered separately in view of the apparent differences in the effect on the CNS of the two hormones.

Adrenocorticotrophic hormone

In the normal animal increases in the levels of circulating ACTH as the result of pituitary release or injection will produce increased secretion of corticosteroids, so tests of the independent effects of ACTH on behaviour must be performed in adrenalectomized animals. In an early study with adrenalectomized rats (Miller and Ogawa, 1962), it was observed that ACTH during avoidance conditioning had no observable effect on the rate of learning, but when the ACTH was administered during extinction, they observed that the frequency of responding was markedly increased. This was the first demonstration that ACTH might have some effects on behaviour independently of effects mediated via the adrenals, and this effect was confirmed by De Wied (1967b) in an experiment in which he tested rats after adrenalectomy, when the ACTH levels were high due to removal of the inhibitory corticosteroid feedback. In these subjects extinction of a shuttlebox avoidance response was retarded, which suggests that the ACTH was either enhancing 'fear', the aversive motivation, or it was interfering with some response-inhibition system. These hypotheses could be tested by means of a passive avoidance situation. If fear were increased then we would expect enhanced passive avoidance performance, but loss of response inhibition would result in impaired performance. As part of a comprehensive study, Weiss, McEwen, Silva and Kalkut (1969) showed that ACTH, secreted in large quantities as a result of adrenalectomy, seemed to enhance the aversive motivation. Fear was measured in terms of the latency for re-entering a compartment where the animal had been shocked severely the previous day. After shocking, both normal and adrenalectomized animals were reluctant to enter the shock compartment, but the adrenalectomized animals had longer latencies and they defecated significantly more· on the first two tests. The experimenters then reduced the number of fear-associated cues by changing the colour of the walls and the colour and texture of the floor of the shock compartment and this reduced the latency for both groups, but adrenalectomized rats still had significantly longer latencies. The explanation for the differences was not general debilitation on the part of the adrenalectomized rats because prior to shocking the two groups had the same latencies for entering the compartment and there were also no differences in a general activity cage. The effects of adrenalectomy on responsiveness to fear-evoking stimuli were seen when normal and adrenalectomized subjects were placed in the passive avoidance chamber, and their 'step out' latency to the larger compartment was measured. Then animals from a separate study were placed in the box and given shock so that they urinated and defecated creating odour in the box which appeared to be aversive to rats. When the adrenalectomized rats were placed in the small chamber they came out signficantly more slowly than normal animals tested under the same conditions and adrenalectomized subjects tested with a clean compartment.

In the second part of this study by Weiss et al., the whole of the pituitary was removed, eliminating ACTH as well as a number of other hormones. In the passive avoidance situation the hypophysectomized rats had the same latencies for initial entry into the compartment as controls, and although these increased after shock they had significantly longer latencies than controls during the fear testing. Their performance during the generalization test was very close to their pre-shock score, and there was no evidence from activity cage scores that the hypophysectomized animals were more active than normals. This study showed that an absence of ACTH resulted in a large reduction in 'fear'.

Corticosteroid hormones

The effects of this hormone can be studied by either injecting ACTH into normal rats, or by using the synthetic corticosteroid, dexamethasone. The advantage of the latter procedure is that ACTH levels are not high at the same time as the corticosterone, enabling separation of the two effects. One study (Wertheim, Conner and Levine, 1967) compared the two sorts of treatment on non-discriminated avoidance, described earlier as a model of human anxiety, in which animals were trained until they had achieved stable performance. Injections of both ACTH and dexamethasone resulted in a marked decrement in shock rate during the avoidance testing, but the greatest decrements occurred with dexamethasone, suggesting that corticosterone's effects were confounded with those of ACTH. The distribution of responses was also modified by dexamethasone, and to a less extent by ACTH, so that there were fewer short-latency responses and concomitantly a slight reduction in response rate; these changes represent an improvement in avoidance responding.

This conclusion was supported by a companion study by the same authors (Wertheim, Conner and Levine, 1969) that correlated the baseline endogenous steroid levels and their steroid response to ether stress with avoidance performance. Initially the rats did not differ in basal levels of corticosterone, but the stress of anaesthetization revealed marked individual differences in responsiveness with a one-to-one correlation with avoidance proficiency of the animal after training. It was also discovered that marked individual differences in baseline steroid levels developed during training, and these correlated with the individual differences in avoidance performance as training progressed.

Another study which demonstrates this relationship was an experiment which examined the effect of the diurnal cycle of ACTH release on shuttlebox avoidance (Pagano and Lovely, 1972). The animals were kept on a 7 a.m.–7 p.m. light on cycle, and they were tested either at the low point of the cycle 8·30–9·30 a.m. or at the crest of the cycle 4·00–5·00 p.m. They found that acquisition was best in groups who received their training in the late afternoon when their ACTH levels were high. Injections of ACTH facilitated acquisition performance significantly in animals trained in the morning, but was without significant effect on those trained in the evening. In this study it was presumed, but not demonstrated, that the effects were mediated by corticosterone. In another study this group (Lovely, Pagano and Paolino, 1972) examined the relationship between endogenous corticosterone levels and avoidance performance. Steroid levels were manipulated by isolating individuals for various lengths of time, 0, 30, and 50 days. After this the animals were trained in a shuttlebox avoidance situation. It was demonstrated that animals housed individually for 30 or 50 days were superior to subjects living in groups in acquisition perform- ance. On the other hand, extinction was retarded in 30- and 50-day isolates, compared with controls. Assay of the basal corticosterone levels in other groups housed for these periods, showed that 30 and 50 days individual housing significantly elevated these levels relative to group-housed rats.

These demonstrate that stress steroids are related to both the ability to learn avoidance and perform the overtrained non-discriminated avoidance response that was discussed as an animal model of human anxiety. This means that either the release of stress steroids and avoidance performance reflects activation of a generalized arousal system (Wertheim et al., 1969), or that the corticosterone feeds back to modulate a generalized arousal system which facilitates avoidance behaviour. In this account the latter position will be taken and it will be argued that the feedback of corticosteroids acting on the neuro- chemical systems in the brain improve avoidance and result in part of the sensations experienced as anxiety in humans. Some evidence for the latter mechanism comes from studies of patients with Cushing's Disease in which there is cortisol hypersecretion; as well as the physical disorder patients display a number of psychiatric symptoms, including agitation and anxiety (Cleghorn, 1957). However, the most convincing evidence for the steroid causing anxiety comes from the effects of administering corticosteroids to man. When cortisone therapy was first introduced there were numerous reports of emotional disturb- ance (Cleghorn, 1957). This was described as an alerted state marked by anxiety, irritability, tension and emotional lability. In these studies anxiety was not necessarily increased although the subjects invariably became more tense. In a more systematic investigation the plasma corticosteroid level was raised rapidly by intravenous injection of hydrocortisone (F) and the subjects' anxiety levels assessed by a battery of tests. It was found that this steroid did not affect the immediate experience of anxiety, but their proneness to anxiety was enhanced (Weiner, Dorman, Persky, Stach, Norton and Levitt, 1963). In order to examine whether corticosteroids play a role in modulating the effects of anxiety-provok- ing situations, subjects were given an injection and an anxiety-provoking suggestion while in a hypnotic trance. These subjects responded to the sugges- tion with a more prolonged anxiety response than subjects receiving a placebo injection (Levitt, Persky, Brady and Fitzgerald, 1963). A later study (Persky, Smith and Basu, 1971) have found slight increases in anxiety measured on the same anxiety scales as Weiner et al. (1963) with small doses of the naturally occurring corticosterone B and corticosterone F given intravenously.

These studies give strong evidence for the idea that anxiety may result from the feedback of corticosteroids to brain and the consequent changes in biochemistry.

D. EVALUATION OF ANTIANXIETY DRUGS

The first situation to be discussed is the Estes and Skinner (1941) conditioned suppression situation which has been extensively used for testing drugs, including the so called 'tranquillizers'. Acute doses of the major tranquillizers like the adrenergic blocking agents, e.g. chlorpromazine, and adrenergic depleting agents, e.g. reserpine, are ineffective in attenuating the response suppression (Kinnard, Aceto and Buckley, 1962; Cicala and Hartley, 1967; Tenen, 1967). Thus blockade of the adrenergic arousal pathways seems to have little effect on the emotional response and in fact there is some evidence that this blockade *enhances* the suppression. In most studies this phenomenon is masked because of overtraining, but Hunt (1956) found that chlorpromazine prevented extinction of the response which would fit in with the suggestion that anxiety depends on the presence of corticosteroids in the blood stream and that there is adrenergic inhibition of corticosteroid release.

The earliest tranquillizers developed were the barbiturates. Tenen (1967) tested the strength of conditioned suppression by measuring the time before resumption of responding, in this case drinking, after presentation of the warning stimulus, and it was found that although chlorpromazine had been ineffective, amobarbital produced faster recovery, suggesting that 'anxiety' had been reduced. This was supported by the findings of Miller (1961) and Lauener (1963) in a study of more conventional conditioned suppression. Lauener (1963) also tested the first of the new generation of minor tranquillizers meprobamate (Miltown) but found small effects. Much larger and consistent results were obtained by Lauener (1963) with the benzodiazepines developed by Hoffman-La Roche, including chlordiazepoxide (Librium) and diazepam (Valium). Tenen (1967) also tested chlordiazepoxide, diazepam and nitrazepam on the recovery responding after onset of the warning signal. He found that there were faster recovery times in spite of their animal's obvious sedated and flaccid condition.

The second test of situations mentioned in Section B were the continuous and discriminated avoidance situations. In tests of the major and minor tranquillizers on performance in continuous avoidance situations (Heise and Boff, 1962), it was found that barbiturates, meprobamate, chlorpromazine, tetrabenazine and the benzodiazepines decreased responding in continuous avoidance. The major difference between the compounds was the dose range ratio which was based on the range of doses over which the drug affected behaviour but did not impair the capacity to press a lever and the incapacitating doses. A compound producing ataxia without any antianxiety effect would have a ratio close to one and so alcohol, the shorter-acting hypnotics and muscle relaxants had values close to one, and the benzodiazepines had values from

3·8 (chlordiaxepoxide) to 19 (nitrazepam) with chlorpromazine and tetra-benazine, intermediate. As Heise and Boff (1962) point out, the boundaries between the three groups were quite arbitrary and so this situation only discriminates between them quantitatively, but the discrete trial avoidance situation does provide some qualitative distinction.

From the discussion of discrete trial avoidance we would not expect anti-anxiety agents to have much effect since there was little uncertainty associated with them. Sure enough, meprobamate and chlordiazepoxide only produced a 25 % loss of avoidance responses at doses which were clearly ataxic (Randall, Schallek, Heise, Keith and Bagdon, 1960). To give a clear comparison of the doses required to modify performance in the two avoidance situations the actual doses will be given from this paper. Thus, 28–115 mg/kg of meprobamate and 49 mg/kg of chlordiazepoxide orally produced a 75 % decrement in continuous avoidance, 640 mg/kg of meprobamate and 60 mg/kg of chlordiazepoxide, orally, decreased locomotor activity by 50 %, while 400–1300 mg/kg of meprobamate and 125–480 mg/kg of chlordiazepoxide orally were required to reduce discrete trial avoidance by 25 %. Chlorpromazine and the other phenothiazines clearly blocked this behaviour at low doses (Irwin, 1961, see Heise and Boff, 1962). In order to investigate this difference, Heise and McConnell (1961) set up a trace conditioning procedure where a 5-sec gap was inserted between the discriminative stimulus and the shock. Typically mepro-bamate and chlordiazepoxide animals responded more during the gap than the normal animals, showing a longer latency to the stimulus due to 'anxiety' reduction, while chlorpromazine animals showed a decrement in avoidance and no 'gap' responses.

The final situation which has proved useful for the preclinical evaluation of antianxiety drugs is the 'conflict' situation developed by Geller, 1962; Geller and Seifter, 1960). Deprived rats were trained to lever-press on a variable interval schedule where liquid food was obtainable once every 2 min on the average. Every 15 min a tone signalled that every lever press during this period would be reinforced, but every reinforcement during this period was paired with electric shock. Since the animals had to balance the positive features of high reward 'payoff' against the negative aspects of receiving shock, the procedure was regarded as 'conflict' producing. This technique was found to be particularly sensitive to minor tranquillizers, so that doses of some barbiturates, meprobamate or chlordiazepoxide produced much more responding in the tone period, even though the animals were ataxic. In contrast, chlorpromazine increased the suppression of responding during the discriminative stimulus. If we argue that the 'conflict' situation results in stress steroid release, because of the uncertainty about which response to make, and so there is 'anxiety' then chlorpromazine by decreasing the adrenergic inhibition on stress steroid release will enhance the 'anxiety' and responses will be suppressed. In the next section evidence will be considered which suggests that the benzodiazepines are acting on the neurochemical systems believed to be involved in stress steroid release.

Neurochemical action of antianxiety drugs

The pattern of action required by the hypothesis is increased functional norepinephrine and decreased functional serotonin and acetylcholine. The effects of the drug on acetylcholine have only been studied in preliminary experiments (Ladinsky, Consolo, Peri and Garattini, 1973) which suggested that the diazepam increased the cortical levels of the transmitter which may represent an inhibition of release, but further studies are required before firm conclusions can be drawn. Thus the rest of the section will be devoted to norepinephrine and serotonin.

Although there are sensitive assay techniques developed for monoamines, there have been very few studies which have studied the effects of the benzodiazepines and other tranquillizers on the synthesis, storage, release, receptor interaction and inactivation of these transmitters. In a pioneering study Taylor and Laverty (1969) found that the major benzodiazepines, chlordiazepoxide, diazepam and nitrazepam, and barbiturates did not change the levels of norepinephrine in any part of the rat brain. In addition there was no evidence for a reduction of re-uptake or inhibition of catechol-0-methyltransferase, the inactivating enzyme, showing that inactivation mechanisms were unaffected by these drugs (Taylor and Laverty, 1969). However, the static levels of transmitter are not good indicators of functional transmitter and so Taylor and Laverty (1973) examined the effect of the three benzodiazepines on norepinephrine turnover in animals stressed with electric shock. This form of stress increased turnover in all regions of the rat brain (Bliss, Aillion and Zwanziger, 1968) as one would expect from the action of corticosteroids on norepinephrine, but the benzodiazepines blocked this effect and maintained the norepinephrine levels close to their control values. It is important for the hypothesis that the human antianxiety potency order was found in these studies, so that nitrazepam was the most effective and chlordiazepoxide was the least effective in maintaining the norepinephrine turnover. From the point of view of the neural pathways involved, this effect was maximal in the midbrain sample containing the hypothalamus, and the cortical sample, and least in the brain stem sample.

The depletion of norepinephrine from the cortical and hypothalamic neurons after injection of a norepinephrine synthesis inhibitor was examined by means of histochemical fluorescence technique (Corrodi et al., 1971). The animals were stressed by restraining them, which accelerated the depletion of the norepinephrine, but chlordiazepoxide and diazepam reduced the rate of depletion. This result shows that these drugs were probably acting to reduce turnover by blocking release of norepinephrine. However, the reduction in turnover was not detectable after six daily injections of benzodiazepine, oxazepam showing the development of tolerance.

A similar sort of result was obtained when the serotonin neurons were examined after the injection of a tryptophan hydroxylase inhibitor, which prevented *de novo* synthesis of serotonin. Benzodiazepines and pentobarbital reduced the rate of depletion of serotonin in the cortex, but the effects may not occur in all parts of the brain (Lidbrink, Corrodi, Fuxe and Olson, 1973). The notion that this result might be due to the benzodiazepines blocking serotonin release was supported by the finding that intracisternally injected radioactive serotonin was retained in the brain longer after diazepam (Chase, Katz and Kopin, 1970). This finding has been confirmed by single and repeated doses of oxazepam (Wise, Berger and Stein, 1972) which reduced serotonin turnover in the midbrain–hindbrain region, and it contrasts with the marked tolerance that developed in the norepinephrine system after repeated doses. This biochemical difference immediately suggests a psychopharmacological experiment examining the behaviour after injections of oxazepam. Margules and Stein (1968) found that the slight non-selective decreased responding disappeared after a few days, but the release of suppressed responding in the Geller and Seifter (1960) conflict situation remained unchanged. This circumstantial evidence points to the involvement of the serotonin system in the antianxiety section of the drug.

Direct tests of the hypothesis that it is either the reduction of functional norepinephrine or the reduced functional serotonin which is responsible for the antianxiety properties of the benzodiazepines have been carried out by workers at Wyeth Laboratories in the U.S.A. They argued that if it was a reduction of adrenergic function then adrenergic blockers should have similar effects as benzodiazepines in the conflict situation of Geller and Seifter (1960), but tests with the alpha-adrenergic blocker, phentolamine, and the beta-adrenergic antagonist propranolol injected into the ventricles were negative. Propranolol had no effect, while phentolamine reduced punished and non-punished responding, showing a non-specific effect of the latter drug (Stein, Wise and Berger, 1973). In the second part of the studies serotonin blockers and a serotonin synthesis inhibitor were tested to produce the reduction of functional serotonin also produced by the benzodiazepines. They injected methysergide, the antagonist, and released the suppressed responding in the conflict situation (Stein et al., 1973). Similar results were obtained with parachlorophenylalanine, the synthesis blocker, and the suppressed behaviour could be reinstated by repletion of serotonin by injecting the precursor 5-hydroxytryptophan (Stein et al., 1973). The serotonin agonist, alpha-methyl-tryptamine and serotonin both enhanced the suppression of responding in the conflict situation (Stein et al., 1973), as one would have predicted from the other results. In a final test of the hypothesis, Stein et al. (1973) injected oxazepam, releasing the suppressed responding, and then introduced serotonin directly into the ventricles which restored the suppression.

The neurochemical pathways on which the oxazepam was acting are not known, although Stein et al. (1973) suggested serotonin fibres arising in the dorsomedial tegmental region of the midbrain. The terminal of these pathways are mainly in the limbic system including the amygdala and septal area (Andén, Dahlstrom, Fuxe, Larsson, Olson and Ungerstedt, 1966). As we saw earlier, these were the regions involved in the control of corticosteroid release. Studies of the antianxiety drugs show that they act on these structures. Barbiturates block hippocampal seizures after fornix stimulation, and suppress post-tetanic potentiation in the hippocampus after amygdaloid and hypothalamic stimulation (Takagi and Ban, 1960). Meprobamate acts on the hypothalamus and septal area and hippocampus (Schallek, Kuehn and Jew, 1962). Barbiturates and meprobamate suppressed hippocampal seizures produced by stimulation of the fornix, but only barbiturates suppressed the hippocampal discharges after stimulation of the amygdala (Kletzkin and Berger, 1959; Schallek et al., 1962). Tests of chlordiazepoxide disclosed some similarities with decreased duration of discharge after stimulation of the septal area, and hippocampus, and decreased amplitude of responses in the amygdala. Chlordiazepoxide also slowed the electrical activity in the septal area, amygdala and hippocampus, but not the cortex, of the unanaesthetized animals just like meprobamate (Schallek et al., 1962).

E. ANTIANXIETY DRUGS IN HUMANS

The debilitating effects of pathological anxiety have been reduced to a remarkable extent by a number of psychoactive drugs. The first drugs used for the treatment of anxiety, besides alcohol, were the barbiturates. The problem of barbiturates is that they tend to induce sleep during the day (see Chapter 6, Section D) and repeated use results in physical dependence (see Chapter 8, Section E). The first modern antianxiety drug was developed at the beginning of the 1950's and released for general use in 1955, and this was meprobamate. It was an instant success and sales went from $7,500 worth in May, 1955, to over $500,000 worth in December of the same year (Ray, 1972).

One of the early clinical trials was by Selling (1955) who found that meprobamate was of considerable value in anxiety and tension states. Of 86 patients complaining of these symptoms, all but 7 showed marked improvement, and these included 29 'pure' anxiety reactives of whom 10 recovered, 17 improved and 2 were not improved. All of the patients had previously been taking a barbiturate, phenobarbital, but all preferred meprobamate. In a later study including acute and chronic anxiety reactives, Gardner (1957) found substantially the same results; of 35 cases of acute anxiety, 60 % showed complete, or almost complete, remission of symptoms with good social and economic adjustment, while another 29 % were satisfactorily improved with easing of the symptoms and considerable adjustment. The chronic anxiety patients responded less often to the drug, and only 3 % were greatly improved with satisfactory improvement in 40 %. There was also some success with other neurotic states in which anxiety and tension were prominent symptoms.

These two studies were not carefully controlled, but a better study compared meprobamate with a placebo, a barbiturate, sodium amobarbital, and a phenothiazine, prochlorperazine (Rickels, Clark, Ewing, Klingensmith, Morris and Smock, 1959). The patients were given a supply of one of these five drugs made up in identical capsules to last 2 weeks, and at the end of 2 weeks the drug was changed according to a randomized block design. The patient's progress was assessed by himself and a doctor on several rating scales and it was found that all four drugs produced an improvement in the patient's condition, showing the importance of the 'placebo' effect. However, if the scores for the first week on the placebo are discounted, i.e. when the placebo followed another sort of treatment, then the scores showed that they did worse on the placebo. A comparison of every drug with the placebo condition showed that meprobamate was the more effective then the barbiturate or the phenothiazine in terms of improvement. This finding was supported by the data on the patient's preference, which showed that all the patients preferred meprobamate as either first choice (60 %) or second choice over the other drugs.

At the same time as the clinical introduction of meprobamate, another group of compounds, the benzodiazepines, were under test, and the first useful compound to be discovered was chlordiazepoxide and more recently diazepam, nitrazepam and oxazepam. The first clinical reports were by Harris (1960) and Tobin and Lewis (1960) in which they collated the results of using chlordiazepoxide with 212 patients whose main symptoms were anxiety and tension. After treatment both the psychiatrists and the patients evaluated the drug, and it was found that over 80 % of the cases demonstrated clinical improvement. The doctors felt that it had a more specific effect on anxiety than meprobamate and the phenothiazines, and did not produce any lowering of mood. On the contrary, the patients reported feelings of wellbeing and increased drive which was reflected in wider social interests and return to work.

More recently study has investigated the adjustment of workers suffering from anxiety and tension for various reasons (Proctor, 1962). It was carried out on fifty women working in a hosiery mill looping thread in the knitting machines and paid by the piece, whose output had declined because of the anxiety-tension symptoms. Use of barbiturates and meprobamate affected the looper's dexterity and co-ordination so chlordiazepoxide was tried with considerable success; all but eight were relieved of their anxiety symptoms and their average production returned to normal in thirty-two of the women. No undesirable side-effects of ataxia and drowsiness were reported and obviously co-ordination was not affected since their production returned to normal.

In a controlled double-blind comparison of chlordiazepoxide with amylobarbital and placebo Jenner, Kerry and Parkin (1961) found that chlordiazepoxide was more effective in relieving anxiety in neurotic patients than a placebo or amylobarbital in the 2-week trial. As a result of the improvement of the symptoms 'many patients were able to live more normal lives and to travel on buses or walk in the streets alone without severe discomfort even after prolonged incapacity' (Jenner et al., 1961). In a comparison of chlordiazepoxide, meprobamate and placebo (McNair, Goldstein, Lorr, Cibelli and Roth, 1965) it was found that chlordiazepoxide was superior to placebo in reducing a patient's anxiety, but meprobamate was not. In a comparison the more recent benzodiazepines like diazepam were better than chlordiazepoxide (Isham, 1966), and of thirty patients diagnosed as anxiety reactives diazepam produced an average improvement of 60 % in the whole group. The patients were also asked to compare diazepam with other drugs which they had been prescribed; two-thirds of the group preferred diazepam and thirteen reported that they found chlordiazepoxide less effective. The author cynically points out that this preference might reflect variables such as the fact that the test drug was supplied free to the patient. In a British study (Jenner and Kerry, 1967) where both drugs were prescribed free, diazepam was preferred by thirty-five of the seventy-five patients, while sixteen selected chlordiazepoxide and thirty felt that the drugs were of equal value from the point of view of relief of symptoms. However, only eight patients preferred diazepam from the point of view of side effects, while thirty-six preferred chlordiazepoxide and thirty-one had no preference. The daytime drowsiness produced by diazepam seems to be one of its principle disadvantages. Pharmacologically oxazepam is considered to be more effective in the suppression of stress hormones released during fear situations than chlordiazepoxide or diazepam. Accordingly, a number of clinical trials of these were carried out. One of these was an extensive survey by Wittenborn (1966) of 613 patients treated with oxazepam and 261 given placebo. In addition a number of chlordiazepoxide patients were used for comparison. The patients given placebo were matched by computer with oxazepam patients on the basis of age, diagnosis, sex, severity of anxiety and the treatment speciality of the prescribing psychiatrist. Over a 5-week period the proportion of patients rated as having severe anxiety decreased in both groups, but the diminution of severe rating was far greater in the oxazepam group. Another way of analysing the data, tried by Wittenborn, was correlating the pre-treatment and post-treatment anxiety scores for the two groups. For the oxazepam, correlation was 0·35 while the same correlation for the placebo patients was 0·58, indicating that the arrangement of individual differences in anxiety has been changed to a greater extent in the oxazepam group. Chlordiazepoxide was also found to have a significant advantage over placebo but no direct comparison between the two drugs was made.

A test of oxazepam and meprobamate on 'normal' subjects who scored on the Taylor Manifest Anxiety Scale was performed by DiMascio and Barrett (1965). It was found that neither drug induced significant changes in any test measure in the group as a whole, but the 'high' anxiety subjects differed markedly from the 'low' anxiety subjects in their responses to the two drugs; oxazepam significantly reduced anxiety in the 'high' anxious subjects but *increased* anxiety measures in the low-scoring subjects, while meprobamate had no effect on anxiety scores. In addition, oxazepam slowed motor activity while meprobamate decreased co-ordination at the doses used. Of particular interest was the finding of opposite effects in different subjects. At the moment the mechanism is unclear, but it does indicate that we do not know the relevant diagnostic measures for predicting the response to drugs, or even manifestations of anxiety that are best treated by the antianxiety drugs. It seems that present clinical practice is to try out the various drugs on the patient until one is found which relieves the symptoms and whose side effects can be tolerated by the patient.

References

Abood, L. G., and Biel, J. H. (1962). Anticholinergic psychotomimetic agents, *Int. Rev. Neurobiol.*, **6**, 218–273.

Abramson, H. A. (1960). *The Use of LSD in Psychotherapy*. New York: Josiah Macy Foundation.

Abramson, H. A., Jarvik, M. E., Kaufman, M. R., Kornetsky, C., Levine, A., and Wagner, M. (1955). Lysergic acid diethylamide (LSD-25): I Physiological and perceptual responses, *J. Psychol.*, **39**, 3–60.

Adam, H. M., and Hye, H. K. A. (1966). Concentration of histamine in different parts of the brain and hypophysis of cat and its modification by drugs, *Br. J. Pharmac. Chemotherap.*, **28**, 137–152.

Adrian, E. D., and Matthews, B. H. C. (1934). Berger rhythm: potential changes from the occipital lobes in man, *Brain*, **57**, 355–385.

Agrawal, H. C., and Himwich, W. A. (1970). Amino acids, proteins and monoamines of the developing brain. In W. A. Himwich (Ed.), *Developmental Neurobiology*. Springfield, Ill.: Charles C. Thomas, 287–310.

Alpern, H. P. (1968). Facilitation of learning by implantation of strychnine sulphate in the central nervous system. Unpublished doctoral dissertation. Univ. of California, Irvine.

Alvord, E. C., Stevenson, L. D., and Vogel, F. S. (1950). Neuropathological findings in phenylpyruvic Oligophrenia, *J. Neuropath. Exp. Neurol.*, **9**, 298–305.

Anand, B. K., Chhina, G. S., and Singh, B. (1962). Effect of glucose on the activity of hypothalamic 'feeding centers', *Science*, **138**, 597–598.

Andèn, N.-E., Corrodi, H., Fuxe, K., and Hökfelt, T. (1968). Evidence for a central 5-hydroxytryptamine receptor stimulation by lysergic acid diethylamide, *Brit. J. Pharmacol.*, **34**, 1–7.

Andèn, N.-E., Dahlstrom, A., Fuxe, K., Larsson, K., Olson, L., and Ungerstedt, U. (1966). Ascending monoamine neurons to the telencephalon and diencephalon, *Acta physiol. Scand.*, **67**, 313–326.

Anderson, E. G., and Bonnycastle, D. D. (1960). A study of the central depressant action of pentobarbital, phenobarbital and diethyl-ether in relationship to increases in brain serotonin, *J. Pharmacol. exp. Therap.*, **130**, 138–146.

Andersson, B. (1952). Polydipsia caused by intrahypothalamic injections of hypertonic NaCl-solutions, *Experientia*, **8**, 157–158.

Andersson, B. (1953). The effect of injections of hypertonic NaCl solution into different parts of the hypothalamus of goats, *Acta physiol. scand.*, **28**, 188–201.

Andersson, B. (1966). The physiology of thirst. In E. Stellar (Ed.), *Progress in Physiological Psychology*, Vol. 1. New York: Academic Press, 191–209.

Andersson, B., and McCann, S. M. (1955). Drinking, antidiuresis and milk ejection from electrical stimulation within the hypothalamus of the goat, *Acta physiol. scand.*, **33**, 333–346.

Anker, J. L., and Milman, D. H. (1972). Patterns of non-medical drug usage among university students. Students attitudes towards drug usage. In Wolfram Keup (Ed.), *Drug Abuse*. Springfield: Charles C. Thomas, 202–214.

Aprison, M. H. (1962). On a proposed theory of the mechanism of action of serotonin in brain, *Rec. Advanc. biol. Psychiat.*, **4**, 133–146.

218

Armitage, A. K., and Hall, G. H. (1967). The effects of nicotine on the electrocorticogram and spontaneous release of acetylcholine from the cerebral cortex of the cat, *J. Physiol. (London)*, **191**, 115–116.

Armitage, A. K., Hall, G. H., and Morrison, C. F. (1968). Pharmacological basis for the tobacco smoking habit, *Nature*, **217**, 331–334.

Aronson, H., Silverstein, A. B., and Klee, G. D. (1959). The influence of lysergic acid diethylamide (LSD-25) on subjective time, *Arch. gen. Psychiat.*, **1**, 469–472.

Aston, R. (1965). Quantitative aspects of tolerance and post-tolerance hypersensitivity to pentobarbital in the rat, *J. Pharmacol. exp. Therap.*, **150**, 253–258.

Atkinson, R. M., and Ditman, K. S. (1965). Tranylcypromine: a review, *Clin. Pharmacol. Therap.*, **6**, 631–655.

Avis, H., and Carlton, P. L. (1968). Retrograde amnesia produced by hippocampal spreading depression, *Science*, **161**, 73–75.

Axelrod, J. (1961). Enzymatic formation of psychotomimetic metabolites from normally occurring compounds, *Science*, **134**, 343.

Ayd, F. J. (1958). Drug-induced depression: fact or fallacy, *N. Y. State J. Med.*, **58**, 354–356.

Ayd, F. J. (1960). Neuroleptics and extrapyramidal reactions. In J. M. Bordeleau (Ed.), *Extrapyramidal System and Neuroleptics*. Montreal: Editions Psychiatriques, 146–157.

Azmitia, F. C., Algheri, S., and Costa, E. (1970). Turnover rate of in vivo conversion of tryptophan into serotonin in brain areas of adrenalectomized rats, *Science*, **169**, 201–203.

Azmitia, E. C., and McEwen, B. S. (1969). Corticosterone regulation of tryptophan hydroxylase in midbrain of the rat, *Science*, **166**, 1274–1276.

Balasz, R., Cocks, W. A., Eayrs, J. R., and Kovacs, S. (1971). Biochemical effects of thyroid hormones on the developing brain. In M. Hamburgh and E. J. W. Barrington (Eds.), *Hormones in Development*. New York: Appleton-Century-Crofts, 357–380.

Baldridge, R. C., Borofsky, L., Baird, H., Reichle, F., and Bullock, D. (1959). Relationship of serum phenylalanine levels and ability of phenylketonurics to hydroxylate tryptophan, *Proc. Soc. exp. biol. Med.*, **100**, 529–531.

Ball, J. R. B., and Kiloh, L. G. (1959). A controlled trial of imipramine in treatment of depressive states, *Brit. med. J.*, **11**, 1052–1055.

Ban, T. (1969). *Psychopharmacology*. Baltimore: Williams and Wilkins.

Banks, A., and Russell, R. W. (1967). Effects of chronic reductions in acetylcholinesterase activity on serial problem solving behavior, *J. comp. physiol., Psychol.*, **64**, 262–267.

Barbeau, A., and MacDowell, F. H. (1970). *L-Dopa and Parkinsonism*. Philadelphia, Penn: F. A. Davis.

Barondes, S. H. (1965). Relationship of biological regulatory mechanisms to learning and memory, *Nature*, **205**, 18–20.

Barondes, S. H., and Cohen, H. W. (1968a). Memory impairment after subcutaneous injection of acetoxycycloheximide, *Science*, **160**, 556–557.

Barondes, S. H., and Cohen, H. W. (1968b). Arousal and the conversion of 'short-term' to 'long-term' memory, *Proc. nat. Acad. Sci.*, **61**, 923–929.

Barnnett, R. J. (1962). The five structural localizations of acetylcholinesterase at the myoneural junction, *J. cell. Biol.*, **12**, 247–262.

Basmajian, J. V. (1972). Electromyography cues of age, *Science*, **176**, 603–609.

Beach, H. D. (1957). Morphine addiction in rats, *Canad. J. Psychol.*, **11**, 104–112.

Beach, G. O., Fitzgerald, R. P., Holmes, R., Phibbs, B., and Stuckenhoff, H. (1964). Scopolamine poisoning, *New Eng. Med.*, **270**, 1354–1355.

Beattie, C. W., Rodgers, C. H., and Soyka, L. F. (1972). Influence of ovariectomy and ovarian steroids on hypothalamic tyrosine hydroxylase (TH) activity in the rat, *Fed. Proc.*, **31**, 211 (Abst.).

Becker, H. S. (1967). History, culture and subjective experience: an exploration of the social bases of drug induced experiences, *J. Health Soc., Beh.*, **8**, 163–176.

Bejerot, N. (1972). *Addiction: an artificially induced drive*. Springfield, Illinois: Charles C. Thomas.

Bell, D. S. (1965). Comparison of amphetamine psychosis and schizophrenia, *Brit. J. Psychiat*, **111**, 701–707.

Benedek, T., and Rubinstein, B. B. (1959). The correlations between ovarian activity and psychodynamic processes 1. the ovulative phase, *Psychosom. Med.*, **1**, 245–270.

Bennett, E. L., and Calvin, M. (1964). Failure to train planarians reliable, *Neuroscience Res. Prog. Bull.*, **2**, (4), 3–24.

Bercel, N. A., Travis, L. E., Obinger, L. B., and Dreikurs, E. (1956). Model psychoses induced by LSD25 in normals, *Arch. Neurol. Psychiat.*, **75**, 588–611.

Berger, R. J. (1963). Experimental modification of dream content by meaningful verbal stimuli, *Brit. J. Psychiat.*, **109**, 722–740.

Berger, B. D., and Stein, L. (1969). An analysis of the learning deficits produced by scopolamine, *Psychopharmacologia (Berl.)*, **14**, 271–283.

Berger, B. D., Wise, C. D., and Stein, L. (1971). Norepinephrine: reversal of anorexia in rats with lateral hypothalamic damage, *Science*, **172**, 281–284.

Berl, S., Puszkin, S., and Nicklas, W. J. (1973). Actomyosin-like protein in brain, *Science*, **179**, 441–446.

Berlet, H. H., Matsumoto, K., Pscheidt, G. R., Spaide, J., Bull, C., and Himwich, H. E. (1965). Biochemical correlates of behavior in schizophrenic patients, *Arch. gen. Psychiat.*, **13**, 521–534.

Berman, P. W., Waisman, H. A., and Graham, F. K. (1966). Intelligence in treated phenylketonuric children: a developmental study, *Child Develop.*, **37**, 731–747.

Bernard, C. (1857). *Leçons sur les effects des substances toxiques et medicamentouse*. Paris: Ballière.

Bernard, C. (1859). *Leçons sur les Proprietés Physiologiques et les Altérations Pathologiques des Liquides de l'Organisme*. Paris: Ballière.

Bernard, C. (1865). *Introduction a l'Étude de la Médécine Expérimentale*. Paris: Ballière.

Bernheimer, H., Birkmayer, W., and Hornykiewicz, O. (1961). Verleilung des 5-Hydroxytryptamine (serotonin) in Jehirn des menschen und sein verhalten bei patrenten mit Parkinson-syndrom, *Klin. Wschr.*, **39**, 1056–1059.

Biase, D. V. (1972). Phoenix Houses: Therapeutic communities for drug addicts. A comparative study of residents in treatment. In Wolfram Keup (Ed.), *Drug Abuse*. Illinois. Charles C. Thomas, 375–380.

Biel, J. H. (1970). Non-monoamine oxidase inhibitor antidepressants structure activity relationships. In W. G. Clark and J. del Giudice (Eds.), *Principles of Psychopharmacology*. London: Academic Press, 289–302.

Bignami, G., and Gatti, G. L. (1966). Neurotoxicity of anticholinesterase agents. Antagonistic action of various centrally acting drugs, *Proc. europ Soc. drug Tox. Vol.* 8, Amsterdam: Exerpta Medica, 93–106.

Bignami, G., and Rosić, N. (1970). The nature of disinhibitory phenomena caused by central cholinergic (muscarinic) blockade. In *Proceedings of the VIIth International Congress of the Collegium Internationale Neuropsychopharmacologicum;* Prague.

Binz, C. (1895). *Lectures on pharmacology*. Trans. Latham London: New Sydenham Society.

Birks, R., and MacIntosh, F. C. (1961). Acetylcholine metabolism of a sympathetic ganglion, *Canad. J. Biochem.*, **39**, 787–827.

Black, A. H. (1958). The extinction of avoidance responses under curare-like drugs, *J. comp. physiol. Psychol.*, **51**, 519–524.

Blaschko, H. (1959). The development of current concepts of catecholamine formation, *Pharmacol. Rev.*, **11**, 307–316.

Bliss, E. J., Ailion, J., and Zwanziger, J. (1968). Metabolism of norepinephrine, serotonin and dopamine in rat brain with stress, *J. pharmacol. exp. Therap.*, **164**, 122–131.

Bloch, V., Denti, A., and Schmaltz, G. (1966). Effets de la stimulation reticulaire sur la phase de consolidation de la trace amnesique, *J. Physiol. (Paris)*, **18**, 469–470.

Bloch, V., and Deweer, B. (1968). Role accelerateur de la phase de consolidation d'un apprentissage en un seul essai, *Compt. Rend. Acad. Sci.*, **266**, 384–387.

220

Bloom, F. E., Costa, E., and Salmoiraghi, E. C. (1965). Anesthesia and the responsiveness of individual neurons of the caudate nucleus of the cat to acetylcholine, norepinephrine and dopamine administered by microelectrophoresis, *J. pharmacol. exp. Therap.*, **150**, 244–252.

Boakes, R. J., Bradley, P. B., Briggs, I., and Dray, A. (1970). Antagonism of 5-hydroxy-tryptamine by LSD-25 in the central nervous system: a possible neuronal basis for the actions of LSD-25, *Brit. J. Pharmacol.*, **40**, 202–218.

Boffey, P. M. (1968). Nerve gas: Dugway accident linked to Utah sheep kill, *Science*, **162**, 1460–1464.

Bohdanecka, M., Bohdancký Z., and Jarvik, M. E. (1967). Amnesia effects of small bilateral brain puncture in the mouse, *Science*, **157**, 334–336.

Bohdancký, Z., and Jarvik, M. E. (1967). Impairment of one trial passive avoidance learning in mice by scopolamine, scopolamine methylbromide and physostigmine, *Int. J. Neuropharmacol.*, **6**, 217–222.

Bolles, R. C. (1971). Species-specific defence reactions. In F. R. Brush (Ed.), *Aversive Conditioning and Learning*. New York: Academic Press, 183–233.

Bonnycastle, D. D., Bonnycastle, M. F., and Anderson, E. G. (1962). The effect of a number of central depressant drugs upon brain 5-hydroxytryptamine levels in the rat, *J. Pharmacol. exp. Therap.*, **135**, 17–32.

Booth, D. A. (1967). Localization of the adrenergic feeding system in the rat diencephalon, *Science*, **158**, 515–517.

Booth, D. A. (1968a). Effects of intrahypothalamic glucose injection on eating and drinking elicited by a single injection of insulin, *J. Comp. physiol. Psychol.*, **65**, 13–16.

Booth, D. A. (1968b). Mechanism of action of norepinephrine in eliciting an eating response on injection into the rat hypothalamus. *J. Pharmac. exp. Therap.*, **160**, 336–348.

Booth, D. A., and Quartermain, D. (1965). Taste sensitivity of eating elicited by chemical stimulation of the rat hypothalamus, *Psychonom. Sci.*, **3**, 525–526.

Bourdillon, R. E., Clarke, C. A., Ridges, A. P., Sheppard, P. M., Harper, P., and Leslie, S. A. (1965). Pink spot in the urine of schizophrenics, *Nature*, **208**, 453–455.

Bowers, M. B., Goodman, E., and Sim, Van N. (1964). Some behavioral changes in man following anticholinesterase administration, *J. ner. ment. Dis.*, **138**, 383–389.

Bowers, M. B., Hartmann, E. L., and Freedman, D. X. (1966). Sleep deprivation and brain acetylcholine, *Science*, **154**, 1416–1417.

Bradley, P. B., and Elkes, J. (1953). The effect of atropine, hyoscyamine, physostigmine and neostigmine on the electrical activity of the conscious cat, *J. Physiol. (Lond.)*, **120**, 13.

Bradley, P. B., and Hance, A. J. (1957). The effect of chlorpromazine and methapromazine on the electrical activity of the brain in the cat, *E. E. G. clin. Neurophysiol.*, **9**, 191–215.

Bradley, P. B., and Key, B. J. (1958). The effects of drugs on arousal responses produced by electrical stimulation of the reticular formation of the brain stem, *E. E. G. Clin. Neurophysiol.*, **10**, 97–110.

Bradley, P. B., and Key, B. J. (1959). A comparative study of the effects of drugs on the arousal system of the brain, *Brit. J. Pharmacol.*, **14**, 340–349.

Brawley, P., and Duffield, J. G. (1972). The pharmacology of hallucinogens, *Pharmacol. Rev.*, **24**, 31–66.

Brazier, M. A. B. (1963). Effects upon physiological systems: the electrophysiological effects of barbiturates on the brain. In W. S. Root and F. G. Hofmann (Eds.), *Physiological Pharmacology*. New York: Academic Press, 219–235.

Brebbia, D. R., and Altshuler, K. Z. (1965). Patterns of energy exchange during sleep and dreams, *Assoc. Physiol. Study Sleep*. Washington, D.C.

Breen, R. A., and McGaugh, J. L. (1961). Facilitation of maze learning with post-trial injections of picrotoxin, *J. comp. physiol. Psychol.*, **54**, 498–501.

Bremer, F. (1954). The neurophysiological problem of sleep. In J. F. Delafresnaye (Ed.), *Brain Mechanisms and Consciousness*. Oxford: Blackwell, 137–157.

Bremer, F. (1961). Neurophysiological mechanisms in cerebral arousal. In G. E. W. Wolstenholme and M. O'Connor (Eds.), *The Nature of Sleep*, Boston: Little, Brown, 30–56.

Briggs, M. H., and Kitto, G. B. (1962). The molecular basis of memory and learning, *Psychol. Rev.*, **69**, 537–541.

Broadbent, D. E., and Gregory, M., (1963). Division of attention and the decision theory of signal detection, *Proc. roy. Soc., B.* **158**, 222–231.

Broadhurst, P. L. (1964). The hereditary base for the action of drugs on animal behaviour. In H. Steinberg (Ed.), *Animal Behaviour and Drug Action*. London: Churchill, 224–236.

Brodie, B. B. (1962). Difficulties in extrapolating data on metabolism of drugs from animal to man, *Clin. Pharmacol. Therap.*, **3**, 374–350.

Brodie, B. B., and Shore, P. A. (1957). A concept for a role of serotonin and norepinephrine as chemical mediators in the brain, *Ann. N. Y. Acad. Sci.*, **66**, 631–642.

Brodie, B. B., and Shore, P. A. (1959). Mechanisms of action of psychotropic drugs. In *Psychopharmacology Frontiers*. Boston: Little, Brown, 413–419.

Brodie, B. B., Sulser, F., and Costa, E. (1961). Theories on mechanisms of action of psychotherapeutic drugs. In J. M. Bordeleau (Ed.), *Extrapyramidal System and Neuroleptics*. Montreal: L'Edition Psychiatriques, 183–189.

Broverman, D. M., Klaiber, E. L. Kobayashi, Y., and Vogel, W. (1968). Role of activation and inhibition in sex differences in cognitive abilities, *Psychol. Rev.*, **75**, 23–50.

Brown, H. M. (1964). *Experimental procedures and state of nucleic acids as factors contributing to 'learning' phenomena in planaria*. Unpublished Ph.D. Thesis. University of Utah.

Brown, K., and Warburton, D. M. (1971). Attenuation of stimulus sensitivity by scopolamine, *Psychon. Sci.*, **22**, 297–298.

Browne-Mayers, A. N., Seelye, E. E., Brown, D., and Fleetwood, M. (1972). Mini delusions. In Keup, W. (Ed.), *Drug Abuse*, Springfield: Charles C. Thomas, 288–295.

Bunney, W. E., Jr., and Davis, J. M. (1965). Norepinephrine in depressive reactions. Review, *Arch. gen. Psychiat.*, **13**, 483–494.

Bureš, J., Bohdanecký, Z., and Weiss, T. (1962). Physostigmine induced hippocampal theta activity and learning in rats, *Psychopharmacologia (Berl.)*, **3**, 254–263.

Burgen, A. S. V., Dickens, F., and Zatman, L. J. (1949). The action of botulinum toxin on the neuromuscular function, *J. Physiol. (Lond.)*, **109**, 10–24.

Burke, A. W., and Broadhurst, P. L. (1966). Behavioural correlates of the oestrous cycle in the rat, *Nature*, **209**, 223–224.

Byrne, W. L., Samuel, D., Bennett, E. L., Rosenzweig, M. R., Wasserman, E., Wagner, A. R., Gardner, R., Galambos, R., Berg, B. W., Margules, D. L., Fenichel, R. L., Stein, L., Corson, J. A., Enesco, H. E., Chorover, S. L., Holt, C. E., III, Schiller, H. P., Chiapetta, L., Jarvik, M. E., Leaf, R. C., Wulder, J. W., Horovitz, Z. P., and Carlton, P. L. (1966). Memory transfer, *Science*, **153**, 658.

Caldwell, D. F., and Domino, E. F. (1967). Electroencephalographic and eye movement patterns during sleep in chronic schizophrenic patients, *E.E.G. Clin. Neurophysiol.*, **22**, 414–420.

Caley, D. W., and Maxwell, D. S. (1971). Developing cerebral cortex in the rat. In M. B. Sterman, D. J. McGinty and A. M. Adiholfi (Eds.), *Brain Development and Behavior*. New York: Academic.

Calloway, E., and Band, R. I. (1958). Some psychopharmacological effects of atropine, *Arch. Neurol. Psychiat. (Chic.)*, **79**, 91–102.

Campbell, B. A., Lytle, L. D., and Fibiger, H. C. (1969). Ontogeny of adrenergic arousal and cholinergic inhibitory mechanisms in the rat, *Science*, **166**, 625–636.

Campbell, B. A., and Spear, N. E. (1972). Ontogeny of memory, *Psychol. Rev.*, **79**, 215–236.

Cancellaro, L. A. (1972). New treatment concepts at the NIMH Clinical Research Centre, Lexington, Kentucky. In Wolfram Keup (Ed.), *Drug Abuse*. Illinois: Charles C. Thomas, 355–367.

Cannon, W. B. (1939). *The Wisdom of the Body* (2nd Ed.). Norton: New York.

222

Carlini, G. R. S., and Carlini, E. A. (1965). Effects of strychnine and cannabis sativa (marijuana) on the nucleic acid content in the brain of the rat, *Med. Pharmacol. Exptl.,* **12**, 21–26.

Carlson, A. J. (1916). *The Control of Hunger in Health and Diseases.* Chicago: University of Chicago Press.

Carlsson A. (1959). The occurrence, distribution and physiological role of catecholamines in the nervous system, *Pharmacol. Rev.,* **11**, 490–493.

Carlsson, A., Lindquist, M., and Magnussen, T. (1957). 4-dihydroxyphenylalanine and 5-hydroxytryptophan as reserpine antagonists, *Nature,* **180**, 1200.

Carlton, P. L. (1963). Cholinergic mechanisms in the control of behavior by the brain, *Psychol. Rev.,* **70**, 19–39.

Carlton, P. L. (1969). Brain acetylcholine and habituation, *Progr. Brain Res.,* **28**, 48–60.

Carr, L. A., and Moore, K. E. (1968). Effects of reserpine and α-methyltyrosine on brain catecholamines and the pituitary-adrenal response to stress, *Neuroendocrinology,* **3**, 285–302.

Cavallito, C. J., White, H. L., Yun, H. S., and Foldes, F. F. (1970). Inhibitions of choline acetyltransferase. In E. Heilbronn and A. Winter (Eds.), *Drugs and Cholinergic Mechanisms in the C.N.S.* Stockholm: Försvarets Forskningsanstalt. Research Institute of National Defence, 97–116.

Chaftez, M. E. (1970). Clinical syndromes of liquor drinkers. In R. E. Popham (Ed.), *Alcohol and Alcoholism.* Toronto: University of Toronto Press, 111–116.

Chambers, C. A., and Russeil Taylor, W. J. (1972). Patterns of 'cheating' among methadone maintenance patients. In Wolfram Keup (Ed.), *Drug Abuse.* Illinois: Charles C. Thomas, 328–336.

Chase, T. N., Katz, R. I., and Kopin, I. J. (1970). Effect of diazepam on fate of intra-cisternally injected serotonin-C^{14}, *Neuropharmacology,* **9**, 103–108.

Chowers, I., Feldman, S., and Davidson, J. M. (1963). Effects of intrahypothalamic crystalline steroids on acute ACTH secretion, *Amer. J. Physiol.,* **205**, 671–673.

Cicala, G. A., and Hartley, D. L. (1967). Drugs and the learning and performance of fear, *J. comp. physiol. Psychol.,* **64**, 175–178.

Clark, F. C., and Steele, B. J. (1966). Effects of d-amphetamine on performance under a multiple schedule in the rat, *Psychopharmacologia (Berl.),* **9**, 157–169.

Clark, L. D., and Beecher, H. K. (1957). Psychopharmacological studies on suppression, *J. nerv. ment. Dis.,* **125**, 316–321.

Cleghorn, R. A. (1957). Steroid hormones in relation to neuropsychiatric disorders. In H. Hoagland (Ed.), *Hormones, Brain Function and Behaviour.* New York: Academic, 3–19.

Clouet, D. H., and Ratner, M. (1970). Catecholamine biosynthesis in brains of rats treated with morphine, *Science,* **168**, 854–856.

Cohen, S. (1970). *Drugs of Hallucination.* London: Paladin.

Cohen, S., and Ditman, K. S. (1963). Prolonged adverse reactions to lysergic acid di-ethylamide, *Arch. gen. Psychiat.,* **8**, 475–480.

Collier, H. O. J. (1969). Humoral transmitters, sensitivity, receptors and dependence. In H. Steinberg (Ed.), *Scientific Basis of Drug Dependence.* London: Churchill, 49–66.

Collier, H. O. J. (1972). Drug dependence: a pharmacological analysis, *Br. J. Addict.* **67**, 277–286.

Connell, P. H. (1958). Amphetamine psychosis. *Maudsley Monograph No. 5.* London: Chapman Hall.

Connor, J. D. (1970). Caudate nucleus neurons: correlation of the effects of substantia nigra stimulation with iontophoretic dopamine, *J. Physiol. (Lond).,* **208**, 691–703.

Connor, J. D., Rossi, G. V., and Baker, W. W. (1966). Analysis of the tremor induced by injection of cholinergic agents into the caudate nucleus, *Int. J. Neuropharmacol.,* **5**, 207–216.

Connor, J. D., Rossi, G. V., and Baker, W. W. (1967). Antagonism of intracaudate carbachol tremor by local injections of catecholamines, *J. pharmacol. exp. Therap.*, **155**, 545–551.

Cools, A. R., and Van Rossum, J. M. (1970). Caudal dopamine and stereotype behaviour of cats, *Arch. int. Pharmacodyn. Thérap.*, **187**, 163–173.

Cooper, J. S. (1965). Surgical treatment of Parkinsonism, *Ann. Rev. Med.*, **16**, 309–333.

Coover, G. D., Goldman, L., and Levine, S. (1971). Plasma corticosterone increases produced by extinction of operant behavior in rats, *Physiol. Behav.*, **7**, 261–263.

Coover, G. D., Ursin, H., and Levine, S. (1972). Plasma corticosterone levels during active avoidance learning in rats, *J. comp. physiol. Psychol.*, **82**, 170–174.

Corrodi, H., Fuxe, K., and Hökfelt, T. (1966). The effect of ethanol on the activity of central catecholamine neurons in rat brain, *J. Pharm. Pharmacol.*, **18**, 821–832.

Corrodi, H., Fuxe, K., Lidbrink, P., and Olson, L. (1971) Minor tranquillizers, stress and central catecholamine neurons, *Brain Res.*, **29**, 1–16.

Cotzias, G. C., Van Woert, M. H., and Schiffer, L. M. (1967). Aromatic amino acids and modification of Parkinsonism, *N. Eng. J. Med.*, **276**, 374–379.

Coury, J. N. (1967). Neural correlates of food and water intake in the rat, *Science*, **156**, 1763–1765.

Cowley, J. J., and Griesel, R. D. (1959). Some effects of a low protein diet on a first filial generation of white rats, *J. genet. Psychol.*, **95**, 187–201.

Cowley, J. J., and Griesel, R. D. (1963). The development of second generation low protein rats, *J. genet. Psychol.*, **103**, 233–242.

Cowley, J. J., and Griesel, R. D. (1964). Low protein diet and emotionality in the albino rat, *J. genet. Psychol.*, **104**, 89–98.

Crancer, A., Jr., Dille, J. M., Delay, J. C., Wallace, J. E., and Haykin, M. D. (1969). Comparison of the effects of marijuana and alcohol on simulated driving performance, *Science*, **164**, 851–854.

Cravioto, J., and Robles, B. (1965). Evolution of adaptive and motor behavior during rehabilitation from Kwashiorkor, *Amer. J. Orthopschiat.*, **35**, 449–464.

Crick, F. H. C. (1959). The present status of the coding problem, *Brookhaven Symp. Biol.*, **12**, 35–39.

Crossland, J. (1960). Chemical transmission in the central nervous system, *J. Pharm. Pharmacol.*, **12**, 1–36.

Culley, W. J., and Lineberger, R. D. (1968). Effect of undernutrition on the size and composition of the brain, *J. Nutr.*, **96**, 375–381.

Curtis, D. R., and Davis, R. (1963). The excitation of lateral geniculate neurons by quaternary ammonium derivatives, *J. Physiol. (Lond.)*, **165**, 62–82.

Curtis, D. R., and Eccles, R. M. (1958). The excitation of Renshaw cells by pharmacological agents applied electrophoretically, *J. Physiol. (Lond.)*, **141**, 435–463.

Curtis, D. R., Phillis, J. W., and Watkins, J. C. (1961). Cholinergic and non-cholinergic transmission in the mammalian spinal cord, *J. Physiol.*, **158**, 296–323.

Curtis, D. R., and Ryall, R. W. (1966). The acetylcholine receptors of Renshaw cells, *Exp. Brain Res.*, **2**, 66–80.

Curzon, G., and Green, A. R. (1968). Effect of hydrocortisone on rat brain 5-hydroxytryptamine, *Life Sci.*, **7**, 657–663.

Dahlström, A., and Fuxe, K. (1964). Evidence for the existence of monoamine-containing neurons in the central nervous system. I. demonstration of monoamines in the cell bodies of the brain stem neurons, *Acta physiol. scand.*, **62**, Suppl. 232 (whole).

Dahlström, A., and Häggendal, J. (1966). Studies on the transport and life span of amine storage granules in a peripheral adrenergic neuron system, *Acta physiol. scand.*, **67**, 278–288.

Dale, H. H. (1935). Pharmacology and nerve endings, *Proc. roy. Soc. Med.*, **28**, 319–332.

Dale, H. H., Feldberg, W., and Vogt, M. (1963). Release of acetylcholine at voluntary motor nerve endings, *J. Physiol. (Lond.)*, **86**, 353–380.

Dalton, K. (1959). Comparative trials of new oral progestogenic compounds in treatment of premenstrual syndrome, *Britt. Med. J.*, **2**, 1307–1309.

Dalton, K. (1960). Effect of menstruation on schoolgirls' weekly work, *Brit. Med. J.*, **1**, 326–328.

Dalton, K. (1968). Menstruation and examinations, *Lancet*, **2**, 1386–1388.

Dalton, K. (1969). *The menstrual cycle*. London: Penguin.

Davis, W. M., and Nichols, J. R. (1962). Physical dependence and sustained opiate directed behavior in the rat, *Psychopharmacologia (Berl.)*, **3**, 139–146.

Davis, J. W., Thomas, R. K., and Adams, H. E. (1971). Interactions of scopolamine and physostigmine with ECS and one trial learning, *Physiol. Behav.*, **6**, 219–222.

Delay, J., and Deniker, P. (1968). Drug-induced extrapyramidal syndromes. In P. J. Vinken and G. W. Bruyn (Eds.), *Handbook of Clinical Neurology*, Vol. 6. 'Diseases of the Basal Ganglia'. Amsterdam: North Holland, 465–508.

del Castillo, J., and Katz, B. (1954). Changes in and plate activity produced by pre synaptic polarization, *J. Physiol. (Lond.)*, **124**, 586–604.

del Castillo, J., and Katz, B. (1957). Curare action studied by an electrical micromethod, *Proc. Roy. Soc. B.*, 146, 339–356.

Delorme, F., Froment, J. L., and Jouvet, M. (1966). Suppression du sommeil par la p-chlorométhamphetamine et la p-chlorophenylalanine, *CR. Séances Soc. Biol.*, **160**, 2347–2351.

Dement, W. (1965). Recent studies on the biological role of rapid eye movement sleep, *Amer. J. Psychiat.*, **122**, 404–408.

Dement, W., Ferguson, J., Cohen, H., and Barchas, J. (1969). Non-chemical methods and data using a biochemical model: the REM quanta. In A. J. Mandell and M. P. Mandell (Eds.), *Psychochemical Research in Man*. London: Academic Press, 275–325.

Dement, W., Henry, P., Cohen, H., and Ferguson, J. (1967). Studies on the effect of REM deprivation in humans and animals. In S. Kety, E. Evarts and H. Williams (Eds.), *Sleep and Altered States of Consciousness*. Baltimore: Williams and Wilkins, 456–468.

Dement, W., and Kleitman, N. (1957). Cyclic variations in EEG during sleep and their relation to eye movements, body motility and dreaming, *EEG clin. Neurophysiol.*, **9**, 673–690.

Deneau, G. A. (1972). The measurement of addiction potential by self-injection experiment in monkeys. In Wolfram Keup (Ed.), *Drug Abuse*. Springfield, Illinois: Thomas, 73–79.

Deneau, G. A., Yanagita, T., and Seevers, M. H. (1969). Self-administration of psychoactive substances by the monkey. A measure of psychological dependence, *Psychopharmacologia (Berl.)*, **16**, 30–48.

Denti, A., McGaugh, J. L., Landfield, P. W., and Shinkman, P. (1970). Facilitation of learning with posttrial stimulation of the reticular formation, *Physiol. Behav.*, **5**, 659–662.

De Robertis, E. (1967). Ultrastructure and cytochemistry of the synaptic region, *Science*, **156**, 907–913.

De Robertis, E. (1971). Molecular biology of synaptic receptors, *Science*, **171**, 963–971.

Desmedt, J. E. (1958). Myasthenic-like features of neuromuscular transmission after administration of an inhibitor of acetylcholine synthesis, *Nature*, **182**, 1673–1674.

Deutsch, J. A., Hamburg, M. D., and Dahl, H. (1966). Anticholinesterase induced amnesia and its temporal aspects, *Science*, **151**, 221–223.

Deweer, B., Hennevin, E., and Block, V. (1969). Nouvelles donnees sur la facilitation réticulairé de la consolidation mnesique, *J. Physiol.*, **60**, 430.

De Wied, D. (1967a). Chlorpromazine and endocrine function, *Pharmacol. Rev.*, **19**, 251–288.

De Wied, D. (1967b). Opposite effects of ACTH and glucocorticoids on extinction of conditioned avoidance behavior. In L. Martini, F. Fraschini and L. Motta (Eds.), *Proceedings Second International Congress on Hormonal Steroids. Milan, 1966*, 945–951.

Dews, P. B. (1958). Studies on behavior. IV Stimulant actions of methamphetamine, *J. pharmacol. exp. Ther.*, **122**, 137–147.

Deza, L., and Eidelberg, E. (1967). Development of cortical electrical activity in the rat, *Exp. Neurol.*, **17**, 425–438.

Diamond, M. C., Krech, D., and Rosenzweig, M. R. (1964). The effects of an enriched environment on the histology of the rat cerebral cortex, *J. comp. Neurol.*, **123**, 111–119.

Dill, R. E., Nickey, W. M., and Little, M. D. (1968). Dyskinesis in rats following chemical stimulation of the neostriatum, *Texas Rep. Biol. Med.*, **265**, 101–106.

DiMascio, A., and Barrett, J. (1965). Comparative effects of oxazepam in 'high' and 'low' anxious student volunteers, *Psychosomatics*, **6**, 298–302.

Dingman, W., and Sporn, M. B. (1961). The incorporation of 8-azaguanine into the brain RNA and its effect on maze-learning by the rat; an inquiry into the biochemical bases of memory, *J. Psychiat. Res.*, **1**, 1–11.

Ditman, K. S., Moss, T., Forgy, E. W., Zunin, L. M., Lynch, R. D., and Funk, W. A. (1969). Dimensions of the LSD, methylphenidate and chlordiazepoxide experiences, *Psychopharmacologia (Berl.)*, **14**, 1–11.

Dobbing, J. (1964). The influence of nutrition on the development and myelination of the brain, *Proc. Roy. Soc. Biol.*, **159**, 503–509.

Dobbing, J. (1968). Effects of Experimental undernutrition on development of the nervous system. In N. S. Scrimshaw and J. E. Gordon (Eds.), *Malnutrition, Learning and Behavior.* Cambridge. Mass.: M.I.T. Press, 181–201.

Dokas, L. A., and Kleinsmith, L. J. (1971). Adenosine $3',5'$-monophosphate increases capacity for RNA synthesis in rat liver nuclei, *Science,* **172**, 1237–1238.

Dole, V. P., and Nyswander, M. (1965). A medical treatment for diacetylmorphine, *J. Amer. med. Assoc.*, **193**, 646–650.

Dole, V. P., Nyswander, M., and Kreek, M. J. (1966). Narcotic blockade, *Arch. int. Med.*, **118**, 304–309.

Dole, V. P., Nyswander, M., and Warner, A. (1968). Successful treatment of 750 criminal addicts, *J.A.M.A.*, **206**(12), 2708–2711.

du Bois-Reymond, E. (1877). Gesammeete adhandt d allgem, *Muskel Nervenphysik,* **2**, 700.

Duffy, E. (1962). *Action and Behavior.* New York: Wiley.

Duritz, G., and Truitt, E. B., Jr. (1966). Importance of acetaldehyde in the action of ethanol on brain norepinepherine and 5-hydroxytryptamine, *Biochem. Pharmacol.*, **15**, 711–715.

Duvoisin, R. C. (1967). Cholinergic anticholinergic antagonism in Parkinsonism, *Arch. Neurol.*, **17**, 124–136.

Eayrs, J. T. (1954). Spontaneous activity in the rat, *Brit. J. Anim. Behav.* **2**, 25–30.

Eccles, J. C. (1957). *The Physiology of Nerve Cells.* Baltimore: John Hopkins Press.

Eccles, J. C. (1964). *The Physiology of Synapses.* New York: Academic Press.

Eccles, J. C., Fatt, P., and Koketsu, K. (1954). Cholinergic and inhibitory synapses in the pathway from motoraxon collaterals to motoneurons, *J. physiol.*, **126**, 524–562.

Edelman, M., Schwartz, I. L., Kronbite, E. P., and Livingston, L. (1965). Studies of ventromedial hypothalamus with autoradiographic techniques, *Ann. N. Y. Acad. Sci.*, **131**, 485–501.

Efron, D., and Kety, S. (1966). *Antidepressant drugs of non-MAO inhibitor type.* Washington D.C.: U.S. Dept. of Health, Education and Welfare.

Eidelberg, E., and Barstow, C. A. (1971). Morphine tolerance and dependence induced by intraventricular injection, *Science,* **174**, 74–76.

Eisner, B. G., and Cohen, S. (1958). Psychotherapy with lysergic acid diethylamide, *J. nerv. ment. dis.*, **127**, 528–539.

Ellenwood, E. H. (1967). Amphetamine psychosis, a description of the individuals and process, *J. nerv. ment. dis.*, **144**, 273–283.

Ellenwood, E. H. (1972). Amphetamine psychosis systems and subjects. In Wolfram Keup (Ed.), *Drug Abuse.* Springfield: Charles C. Thomas, 302–306.

Elmqvist, D., and Quastel, D. J. M. (1965). Presynaptic action of hemicholinium at the neuromuscular junction, *J. Physiol. (Lond.)*, **177**, 463–482.

Endročzi, E., Hartmann, G., and Lissák, K. (1963). Effect of intracerebrally administered cholinergic and adrenergic drugs on neocortical and archicortical electrical activity, *Acta Physiol. Hung.*, **24**, 200–209.

Epstein, A. N. (1960). Reciprocal changes in feeding behavior produced by intra-hypothalamic chemical injections, *Amer. J. Physiol.*, **199**, 969–974.

Epstein, A. N. (1971). The lateral hypothalamic syndrome. Its implication for the physiological psychology of hunger and thirst. In E. Stellar and J. Sprague (Eds.), *Progress in Physiological Psychology*, Vol. 4. New York: Academic Press, 263–317.

Erickson, C. K., and Chalmers, R. K. (1966). Hippocampal theta rhythm involvement in cholinergic-induced blockade of discriminated avoidance responding in rats, *Arch. Int. Pharmacodyn. Therap.*, **163**, 70–78.

Erickson, C. K., and Patel, J. B. (1969). Facilitation of avoidance learning by post-trial hippocampal electrical stimulation, *J. comp. physiol. Psychol.*, **68**, 400–406.

Ersner, J. S. (1940). The treatment of obesity due to dietary indiscretion (over eating) with benzedrine sulfate, *Endocrinology*, **27**, 774–780.

Estes, W. K., and Skinner, B. F. (1941). Some quantitative properties of anxiety, *J. exp. Psychol.*, **29**, 390–400.

Everett, G. M., and Toman, J. E. P. (1959). Mode of action of Rauwolfia Alkaloids and motor activity. In Masserman, J. (Ed.), *Biological Psychiatry*. New York: Grune and Stratton, 75–81.

Everett, G. M., Wiegand, R. G., and Rinaldi, F. U. (1963). Pharmacologic studies of some non-hydrazine MAO inhibitors, *Ann. N.Y. Acad. Sci.*, **107**, 1068–1080.

Farrell, J. P., and Sherwood, S. L. (1956). An alpha correlate to behavior changes produced in psychotics by intraventricular injections, *E.E.G. clin. neurophysiol.*, **8**, 713.

Fatt, P. (1959). Skeletal neuromuscular transmission. In J. Field (Ed.), *Handbook of Neurophysiology*, Vol. 1. Washington, D.C.: American Physiological Society, 199–213.

Fazekas, J. F., Ehrmantraut, W. R., and Kleh, J. (1958). A study of the effectiveness of certain anorexigenic agents, *Am. J. Med. Sci.*, **236**, 692–699.

Feldberg, W. (1963). *A Pharmacological Approach to the Brain from its Inner and Outer Surface*. Baltimore: Williams and Wilkins.

Feldberg, W. (1964). Discussion on extrapolation from animals to man: catatonia. In H. Steinberg (Ed.), *Animal Behaviour and Drug Action*. London: Churchill, 429–439.

Feldberg, W., and Sherwood, S. L. (1954). Injection of drugs into the lateral ventricle of the cat, *J. Physiol.*, **123**, 148–167.

Feldberg, W., and Sherwood, S. L. (1955). Recent experiments with injections of drugs into the ventricular system of the brain, *Proc. roy. soc. Med.*, **48**, 853–863.

Fink, M. (1960). Effect of anticholinergic compounds on post convulsive electroencephalogram and behavior of psychiatric patients, *E.E.G. clin. Neurophysiol.*, **12**, 359–369.

Fink, M. (1970). Narcotic antagonists in opiate dependence, *Science*, **169**, 1005–1006.

Fink, M., Klein, D. F., and Kramer, J. C. (1965). Clinical efficacy of chlorpromazine–procyclidine combination, imipramine and placebo in depressive disorders, *Psychopharmacologia (Berl.)*, **7**, 27–36.

Fisher, A. E., and Coury, J. N. (1962). Cholinergic tracing of a central neural circuit underlying the thirst drive, *Science*, **138**, 691–693.

Fisher, A. E., and Levitt, R. A. (1967). Drinking induced by carbachol: thirst circuit or ventricular modifications, *Science*, **157**, 839–841.

Fitzsimmons, J. T. (1961). Drinking by rats depleted by body fluid without increase in osmotic pressure, *J. Physiol. (Lond.)*, **159**, 297–309.

Fitzsimmons, J. T. (1971). The physiology of thirst. In E. Stellar and J. M. Sprague (Eds.), *Progress in Physiological Psychology*, Vol. 4. New York: London: Academic Press, 119–201.

Fjerdingstad, E. J., Nissen, Th., and Røigaard-Petersen, H. H. (1965). Effect of ribonucleic acid (RNA) extracted from the brain of trained animals on learning in rats, *Scand. J. Psychol.*, **6**, 1–6.

Flechsig, P. (1896 repr. 1962). Extract from Gehrirn and Seele. In M. C. H. Dodgson

227

(Ed.), *The Growing Brain. An Essay in Developmental Biology.* Bristol, England: Wright, 160–168.

Flexner, L. B., Flexner, J. B., and Roberts, R. B. (1967). Memory in mice analysed with antibiotics, *Science,* **155,** 1377–1383.

Flexner, L. B., Flexner, J. B., Stellar, E., Roberts, R. B., and de la Haba, G. (1964). Loss of recent memory in mice as related to regional inhibition of cerebral protein synthesis, *Proc. nat. Acad. Sci.,* **52,** 1165–1169.

Flexner, J. B., Flexner, L. B., and Stellar, E. (1963). Memory in mice as affected by intra-cerebral puromycin, *Science,* **141,** 57–59.

Flourens, P. (1824). *Recherches expérimentales sur les propriétés et les fonctions du système nerveux dans les animaux vertébrés.* Paris: Balliére.

Forrer, G. R. (1956). Symposium on atropine toxicity therapy, *J. nerv. men. dis.,* **124,** 257–283.

Forrer, G. R., and Miller, J. J. (1958). Atropine coma: a somatic therapy in psychiatry, *Amer. J. Psychiat.,* **115,** 455–458.

Forssberg, A., and Larsson, S. (1954). Studies of isotope distribution and chemical composition in the hypothalamic region of hungry and fed rats, *Acta physiol. scand.,* **32,** *Suppl.* 115, 41–63.

Frankenhaeuser, M., Myrsten, A. L., Johansson, G., and Post, B. (1971). Behavioral and physiological effects of cigarette smoking in a monotonous situation, *Psychopharmacologia (Berl.),* **22,** 1–7.

Frey, P. N., and Polidora, V. J. (1966). Magnesium Pemoline: effect on avoidance conditioning in rats, *Science,* **155,** 1281–1282.

Freyhan, F. A. (1956). Comments on the biological and psychopathological bases of individual variation in chlorpromazine therapy, *Encéphale, Colloq. Spécial,* 613–619.

Friedhoff, A. J. (1969). Strategies for investigating biochemical aberrations in mental dysfunction. In A. J. Mandell and M. P. Mandell (Eds.), *Psychochemical Research in man.* New York: Academic, 161–174.

Friedhoff, A. J., and Van Winkle, E. (1962). Isolation and characterization of a compound from the urine of schizophrenics, *Nature,* **194,** 897–898.

Friedhoff, A. J., and Van Winkle, E. (1965). A neurotropic compound identified in urine of schizophrenic patients. In P. Hoch and J. Zubin (Eds.), *Psychopathology of Schizophrenia.* New York: Grune and Stratton.

Frosch, W. A., Robbins, E. S., and Stern, M. (1965). Untoward reactions to lysergic acid diethylamide (LSD) resulting in hospitalization, *New Eng. J. Med.,* **273,** 1235–1248.

Funderburk, W. H., and Case, T. (1947). Effect of parasympathetic drugs on the conditioned response, *J. Neurophysiol.,* **10,** 179–188.

Funderburk, W. H., and Case, T. J. (1951). The effect of atropine on cortical potentials, *E.E.G. clin. Neurophysiol.,* **3,** 213–223.

Fuxe, K., Corrodi, H., Hokfelt, T., and Jonsson, G. (1970). Central monoamine neurons and pituitary-adrenal activity, *Prog. Brain. Res.,* **32,** 42–56.

Gaddum, J. H. (1957). Serotonin–LSD interaction, *Ann. N.Y. Acad. Sci.,* **66,** 643–648.

Gardner, A. (1957). Meprobamate—a clinical study, *Amer. J. Psychiat.,* **114,** 524–526.

Garrod, A. E. (1908). *Inborn Errors of Metabolism.* London: Oxford University.

Geffen, L. B., and Rush, R. A. (1968). Transport of noradrenaline in sympathetic nerves and the effect of nerve impulses on its contribution to transmitter stores, *J. Neurochem.,* **15,** 925–930.

Geller, E., Yuwiler, A., and Zolman, J. F. (1966). Effects of environmental complexity on constituents of brain and liver, *J. Neurochem.,* **12,** 949–955.

Geller, I. (1962). Use of approach avoidance behaviour (conflict) for evaluating depressant drugs. In J. H. Nodine and J. H. Moyer (Eds.), *Psychosomatic Medicine.* Philadelphia: Lea and Febiger, 267–274.

Geller, I., and Seifter, J. (1960). The effects of meprobamate, barbiturates d-amphetamine and promazine on experimentally induced conflict in the rat, *Psychopharmacologia (Berl.),* **1,** 482–492.

228

George, R., Haslett, U. W. L., and Jenden, D. J. (1964). A cholinergic mechanism in the brainstem reticular formation: induction of paradoxical sleep, *Int. J. Neuropharmacol.*, **3**, 541–552.

Gerard, R. W. (1955). Biological roots of psychiatry, *Science*, **122**, 225–230.

Gerlach, J. L. and McEwen, B. S. (1972). Rat brain binds adrenal steroid hormone radiography of hippocampus with corticosterone, *Science*, **175**, 1133–1136.

Gershon, S., and Shaw, F. H. (1961). Psychiatric sequelae of chronic exposure to organophosphorus insecticides, *Lancet*, **1**, 1371–1374.

Gerstl, B., Malamud, N., Eng, L. F., Hayman, R. B. (1967). Lipid alterations in human brains in phenylketonuria, *Neurology*, **17**, 51–58.

Giachetti, A., and Shore, P. A. (1966). Studies in vitro of amine uptake mechanisms in heart, *Biochem. Pharmacol.*, **15**, 607–614.

Giarman, N. J., and Pepeu, G. (1964). The influence of centrally acting cholinolytic drugs on brain acetylcholine levels, *Brit. J. Pharmacol. Chemother.*, **23**, 123–130.

Gioscia, V. (1972). Drugs as chronetic agents. In W. Keup (Ed.), *Drug Abuse*. Springfield, Ill.: Charles C. Thomas, 164–170.

Glasky, A. J., and Simon, L. N. (1966). Magnesium pemoline: enhancement of brain RNA polymerases, *Science*, **151**, 702–703.

Glickman, L., and Blumenfield, M. (1967). Psychological determinants of LSD reactions, *J. nerv. ment. Dis.*, **145**, 79–83.

Glowinski, J., and Axelrod, J. (1966). Effects of drugs on disposition of H^3 norepinephrine in rat brain, *Pharmacol. Rev.*, **18**, 775–785.

Glowinski, J., Axelrod, J., and Iversen, L. L. (1966). Regional studies of catecholamines in the rat brain. IV—Effects of drugs on the disposition and metabolism of H^3 norepinephrine and H^3-dopamine, *J. Pharmac. exp. Ther.*, **153**, 30–41.

Gold, R. M. (1967). Aphagia and adipsia following unilateral and bilaterally asymmetrical lesions in rats, *Physiol. Behav.*, **2**, 211–220.

Goldberg, L. (1970). Effects of ethanol in the central nervous system. In R. E. Popham (Ed.), *Alcohol and Alcoholism*. Toronto: University of Toronto Press, 42–56.

Goldsmith, S. R., Frank, I., and Ungerleider, J. T. (1968). Poisoning from ingestion of a stramonium belladonna mixture. Flower power gone sour, *J. amer. med. Assoc.*, **204**, 2, 168–169.

Goldstein, A. (1973). Comments on the isolation, identification and synthesis of a specific behaviour inducing brain peptide, *Nature*, **242**, 60–62.

Goldstein, M., Battista, A. F., Ohmoto, T., Anagnoste, B., and Fuxe, K. (1973). Tremor and involuntary movements in monkeys: effect of l-dopa and of a dopamine receptor stimulating agent, *Science*, **79**, 816–817.

Goodman, L. S., and Gilman, A. (1955). The pharmacological basis of therapeutics. New York: Macmillan.

Goodwin, D. W., Powell, B., Bremer, D., Hoine, H., and Stern, J. (1969). Alcohol and recall. State dependent effects in man, *Science*, **163**, 1358–1360.

Gordon, E. (1972). Methadone maintenance treatment: an overview. In W. Keup (Ed.), *Drug Abuse*. Springfield, Ill.: Charles C. Thomas, 418–423.

Gordon, P. (1971). Molecular approaches to the drug enhancement of deteriorated functioning in the aged, *Adv. gerontol. Res.*, **3**, 199–248.

Gottschalk, L. A. (1969). Phasic circulating biochemical reflections of transient mental content. In A. Mandell and M. P. Mandell (Eds.), *Psychochemical Research in Man*. New York: Academic, 357–378.

Granit, R., Holmgren, B., and Merton, P. A. (1955). The two routes for excitation of muscle and their subservience to the cerebellum, *J. Physiol.*, **130**, 213–224.

Gray, J. A., and Levine, S. (1964). Effects of induced oestrus on emotional behaviour in selected strains of rats, *Nature*, **201**, 1198–1200.

Greene, R., and Dalton, K. (1953). The premenstrual syndrome, *Brit. med. J.*, **1**, 1007–1013.

Grisell, J. L., and Bynum, H. J. (1956). A study of the relationship between anxiety level, ego strength and response to atropine toxicity therapy, *J. nerv. ment. Dis.*, **124**, 265–268.

Grob, D. (1956). Uses and hazards of the organic phosphate and cholinesterase compounds, *Ann. int. Med.*, **32**, 1229–1234.

Grob, D., Garlick, W. L., and Harvey, A. M. (1950). The toxic effects in man of the anticholinersterase insecticide parathion (*p*-nitrophenyl diethyl thionophosphate), *Johns. Hopk. Hosp. Bull.*, **87**, 106–115.

Grob, D., Garlick, W. L., Merrill, G. G., and Freimuth, H. C. (1949). Death due to parathion an anticholinesterase insecticide, *Ann. int. Med.*, **35**, 899–904.

Grob, D., and Harvey, A. M. (1958). Effects on man of the anticholinesterase compound sarin (isopropl methyl phosphorofluoridate), *J. clin. Invest.*, **37**, 350–368.

Grob, D., Harvey, A. M., Langworthy, O. R., and Lilienthal, J. L. (1947). Administration of di-isopropylfluorophosphate to man. III—The effect on the central nervous system with special reference to the electrical activity of the brain, *Bull. John Hopkins Hosp.*, **81**, 257–266.

Grossman, S. P. (1960). Eating or drinking elicited by direct adrenergic or cholinergic stimulation of hypothalamus, *Science*, **132**, 301–302.

Grossman, S. P. (1962). Direct adrenergic and cholinergic stimulation of hypothalamic mechanisms, *Amer. J. Physiol.*, **202**, 872–882.

Grossman, S. P. (1964a). Behavioral effects of chemical stimulation of the ventral amygdala, *J. comp. physiol. Psychol.*, **57**, 29–36.

Grossman, S. P. (1964b). Effects of chemical stimulation of the septal area on motivation, *J. comp. physiol. Psychol.*, **58**, 194–200.

Grossman, S. P. (1967). *A Textbook of Physiological Psychology*, New York: Wiley.

Grossman, S. P. (1969). Neuropharmacology of central mechanisms contributing to control of food and water intake. In J. Field (Ed.), *Handbook of Physiology—alimentary canal*, 287–301.

Gruber, R. P., Stone, G. C., and Reed, D. R. (1967). Scopolamine induced anterograde amnesia, *Int. J. Neuropharm.*, **6**, 187–190.

Gurdjian, E. S. (1926). The hypothalamus in the rat, *Anat. Rec.*, **32**, 208.

Gursey, D., Vester, J. W., and Olsen, R. E. (1959). Effect of ethanol administration upon serotonin and norepinephrine levels in rabbit brain, *J. clin. Invest.*, **38**, 1008–1017.

Gursey, D., and Olsen, R. E. (1960). Depression of serotonin and norepinephrine levels in brain stem of rabbit by ethanol, *Proc. Soc. exp. biol. Med.*, **104**, 280–290.

Haggendahl, J., and Lindquist, M. (1961). Ineffectiveness of ethanol on noradrenalin dopamine or 5-hydroxytryptamine levels in brain, *Acta pharmacol. toxicol.*, **18**, 278–284.

Haggendahl, J., and Lindquist, M. (1964). Disclosure of labile monoamine fractions in brain and their correlation to behaviour, *Acta. Physiol. scand.*, **60**, 351–357.

Halas, E. S., Bradfield, K., Sandlie, M. E., Theye, F., and Beardsley, J. (1966). Changes in rat behavior due to RNA injection, *Physiol. Behav.*, **1**, 281.

Hamburg, D. A., Moos, R. H., and Yalom, I. D. (1968). Studies of distress in the menstrual cycle and the postpartum period. In R. P. Michael (Ed.), *Endocrinology and Human Behaviour*. London: Oxford Univ. Press, 94–116.

Hamilton, L. W., and Grossman, S. P. (1969). Behavioral changes following disruption of central cholinergic pathways, *J. comp. physiol. Psychol.*, **69**, 76–82.

Harris, G. W., and Naftolin, F. (1970). The hypothalamus and control of ovulation, *Brit. Med. Bull.*, **26**, 3–8.

Harris, T. H. (1960). Methaminodiazepoxide, *J. Amer. med. Assoc.*, **172**, 1162–1163.

Hartmann, E. (1966). Mechanism underlying the sleep dream cycle, *Nature*, **212**, 648–650.

Harvey, A. M., Lilienthal, J. L., Grob, D., Jones, B. F., and Talbot, S. A. (1947). The administration of di-isopropylfluorophosphate to man. IV—The effects on neuromuscular function in normal subjects and in myasthenia gravis, *Bull. John Hopkins Hosp.*, **81**, 267–278.

Hashim, S. A., and van Itallie, T. B. (1965). Studies in normal and obese subjects with a monitored food dispensing device, *Ann. N.Y. Acad. Science*, **131**, 654–661.

Hauser, H., Schwartz, B. F., Ross, G., and Bickford, R. G. (1958). Electroencephalographic changes related to smoking, *Electroenceph. clin. Neurophysiol.*, **10**, 576.

Hearst, E. (1959). Effects of scopolamine on discriminated responding in the rat, *J. Pharmacol. exp. Ther.*, **126**, 349–358.

Hebb, D. O. (1949). *The Organization of Behavior: A Neuropsychological Theory*. London: Chapman and Hall.

Hedge, G. A., and Smelik, P. G. (1968). Corticotrophin release: inhibition by intrahypothalamic implantation of atropine, *Science*, **159**, 891–892.

Heimstra, N. W., Bancroft, N. R., and DeKoch, A. R. (1967). Effects of smoking upon sustained performance in a simulated driving task, *Ann. N.Y. Acad. Sci.*, **142**, 295–300.

Heinrich, U., Lichtensteiger, W., and Largemann, H. (1971). Effect of morphine on the catecholamine content of midbrain nerve cell groups in rat and mouse, *J. Pharmacol. exp. Therap.*, **179**, 259–267.

Heise, G. A., and Boff, E. (1960). Behavorial determination of time and dose parameters of monoamine oxidase inhibitors, *J. pharmacol. exper. Therap.*, **129**, 155–161.

Heise, G. A., and Boff, E. (1962). Continuous avoidance as a baseline for measuring behavioral effects of drugs, *Psychopharmacologia (Berl.)*, **3**, 264–282.

Heise, G. A., and Boff, E. (1971). Stimulant action of d-amphetamine in relation to test compartment dimesions and behavioral measure, *Neuropharmacology*, **10**, 259–266.

Heise, G. A., Keller, C., Khavari, K. A., and Laughlin, N. (1969). Learning of discrete trial, go–no go alternation patterns by the rat, *J. exp. anal. Beh.*, **12**, 609–622.

Heise, G. A., and Lilie, N. L. (1970). Effects of scopolamine, atropine and amphetamine on internal and external control of responding and non-reinforced trials, *Psychopharmacologia (Berl.)*, **18**, 38–49.

Heise, G. A., Laughlin, N., and Keller, C. (1970). Behavioral analysis of reinforcement withdrawal, *Psychopharmacologia (Berl.)*, **16**, 345–368.

Heise, G. A., and McConnell, H. (1961). Differences between chlordiazepoxide-type and chlorpromazine type action in 'trace' avoidance. In R. A. Cleghorn (Ed.), *Proceedings of the Third World Congress of Psychiatry*, Vol. 2. Toronto: University of Toronto, McGill University, 917–921.

Hernandez-Peon, R. (1963). Sleep induced by localized electrical or chemical stimulation of the forebrain, *E.E.G. Clin. Neurophysiol. Suppl.*, **24**, 188–198.

Hernandex-Peon, R. (1965). Physiological mechanisms in attention. In R. W. Russell (Ed.), *Frontiers in Physiological Psychology*. New York: Academic, 121–147.

Hernandez-Peon, R. (1969). A neurophysiological and evolutionary model of attention. In C. R. Evans and T. B. Mulholland (Eds.), *Attention in Neurophysiology*. London: Butterworths, 417–425.

Hernandez-Peon, R., Chavez-Ibarra, G., Morgane, P. J., and Timo-Iaria, C. (1963). Limbic cholinergic pathways involved in sleep and emotional behavior, *Exp. Neurol.*, **8**, 93–111.

Herz, A. (1968). Some actions of cholinergic and anticholinergic drugs on reactive behavior, *Progr. Brain Res.*, **28**, 78–85.

Herzberg, B. N., Johnson, A. L., and Brown, S. (1970). Depressive symptoms and oral contraceptives, *Brit. Med. J.*, **4**, 142–145.

Hess, R. (1954). Group discussion of diencephalic sleep centre. In J. F. Delafresnaye (Ed.), *Brain Mechanisms and Consciousness*. Oxford: Blackwell, 125–136.

Hess, W. R. (1954a). The diencephalic sleep centre. In J. F. Delafresnaye (Ed.), *Brain and Consciousness*. Oxford: Blackwell, 117–125.

Hess, W. R. (1954b). *Das Zwischenhirn Syndrome Lokalisationen Functionen*. Basel: Benno Schwabe.

Hillarp, N. Ä., Fuxe, K., and Dahlstrom, A. (1966). Demonstration and mapping of central neurons containing dopamine noradrenaline and 5-hydroxytryptamine and their reactions to psychopharmaca, *Pharmacol. Rev.*, **18**, 727–741.

Hillarp, N. Ä., and Malmfors, T. (1964). Reserpine and cocaine blocking of the uptake and storage mechanism in adrenergic nerves, *Life Sci.*, **3**, 703–708.

Himwich, H. E., Narasimhachari, N., Heller, B., Spaide, J., Haškovec, L., Fujimori,

M., and Tabushi, K. (1970). Comparative behavioral and urinary studies on schizophrenics and normal controls. In R. E. Bowman and S. P. Datta (Ed.), *Biochemistry of Brain and Behavior*. New York: Plenum, 207–221.

Hoch, P. H. (1946). The present stasus of narcodiagnosis and therapy, *J. Nerv. and Ment. Dis.*, **103**, 248–259.

Hoch, P. H., and Polatin, P. (1952). Narcodiagnosis and narcotherapy. In G. Bychowski and J. L. Despert (Eds.), *Specialized Techniques in Psychotherapy*. New York: Grove Press, 1–23.

Hodos, W., and Valenstein, E. S. (1960). Motivational variables affecting the role of behavior maintained by intracranial stimulation, *J. comp. physiol. Psychol.*, **53**, 502–508.

Hodos, W., and Valenstein, E. (1962). An evaluation of response rate as a measure of rewarding intracranial stimulation, *J. comp. physiol. Psych.*, **55**, 80–84.

Hoffer, A. (1958). Adrenochrome in blood plasma, *Amer. J. Psychiat.*, **114**, 752–753; *Amer. J. Psychiat.*, **114**, 752.

Hoffer, A., and Osmond, H. (1960). *The Chemical Basis of Clinical Psychiatry*. Springfield, Ill.: Charles Thomas.

Hoffer, A., and Osmond, H. (1966). Some psychological consequences of perceptual disorder and schizophrenia, *Int. J. Neuropsychiat.*, **2**, 1–19.

Hoffer, A., and Osmond, H. (1968). *The Hallucinogens*. New York: Academic.

Hoffer, A., Osmond, H., and Smythies, J. R. (1954). Schizophrenia: A new approach II, *J. ment. Sci.*, **100**, 29–45.

Hoffer, A., and Payza, A. N. (1960). The presence of adrenochrome in blood. *Amer. J. Psychiat.*, **116**, 664.

Hoffman, G. M., and Gay, J. R. (1959). Accidental atropine poisoning, *Pennsylvania Med. J. Sept.*, 1340–1341.

Hofmann, A. (1958). Lysergic acid diethylamide and related compounds. Relationships between spatial arrangement and mental effects. In M. Rinkel and H. C. B. Denber (Ed.), *Chemical Concepts of Psychosis*. New York: Oblensky.

Holden, C. (1972). Methadone. New F.D.A. guidelines would tighten distribution, *Science*, **177**, 502–504.

Hollinger, M. (1969). Effect of reserpine, a methyl-*p*-tyrosine,*p*-chlorophenylalanine and pargyline on levorphanol-induced running activity in mice, *Arch. int. pharmacodyn. therap.*, **179**, 419–424.

Hollister, L. E. (1971). Actions of various marijuana derivatives in man, *Pharmacol. Rev.*, **23**, 349–357.

Hollister, L. E., and Friedhoff, A. J. (1966). Effects of 3,4-dimethoxyphenylethylamine in man, *Nature*, **210**, 1377–1378.

Holmstedt, B. (1963). Structure-activity relationships of the organophosphorus anticholinesterase agents. In G. B. Koelle (Ed.), *Handbuck Der Experimentellen Pharmachologie Ergänzungswerk*. Berlin: Springer-Verlag, 429–485.

Holmstedt, B. (1967), Mobilization of acetylcholine by cholinergic agents, *Ann. N. Y. Acad. Sci.*, **144**, 433–458.

Holmstedt, B., and Lundgren, G. (1967). Arecoline, nicotine and related compounds. Tremorgenic activity and effect upon brain acetylcholine, *Ann. N. Y. Acad. Sci.*, **142**, 126–142.

Holtzman, D., Lovell, R. A., Jaffe, J. H., and Freedman, D. X. (1969). 1- Δ^a-tetrahydrocannabinol: neurochemical and behavioral effects in the mouse, *Science*, **163**, 1464–1467.

Hornykiewicz, O. (1966). Dopamine (3-hydroxytyramine) and brain function, *Pharmacol. Rev.*, **18**, 927–963.

Hornykiewicz, O. (1973). Dopamine in the basal ganglia, *Brit. med. Bull.*, **29**, 172–178.

Hornykiewicz, O., Markham, C. H., Clark, W. G., and Fleming, R. M. (1970). Mechanisms of extrapyramidal side effects of therapeutic agents. In W. G. Clark and J. del Guidice (Eds.), *Principles of Psychopharmacology*. New York: Academic, 585–596.

Horsley, J. S. (1936). Narcoanalysis, *Lancet*, **1**, 55.

Hosein, E. A., and Koh, T. Y. (1966). Acetylcholine like activity of acetyl-L-carnityl C. A. in subcellular particles of narcotized brain homogenates, *Arch Biochem. Biophys*, **114**, 94–99.

Hoskins, R. G. (1925). Studies on vigor II. The effect of castration on voluntary activity, *Amer. J. Physiol.*, **72**, 324–330.

Hoskins, R. G., and Bevin, S. (1941). The effect of fractionated chorionic gonadotropic extract on spontaneous activity and weight of elderly male rats, *Endocrinology*, **27**, 929–931.

House, R. C. (1931). The use of scopolamine in criminology, *Amer. J. police Sci.*, **2**, 328–336.

Hsia, D. Y. Y. (1967). The hereditary metabolic diseases. In J. Hirsch (Ed.), *Behavior—Genetic Analysis*. New York: McGraw-Hill, 176–210.

Hunt, H. F. (1956). Some effects of drugs on classical (type S) conditioning, *Ann. N. Y. Acad. Sci.*, **65**, 258–267.

Huxley, H. E. (1969). The mechanism of muscular contraction. *Science*, **164**, 1356–1366.

Hydén, H. (1959). Biochemical changes in glial cells and nerve cells at varying activity. In O. Hoffman Ostenhof (Ed.), *Biochemistry of the Central Nervous System*. New York: Pergamon, 64–88.

Hydén, H. (1961). Satellite cells in the nervous system, *Sci Amer.*, **205**, No. 6, 62–70.

Hydén, H., and Egyházi, E. (1962). Nuclear RNA changes of nerve cells during a learning experiment with rats, *Proc. nat. Acad. Sci.*, **48**, 1366–1373.

Hydén, H., and Egyházi, E. (1963). Glial RNA changes during a learning experiment in rats, *Proc. nat. Acad. Sci.*, **49**, 618–624.

Hydén, H., and Lange, P. W. (1964). Changes in RNA control and base composition in cortical neurons of rats in a learning experiment involving transfer of handedness, *Proc. nat. Acad. Sci.*, **52**, 1030–1035.

Hydén, H., and Lange, P. W. (1965). A differentiation in RNA response in neurons early and late in learning, *Proc. nat. Acad. Sci.*, **53**, 946–952.

Hydén, H., and Lange, P. W. (1968). Protein synthesis in the hippocampal pyramidal cells of rats during a behavioral test, *Science*, **159**, 1370–1373.

Hydén, H., and Lange, P. W. (1970). Protein changes in nerve cells related to learning and conditioning. In F. O. Schmitt (Ed.), *The Neurosciences: Second Study Program*. New York: Rockefeller University Press, 278–288.

Hyppä, M. (1969). Differentiation of the hypothalamic nuclei during ontogenetic development in the rat, *Z. Anat. Entwickl.-Gesch.*, **129**, 41–52.

Il'yutchenok, R. Yu, and Gilinskii, M. A. (1969). Deĭstvie Kholinoliticheskikh veschestv na spontannuyun i vyzannuyn aktivnost, Karkovykh neĭronov, *Framakologiya i Iskisiskologiya*, **32**, 515–519.

Im, H. S., Barnes, R. H., and Levitsky, D. A. (1971). Postnatal malnutrition and brain cholinesterase in rats, *Nature*, **233**, 269–270.

Irwin, R. L., and Hein, M. N. (1962). The inhibition of rat brain cholinesterase after administration of the dimethyl-carbamates of deoxy-demethyl lynoramine neostigmine or physostigmine, *J. Pharm. exp. Therap.*, **136**, 20–25.

Isaacson, R. L. (1972). Neural systems of the limbic brain and behavioural inhibition. In R. A. Boakes and M. S. Halliday (Eds.), *Inhibition and Learning*. London; Academic Press, 497–528.

Isbell, H. (1970). Experimental physical dependence on alcohol in humans. In R. E. Popham (Ed.), *Alcohol and Alcoholism*. Toronto: University of Toronto Press, 106–110.

Isbell, H., and Gorodetzky, C. W. (1966). Effect of alkaloids in ololingui in man, *Psychopharmacologia (Berl.)*, **8**, 331–339.

Isham, A. C. (1966). Office evaluation of diazepam for psychoneurotic anxiety and depressive reactions, *Int. J. Neuropsychiat.*, **2**, 111–121.

Iversen, L. L. (1967). *The Uptake and Storage of Noradrenaline in Sympathetic Nerves.* Cambridge: Cambridge University Press.

Iversen, L. L. (1970). Neurotransmitters. Neurohormones and other small molecules in neurons. In F. O. Schmitt (Ed.), *The Neurosciences. Second study Program.* New York: Rockfeller University Press.

Iversen, L. L., and Salt, P. J., (1970). Inhibition of catecholamine uptake by steroids in the isolated rat heart, *Brit. J. Pharmacol.,* **40,** 528–530.

Ivey, M. E., and Bardwick, J. M. (1968). Patterns of affective fluctuation in the menstrual cycle, *Psychosomat. Med.,* **30,** 336–345.

Jackson, C. O. (1971). The amphetamine inhaler. A case study of medical abuse, *J. Hist. Med.,* **25,** 187–196.

Jacobs, H. L. (1962). Some physical metabolic and sensory components in the appetite for glucose, *Amer. J. Physiol.,* **203,** 1043–1054.

Jacobsen, A. L., Babich, F. R., Bubash, S., and Jacobson, A. (1965). Differential approach tendencies produced by injection of RNA from trained rats, *Science,* **150,** 636–737.

Jacobsen, E. (1968). The hallucinogens. In C. R. B. Joyce (Ed.), *Psychopharmacology.* London: Tavistock, 175–213.

Jaffe, J. H. (1970). Treatment of drug abusers. In W. G. Clark and J. del Guidice (Eds.), *Principles of Psychopharmacology.* New York: Academic, 547–570.

Jaffe, J. H., Zacks, M. S., and Washington, E. N. (1969). Experience with the use of methadone in multi-modality programs for the treatment of narcotic users, *Int. J. Addict.,* **4,** 481–490.

Janowitz, H. D., and Hollander, F. (1953). Effect of prolonged intragastric feeding on oral digestion, *Fed. Proc.,* **12,** 72.

Janowsky, D. S., Gorney, R., and Mandell, A. J. (1967). The menstrual cycle: psychiatric and ovarian-adrenocortical hormone correlates: case study and literature review, *Arch. gen. Psychiat.,* **17,** 459–469.

Jarvik, M. E. (1964). The influence of drugs upon memory. In H. Steinberg (Ed.), *Animal Behavior and Drug Action.* London: Churchill, 44–61.

Jenner, F. A., and Kerry, R. J. (1967). Comparison of diazepan, cholordiazepoxide and amylobarbitone (a multidose double-blind cross-over study), *Dis. Nerv. Syst.,* **28,** 245–249.

Jenner, F. A., Kerry, R. J., and Parkin, D. (1961). A controlled trial of methamino-diazepoxide (chlordiazepoxide, Librium) in the treatment of anxiety in neurotic patients, *J. ment. Sci.,* **107,** 575–582.

Jensen, D. D. (1965). Paramecia, planaria and pseudo-learning, *Anim. Behav. Suppl.,* **7,** 9–20.

Jéquier, E., Lovenberg, W., and Sjoerdsma, A. (1967). Tryptophan hydroxylase inhibition: the mechanism by which p-chlorophenylalanine depletes rat brain serotonin, *Mol. Pharmacol.,* **3,** 274–278.

John, E. R. (1967). *Mechanisms of Memory.* New York: Academic Press.

Jones, R. T. (1971). Marijuana-induced 'high': influence of expectation, setting and previous drug experience, *Pharmacol. Rev.,* **23,** 359–369.

Jönsson, L. E., Änggärd, E., and Gunne, L. M. (1971). Blockade of intravenous amphetamine euphoria in man, *Clin. Pharmacol. Therap.,* **12,** 889–896.

Jönsson, L. E., and Sjöström, K. (1970). A rating scale for evaluation of the clinical course and symptamatology in amphetamine psychosis, *Br. J. Psychiat.,* **117,** 661–665.

Jouvet, M. (1962). Recherches sur les structures nerveuses et les mécanismes responsables des différentes phases du sommeil physiologique, *Arch. ital. Biol.,* **100,** 125–206.

Jouvet, M. (1967). Mechanisms of the states of sleep. a neuropharmacological approach, *Ass. Res. nerv. ment. Dis. Res. Publ.,* **45,** 86–126.

Jouvet, M. (1969). Biogenic amines and the states of sleep, *Science,* **163,** 32–40.

Jouvet, M., Bobillier, P., Pujol, J. F., and Renault, J. (1966). Effets des lésions due systeme du raphé sur le sommiel et al sérotonine cérébrale, *Compt. Rend. Soc. biol.* **160,** 2343–2346.

Jouvet, M., and Renault, J. (1966). Insomnie persistante apres lesions des moyaux du raphe chez le chat, *Comp. Rend. Soc. biol.*, **160**, 1461–1465.

Jouvet, M., Vimont, P., and Delorme, F. (1965). Suppression elective du sommeil paradoxical chez le chat par les inhibiteurs monoamine oxydase, *Compt. Rend. Soc. Biol.*, **159**, 1595–1599.

Jung, R., and Hassler, R. (1959). The extrapyramidal motor system. In J. Field (Ed.), *Handbook of Neurophysiology*, Vol. 2. Washington, D. C.: Americal Physiological Society, 863–927.

Kaito, A. A., and Goldberg, A. M. (1969). Control of acetylcholine synthesis—the inhibition of choline acetyltransferase by acetylcholine, *J. Neurochem.*, **16**, 1185–1191.

Kalant, H., and Grose, W. (1967). Effects of ethanol and pentobarbital on release of acetylcholine from cerebral cortex slices, *J. Pharmacol. Exp. Therap.*, **158**, 386–936.

Kalant, H., Israel, Y., and Mahon, M. A. (1967). The effect of ethanol on acetylcholine synthesis release and degradation in brain, *Canad. J. Physiol. Pharmacol.*, **45**, 172–177.

Kalant, H., Leblanc, A. E., and Gibbins, R. J. (1971). Tolerance to, and dependence on, ethanol. In Y. Israel and J. Mardones (Eds.), *Biological Basis of Alcoholism*. New York: Wiley, 235–270.

Kamberi, J. A., and Kobayashi, Y. (1970). Monoamine oxidase activity in the hypothalamus and various other brain areas and in some endocrine glands of the rat during the oestrous cycle, *J. Neurochem.*, **17**, 261–268.

Kanai, T., and Szerb, J. C. (1965). Mesencephalic reticular activating system and cortical acetylcholine output, *Nature*, **205**, 81–88.

Kanijo, K., Koelle, G. B., and Wagner, H. H. (1955). Modification of the effects of sympathomimetic amines and of adrengergic nerve stimulation by 1-isonicotinyl 2-isopropythydrazine and isonicotinic acid hydrazide, *J. Pharmacol. exp. Therap.*, **117**, 213–227.

Kare, M. R., Schechter, P. J., Grossman, S. P., and Roth L. J. (1969). Direct pathway to the brain, *Science*, **163**, 952–954.

Karki, N., Kuntzman, R., and Brodie, B. B. (1962). Storage, synthesis, and metabolism of monoamines in the developing brain, *J. Neurochem.*, **9**, 53–58.

Karlson, P. (1963). *Introduction to Modern Biochemistry*. New York: Academic.

Karrer, R., and Cahilly, G. (1965). Experimental attempts to produce phenylketonuria in animals. A critical review, *Psychol. Bull.*, **64**, 52–64.

Katz, B. (1958a). Microphysiology of the neuromuscular junction. A physiological 'quantum of action' at the myoneural junction, *Bull. John Hopkins Hosp.*, **102**, 275–295.

Katz, B. (1958b). Microphysiology of the neuromuscular junction. The chemoreceptor function of the motor end-plate, *Bull. John Hopkins Hosp.*, **102**, 296–321.

Katz, M. M., Waskow, I. E., and Olsson, J., (1968). Characterising the psychological state produced by LSD, *J. abn. Psychol.*, **73**, 1–14.

Kaufman, E. (1972). Reality House, a self-help day care centre for narcotic addicts. In Wolfram Keup (Ed.), *Drug Abuse*. Illinois: Charles C. Thomas, 381–386.

Kelleher, R. T., and Morse, W. H. (1968). Determinants of the specificity of behavioral effects of drugs. *Ergebn. Physiol.*, **60**, 1–56.

Kelsey, J. E., and Grossman, S. P. (1969). Cholinergic blockade and lesions in the ventromedial septum of the rat, *Physiol. Behav.*, **4**, 837–845.

Kennedy, C. C. (1966). Food intake, energy, balance and growth, *Brit. med. Bull.*, **23**, 216–219.

Kessel, N., and Walton, H. (1969). *Alcoholism* (Revised Edition). London: Penguin.

Kety, S. S. (1969). The precursor-load strategy in psychochemical research. In A. J. Mandell and M. P. Mandell (Ed.), *Psychochemical Research in Man*. New York: Academic, 127–131.

Kety, S. S. (1970). The biogenic amines in the central nervous system: their possible roles in arousal, emotion and learning. In F. O. Schmitt (Ed.), *The Neurosciences. Second Study Program*. New York: Rockefeller University Press, 324–336.

Key, B. J. (1965). Effect of lysergic acid diethylamide on potentials evoked in the specific sensory pathways, *Brit. med. Bull.*, **21**, 30–35.

Khavari, K. A., Heebink, P., and Traupman, J. (1968). Effects of intraventricular carbachol and eserine on drinking, *Psychonom. Sci.*, **11**, 93–94.

Khavari, K .A., and Russell, R. W. (1966). Acquisition, retention and extinction under conditions of water deprivation and of central cholinergic stimulation, *J. comp. physiol. Psychol.*, **61**, 339–345.

Kiloh, L. G., Ball, J. R. B., and Garside, R. F. (1962). Prognostic factors in treatment of depressive states with imipramine, *Brit. Med. J.*, **1**, 1225–1227.

Kiloh, L. G., and Brandon, S. (1962). Habituation and addiction to amphetamines, *Brit. Med. J.*, **2**, 40–43.

Kiloh, L. G., Child, J. P., and Latner, G. (1960). A controlled trial of ipromiazid in the treatment of endogenous depression, *J. Ment. Sci.*, **106**, 1139–1143.

Kimble, G. A. (1961). Hilgard and Marquis (Eds.). *Conditioning and Learning*. New York: Appleton Century-Crofts.

Kinnard, W. J., Aceto. M. D. G., and Buckley, J. P. (1962). The effects of certain psychotropic agents on the conditioned emotional response behavior pattern of the albino rat, *Psychopharmacologia (Berl.)*, **3**, 227–230.

Kissin, B. (1972). Alcohol as it compares to other addictive substances. In W. Keup (Ed.), *Drug Abuse*. Springfield: Charles C. Thomas, 251–262.

Klawans, H. L. (1970). A pharmacological analysis of Huntington's Chorea, *Europ. Neurol.*, **4**, 148–163.

Klawans, H. L., and Rubovits, R. (1972). Central cholinergic-anticholinergic antagonism in Huntington's Chorea, *Neurology*, **22**, 107–116.

Kleitman, N. (1963). *Sleep and Wakefulness*. Chicago: University of Chicago Press.

Klerman, G. L. (1966). Comments from the viewpoint of a clinical psychiatrist. In J. R. Wittenborn and P. R. A. May (Eds.), *Prediction of Response to Pharmacotherapy*. Springfield, Ill: Thomas, 135–138.

Kletzkin, M., and Berger, F. M. (1959). Effect of meprobamate on limbic system of the brain, *Proc. Soc. exp. biol. Med.*, **100**, 681–683.

Kline, N. S. (1963). Use of Pargyline (Eutonyl) in private practice, *Ann. N. Y. Acad. Sci.*, **107**, 1090–1106.

Knoll, M., Kugler, J., Höfer, O., and Lawder, S. D. (1963). Effects of chemical stimulation of electrically induced phosphenes on their bandwidth, shape, number and intensity, *Confin. Neurol.*, **23**, 201–226.

Knowles, J. B., Laverty, S. G., and Kuechler, H. A. (1968). Effects of alcohol on R. E. M. sleep, *Quart. J. Stud. Alc.*, **29**, 342–349.

Kobayashi, T., Kobayashi, T., Kato, J., and Minaguchi, H. (1966). Cholinergic and adrenergic mechanisms in the female rat hypothalamus with special reference to feedback of ovarian steroid hormones. In G. Pincus, T. Nakao and J. Tait (Eds.), *Steroid Dynamics*. New York: Academic, 303–339.

Koe, B. K., and Weissman, A. (1966). *p*-Chlorophenylalanine: a specific depletor of brain serotonin, *J. Pharmacol. exp. Therap.*, **154**, 499–516.

Koella, W. P., Feldstein, A., and Czieman, J. S. (1968). The effect of *para*-chlorphenylalanine on the sleep of cats, *E.E.G. clin. Neurophysiol.*, **25**, 481–490.

Kopin, I. J. (1968). False adrenergic transmitters, *Ann. Rev. Pharmac.*, **8**, 377–394.

Krech, D. (1968). Brain chemistry and anatomy: Implications for behavior therapy. In C. Rupp (Ed.), *Mind as Tissue*. New York: Hoeber (Harper and Row), 39–54.

Krechevsky, I. (1933). Hereditary nature of 'hypotheses', *J. comp. Psychol.*, **16**, 99–116.

Krieger, H. P., and Krieger, D. T. (1970). Chemical stimulation of the brain: effect on adrenal corticoid release, *Amer. J. Physiol.*, **218**, 1632–1641.

Krnjević, K., and Phillis, J. W. (1963a and b). Acetylcholine sensitive cells in the cerebral cortex, *J. Physiol (Lond.)*, **166**, 296–327.

Krnjević, K., and Phillis, J. W. (1963b). Pharmacological properties of acetylcholine sensitive cells in the cerebral cortex, *J. Physiol. (Lond.)*, **166**, 328–350.

Krnjević, K., Pumain, R., and Renaud, L. (1971). The mechanism of excitation by acetylcholine in the cerebral cortex, *J. Physiol (Lond).*, **215**, 247–268.

Kuffler, S. W. (1949). Transmitter mechanism at the nerve-muscle junction, *Arch. Sci. physiol.*, **3**, 585–601.

Kuhn, R. (1958). The treatment of depressive states with G. 22355 (imipramine hydrochloride), *Amer. J. Psychiat.*, **115**, 459–464.

Ladinsky, H., Consolo, S., Peri G., and Garattini, S. (1973). Increase in mouse and rat brain acetylcholine levels by diazepam. In S. Garattini, E. Mussini and L. O. Randall (Eds.), *The Benzodiazepines*. New York: Raven, 241–242.

Ladisich, W., and Baumann, P. (1971). Influence of progesterone on norepinephrine metabolism of the rat brain in connection with amphetamine and stress, *Neuroendocrinology*, **7**, 16–24.

Landauer, T. K. (1969). Reinforcement as consolidation, *Psychol. Rev.*, **76**, 82–96.

Landfield, P. W., McGaugh, J. L., and Tusa, R. J. (1972). Theta rhythm: a correlate of posttrial memory storage processes in rats, *Science*, **175**, 87–88.

Langley, J. N. (1905). On the reaction of cells and of nerve-ending to certain, poisons, chifely as regards the reaction of striated muscle to nicotine and to curare, *J. Physiol. (Lond.)*, **33**, 374–392.

Larson, J. A. (1932). *Lying and its Detection*. Chicago: University of Chicago Press.

Laties, V., and Weiss, B. (1966). Influence of drugs on behavior controlled by internal and external stimuli, *J. pharmacol. exp. Therap.*, **152**, 388–396.

Lauener, H. (1963). Conditioned suppression in rats and the effect of pharmacological agents thereon, *Psychopharmacologia (Berl.)* **4**, 311–325.

Laverty, R., Michaelson, I. A., Sharman, D. F., and Whittaker, V. P. (1963). The subcellular localization of dopamine and acetylcholine in the dog caudate nucleus, *Brit. J. Pharmacol.*, **21**, 482–490.

Lee. D. K., Markham, C. H., and Clark, W. G. (1968). Serotonin (5-HT) metabolism in Huntington's Chorea, *Life Sci.*, **7**, 707–715.

Lee, D. K., Markham, C. H., and Clark, W. G. (1969). Serotonin metabolism in Huntington's Chorea. In A. Barbeau and C. Brunnette (Eds.), *Progress in Neurogenetics*. Amsterdam: Exerpta Medica, 577–588.

Lehmann, H. E. (1966). Depression: categories, mechanisms and phenomena. In J. O. Cole and J. R. Wittenhorn (Eds.), *Pharmacotherapy of Depression*. Springfield, Ill.: Charles C. Thomas.

Leibowitz, S. F. (1968). *Memory and emotionality after anticholinesterase in the hippocampus: inverse function of prior learning level.* Unpublished Doctoral Dissertation, New York University.

Leibowitz, S. F. (1970a). Hypothalamic β-adrenergic 'satiety' system antagonizes an α-adrenergic 'hunger' system in the rat, *Nature*, **226**, 963–964.

Lebowitz, S. F. (1970b) Reciprocal hunger regulating circuits involving alpha-and beta-adrenergic receptors located respectively, in the ventromedial and lateral hypothalamus, *Proc. nat. Acad. Sci.*, **67**, 1063–1070.

Leibowitz, S. F. (1971). Hypothalamic alpha- and beta-adrenergic systems regulate both thirst and hunger in the rat, *Proc. nat. Acad. Sci.*, **68**, 2, 332–334.

Le Moal, M., Stinus, J., and Cardo, B. (1969). Influence of ventral mesencephalic lesions on various spontaneous and conditioned behaviors in the rat, *Physiol. Behav.*, **4**, 567–573.

Lennard, H. L., Epstein, L. J., and Rosenthal, M. S. (1972). The methadone illusion, **176**, 881–884.

Leonard, B. E., and Tonge, S. R. (1969). The effects of some hallucinogenic drugs upon the metabolism of noradrenaline, *Life Sci.*, **8**, 815–825.

Lester, B. K., and Guerrero-Figueroa, R. (1966). Effects of some drugs on electoencephalographic fast activity and dream time, *Psychophysiology*, **2**, 224–236.

Levi, R., and Maynert, E. W. (1974). The subcellular localization of brain stem norepinephrine and 5-hydroxytryptamine in stressed rats, *Biochem. Pharmacol.*, **3**, 615–621.

Levine, J., and Ludwig, A. M. (1965). Alternations in consciousness produced by combinations of LSD, hypnosis and psychotherapy, *Psychopharmacologia (Berl.)*, **7**, 123–137.

Levine, S., and Broadhurst, P. L. (1963). Genetic and ontogenetic determinants of behaviour. Effects of infantile stimulation on adult emotionality and learning in selected strains of rats, *J. comp. physiol. Psychol.*, **56**, 423–428.

Levine, S., Goldman, L., and Coover, G. D. (1972). Expectancy and the pituitary-adrenal system. In R. Porter and J. Knight (Eds.), *Physiology, Emotion and Psychosomatic Illness*. Amsteradam: Elsevier, 281–291.

Levitsky, D. A., and Barnes, R. H. (1970). Effect of early malnutrition on the reaction of adult rats to aversive stimuli, *Nature*, **225**, 468–469.

Levitt, E. E., Persky, H., Brady, J. P., and Fitzgerald, J. A. (1963). The effect of hydrocortisone infusion on hypnotically induced anxiety, *Psychosom. Med.*, **25**, 158–161.

Levitt, R. A., and Fisher, A. E. (1966). Anticholinergic blockade of centrally induced thirst, *Science*, **154**, 520–522.

Levitt, R. A., White, C. S., and Sander, D. M. (1970). Dose response analysis of carbachol elicited drinking in the rat limbic system, *J. comp. physiol.*, **72**, 345–350.

Lewin, L. (1964). *Phantastica, narcotic and stimulating drugs*. London: Routledge, Kegan Paul.

Lewis, P. R., and Shute, C. C. D. (1967). The cholinergic limbic system: projections to hippocampal formation medial cortex, nuclei of the ascending cholinergic reticular system and the subfornical organ and supra-optic crest, *Brain*, **90**, 521–540.

Lewis, P. R., Shute, C. C. D., and Silver, A. (1964). Confirmation from cholin acetylase analyses of a massive cholinergic innervation to the rat hippocampus, *J. Physiol. (Lond.)*, **191**, 215–224.

Libet, B. (1965). Cortical activation in conscious and unconscious experience, *Pers. Biol. Med.*, **9**, 77–86.

Libet, B. (1966). Brain stimulation and the threshold of conscious experience. In J. C. Eccles, *Brain and Conscious Experience*, New York. Springer-Verlag, 165–181.

Lidbrink, P., Corrodi, H., Fuxe, K., and Olson, L. (1973). The effects of benzodiazepines, meprobamate and barbiturates on central monoamine neurons. In S. Garattini, E. Mussini, and L. O. Randall (Eds.), *The Benzodiazepines*. New York: Raven, 203–220.

Lindemann, E. (1932). Psychological changes in normal and abnormal individuals under the influence of sodium amytal, *Amer. J. Psychiat.*, **11**, 1083–1091.

Lindesmith, A. (1970). Psychology of addiction. In W. G. Clarm and J. del Guidice (Eds.), *Principles of Psychopharmacology*. Academic Press: New York and London, 471–476.

Lindsley, D. B., Shreiner, L. H., and Magoun, H. W. (1949). An electromyographic study of spasticity, *J. Neurophysiol.*, **12**, 197–205.

Linton, H. B., and Langs, R. J. (1962). Subjective reactions to lysergic acid diethylamide, LSD-25, *Arch. gen. Psychiat.*, **6**, 352–368.

Lloyd, K., and Hornykiewicz, O. (1970). Parkinson's disease: activity of *l*-dopa decarboxylase in discrete brain regions, *Science*, **170**, 1212–1213.

Lovely, R. B., Pagano, R. R., and Paolino, R. M. (1972). Shuttlebox avoidance performance and basal corticosterone levels as a function of duration of individual housing in rats, *J. comp. physiol.*, **81**, 331–335.

Lubin, A. (1967). Performance under sleep loss and fatigue, *Assoc. Res. nerv. ment. Dis.*, **45**, 506–513.

Luby, E. D., Grisell, J. L., Frohman, C. E., Lees, H., Cohen, B. D., and Gottlieb, J. S. (1961). Biochemical, psychological and behavioral responses to sleep deprivation, *Ann. N.Y. Acad. Sci.*, **96**, 71–78.

Luttges, M. (1968). *Electrophysiological dose response effects of megamide and strychnine*, Unpublished doctoral dissertation. University of California, Irvine.

Luttges, M., Johnson, T., Buck, C., Holland, J., and McGaugh, J. (1966). An examination of 'transfer of learning' by nucleic acid, *Science*, **151**, 834–837.

McAfee, D. A., Schonderet, M., and Greengard, P. (1971). Adenosine $3',5'$-monophosphate in nervous tissue: increase associated with synaptic transmission, *Science*, **171**, 1156–1158.

McConnell, J. V., Jacobsen, A. L., and Kimble, D. P. (1959). The effects of regeneration upon retention of a conditioned response in the planarian, *J. comp. physiol. Psychol.,* **52,** 1–5.

McConnell, W. B. (1963). Amphetamine substances in mental illness in Northern Ireland, *Brit. J. Psychiat.,* **109,** 218–224.

McEwen, B. S., Zigmond, R. E., Azmitia, E. C., and Weiss, J. M. (1970). Steroid hormone interaction with specific brain regions. In R. E. Bowman and S. P. Datta (Eds.), *Biochemistry of Brain and Behavior.* New York: Plenum, 123–167.

McGaugh, J. L. (1959). *Some neurochemical factors in learning.* Unpublished doctoral dissertation University of California, Berkley.

McGaugh, J. L. (1966). Time-dependent processes in memory storage, *Science,* **153,** 1351–1358.

McGaugh, J. L. (1968). Drug facilitation of memory and learning. In Efron, D. H. (Ed.), *Psychopharmacology. A Review of Progress.* Washington, D. C. Public Health Service. Publ. No. 1836, 891–904.

McGaugh, J. L., and Dawson, R. G. (1971). Modification of memory storage processes, *Behav. Sci.,* **11,** 45–63.

McGaugh, J. L., and Krivanek, J. (1970). Strychnine effects on discrimination learning in mice. Effects of dose and time of administration, *Physiol. Behav.,* **5,** 1437–1442.

McGaugh, J. L., and Petrinovich, L. (1965). Effects of drugs on learning and memory, *Int. Rev. Neurobiol.,* **8,** 139–191.

McGaugh, J. L., Thomson, C. W., Westbrook, W. H., and Hudspeth, W. J. (1962). A further study of learning facilitation with strychnine sulphate, *Psychopharmacologia (Berl.),* **3,** 352–360.

McGaugh, J. L., Westbrook, W. H., and Burt, G. (1961). Strain differences in the facilitative effects of 5-7-diphenyl-1-3-diazadamantan-6-OL (1757 I. S.) on maze learning, *Psychopharmacologia (Berl,),* **54,** 502–505.

McGeer, P. L., Boulding, J. E., Gibson, W. C., and Foulkes, R. G. (1961). Drug-induced extrapyramidal reactions, *J. amer. med. Assoc.,* **177,** 665–670.

McIlwain, H. (1955). *Biochemistry and the Central Nervous System,* London: Churchill.

McKean, C. M., Schanberg, S. M., and Giarman, N. J. (1962). A mechanism of the indole defect in experimental phenylketonuria, *Science,* **137,** 604–605.

McKennee, C. T., Timiras, P. S., and Quay, W. B. (1968). Concentrations of 5-hydroxy-tryptamine in the rat brain and pineal after adrenelectomy and cortisol administration, *Neuroendocrinology,* **1,** 251–256.

McKinstry, D. N., and Koelle, G. B. (1967). Acetylcholine release from the cat superior cervical ganglion by carbachol, *J. Pharmac. exp. Therap.,* **157,** 319–327.

MacLean, P. D. (1957). Chemical and electrical stimulation of hippocampus in unrestrained animals I and II, *A. M. A. Arch. Neurol. Psychiat.,* **78,** 113–142.

MacLean, P. D., Flanigan, S., Flynn, J. P., Kim, C., and Stevens, J. R. (1956). Hippocampal functions tentative correlations of conditioning E.E.G., drug, and radio-autographic studies, *Yale J. biol. med.,* **28,** 380–395.

McLennan, H. (1964). The release of acetylcholine and of 3-hydroxytyramine from the caudate neucleus, *J. Physiol.,* **174,** 152–161.

McLennan, H., and Elliott, K. A. C. (1951). Effects of convulsant and narcotic drugs on acetylcholine synthesis, *J. Pharmacol. exp. Therap.,* **103,** 35–43.

McLennan, H., and York, D. H. (1966). Cholinergic mechanisms in the caudate nucleus, *J. Physiol. (Lond.),* **187,** 163–175.

McLennan, H., and York, D. H. (1967). The action of dopamine on neurons of the caudate nucleus, *J. Physiol.,* **189,** 393–402.

McNair, D. M., Goldstein, A. P., Lorr, M., Cibelli, L. A., and Roth, I. (1965). Some effects of chlordiazepoxide and meprobamate with psychiatric outpatients, *Psychopharmacologia (Berl.),* **1,** 256–265.

Magoun, H. W. (1950). Caudal and cephalic influences of the brain stem reticular formation, *Physiol. Rev.,* **30,** 459–474.

Mahler, H. R., and Cordes, E. H. (1966). *Biological Chemistry*, Harper Row: New York.

Maickel, R. P., Westermann, E. O., and Brodie, B. B. (1961). Effects of reserpine and cold exposure on pituitary-adrenocortical function in rats, *J. pharmacol. exp. Therap.*, **134**, 167–175.

Malin, D. H., and Guttman, H. N. (1972). Synthetic rat scotophobin induces dark avoidance in mice, *Science*, **178**, 1219–1220.

Mandel, P., Rein, H., Harth-Edel, S., and Mardell, R. (1964). Distribution and metabolism of ribonucleic acid in the vertebrate central nervous system. In D. Richter (Ed.), *Comparative Neurochemistry*. New York: Pergamon, 23–41.

Mandell, A. J. (1970). Drug induced alterations in brain biosynthetic enzyme activity—a mode for adaptation to the environment by the central nervous system. In R. E. Bowman and S. Datta (Eds.), *Biochemistry of brain and behavior*. New York: Plenum Press, 97–121.

Mandell, A. J., and Mandell, M. P. (1967). Suicide and the menstrual cycle, *J. Am. med. Assoc.*, **200**, 792–793.

Mandell, A. J., and Morgan, M. (1971). Indole(ethyl)amine *N*-methyltransferase in human brain, *Nature, New Biol.*, **230**, 85–87.

Mandell, A. J., and Spooner, C. E. (1968). Psychochemical research studies in man, *Science*, **162**, 1442–1453.

Mandell, M. P., Mandell, A. J., and Jacobson, A. (1965). Biochemical and neurophysiological studies of paradoxical sleep, *Rec. Advanc. biol. Psychiat.*, **7**, 115–122.

Mandler, G. (1967). The conditions for emotional behavior. In D. C. Glass (Ed.), *Neurophysiology and Emotion*. New York: Rockefeller, 96–102.

Marantz, R., and Rechtschaffen, A. (1967). Effect of alpha-methyltyrosine on sleep in the rat, *Perc. mot. Skills*, **25**, 805–808.

Marczynski, T. J. (1969). Postreinforcement synchronization and the cholinergic system, *Fed. Proc.*, **28**, 132–134.

Marczynski, T. J. (1971). Cholinergic mechanism determines the occurrence of reward contingent positive variation (RCPV) in cat, *Brain Res.*, **28**, 71–83.

Marczynski, T. J., and Hackett, J. T. (1969). Post-reinforcement electrocortical synchronization and facilitation of cortical somato-sensory evoked potentials during instrumentally conditioned appetitive behavior in the cat, *E.E.G. Clin. Neurophysiol.*, **26**, 41–49.

Marczynski, T. J., Hackett, J. T., Sherry, C. J., and Allen, S. L. (1971). Diffuse light input and quality of reward determine the occurrence of 'reward contingent positive variation' (RCPV) in cat, *Brain Res.*, **28**, 57–70.

Margules, D. L., and Stein, L. (1968). Increase of anxiety activity and tolerance of behavioural depression during chronic administration of oxazepam, *Psychopharmacologia (Berl.)*, **13**. 74–80.

Marshall, N. B., Barnett, R. J., and Mayer, J. (1955). Hypothalamic lesions in goldthioglucose injected mice, *Proc. Soc. exp. Biol. Med.*, **90**, 240–246.

Martin, W. R. (1970). Pharmacological redundancy as an adaptive mechanism in the central nervous system, *Fed. Proc.*, **29**, 13–18.

Masaki, T. (1956). The amphetamine problem in Japan, *W.H.O. Tech. Ref. Ser.*, **102**, 14–21.

Mason, J., Brady, J. P., and Sidman, M. (1957). Plasma 17-hydroxycorticosteroid levels and conditioned behavior in the rhesus monkey, *Endocrinology*, **60**, 741–752.

Masters, R. E. L., and Huston, J. (1966). *The Varieties of Psychedelic Experiences*. New York: Rinehart and Winston.

Matsumoto, J., and Jouvet. M. (1964). Effets de reserpine, DOPA et 5HTP sur les deux etats de sommeil, *Compt. Rend. Soc. Biol.*, **158**, 2137–2140.

Mayer, J. (1953). Glucostatic mechanisms of regulation of food intake, *New Engl. J. Med.*, **249**, 13–16.

Mayer, J. (1955). Regulation of energy intake and the body weight. The glucostatic theory and the lipostatic hypothesis, *Ann. N.Y. Acad. Sci.*, **63**, 15–43.

Mayer, J., and Thomas, D. W. (1967). Regulation of Food Intake and Obesity, *Science*, **156,** 328–337.

Mayer-Gross, W., Slater, E., and Roth, M. (1954). *Clinical Psychiatry*, London: Cassell.

Maynert, E. W., and Levi, R. (1964). Stress-induced release of brain norepinephrine and its inhibition by drugs, *J. pharmacol. exp. therap.*, **143,** 90–95.

Mechoulam, R. (1970). Marijuana chemistry, *Science*, **168,** 1159–1166.

Medical Research Council (1965). Clinical trial of the treatment of depressive illness, *Brit. med. J.*, **1,** 881–886.

Meeter, E. (1969). Desensitization of the end-plate membrane following cholinesterase inhibition an adjustment to a new working situation, *Acta physiol. Pharmacol. neerland.*, **15,** 243–258.

Melges, F. T., Tinklenberg, J. R., Hollister, L. E., and Gillespie, H. K. (1970). Marijuana and temporal disintegration, *Science*, **168,** 1118–1120.

Mello, N. K., and Mendelson, J. H. (1971). Experimentally induced intoxication in alcoholics: a comparison between programmed and spontaneous drinking. In Y. Israel (Ed.) and J. Mardones, *Biological Basis of Alcoholism*. New York: Wiley, 271–298.

Mendelson, J. H., and Mello, N. K. (1970). Mechanisms of physical dependence in alcoholism. In R. E. Popham (Ed.), *Alcohol and Alcoholism*. Toronto: University of Toronto Press, 126–129.

Meselson, M. S. (1970). Chemical and biological weapons, *Sci. Amer.*, **222,**(5), 15–25.

Metz, B. (1958). Brain acetylcholinesterase and a respiratory reflex, *Amer. J. Physiol.*, **192,** 101–105.

Meyers, B., and Domino, E. F. (1964). The effect of cholinergic blocking drugs on spontaneous alternation in rats, *Arch. Int. Pharmadynam. Therap.*, **150,** 525–529.

Meyerson, B. J., and Sawyer, C. H. (1968). Monoamines and ovulation in the rat, *Endocrinology*, **83,** 170–176.

Michelson, M. J. (1961). Pharmacological evidences of the role of acetylcholine in the higher nervous activity of man and animals, *Activas Nerv. Sup. (Prague)*, **3,** 140–147.

Miles, S. (1955). *Some effects of injection on atropine sulphate in healthy young men*. U.K. Ministry of Defence Unpublished Report.

Miller, N. E. (1960). Motivational effects of brain stimulation and drugs. *Fed. Proc.*, **19,** 846–853.

Miller, N. E. (1961). Some recent studies of conflict behavior and drugs, *Amer. Psychologist*, **16,** 12–24.

Miller, N. E. (1963). Some reflections on the law of effect produce a new alternative to drive reduction, *Nebraska Symp. Motiv.*, **11,** 65–112.

Miller, N. E. (1965). Chemical coding of behavior in the brain, *Science*, **148,** 328–338.

Miller, N. E., Gottesman, Kay, S., and Emery, N. (1964). Dose response to carbachol and norepinephrine in rat hypothalamus, *Amer. J. Physiol.*, **206,** 1384–1388.

Miller, R., and Ogawa, N. (1962). The effect of adrenocorticotrophic hormone (ACTH) on avoidance conditioning in the adrenalectomised rat, *J. comp. physiol. Psychol.*, **55,** 211–213.

Millman, D. H. (1967). An untoward reaction to accidental ingestion of LSD in a 5-year old girl, *J. amer. med. Assoc.*, **201,** 821.

Millman, D. H., and Anker, J. L. (1972). Patterns of drug usage among university student's multiple drug usage. In Wolfram Keup (Ed.), *Drug Abuse*. Springfield, Ill.: Charles C. Thomas, 190–201.

Mishkin, M. (1964). Preseveration of central sets after frontal lesions in monkeys. In J. M. Warren and K. Akert (Eds.), *The Frontal Granular Cortex and Behavior*. New York: McGraw-Hill, 219–241.

Mitoma, C., Auld, R. M., and Udenfriend, S. (1957). On the nature of enzymatic defect in phenylpyruric oligophrenia, *Proc. Soc. expt. Biol. Med.* , **94,** 632–638.

Mizner, G. L., Barter, J. T., and Wermer, P. H. (1970). Patterns of drug use among college students: a preliminary report, *Amer. J. Psychiat.*, **127,** 15–24.

Mönckeberg, F. (1968). Effect of early marasmic malnutrition on subsequent physical and psychological development. In Scrimshaw N. S. and Gordon, J. E. (Eds.), *Malnutrition, Learning and Behavior:* Cambridge, Mass. and London, England: M.I.T. Press, 269–278.

Monnier, M., and Hosli, L. (1964). Dialysis of sleep and waking factors in blood of the rabbit, *Science*, **146**, 796–797.

Moore, B. W., and McGregor, D. (1965). Chromatographic and electrophoretic fractionation of soluble proteins of brain and liver, *J. biol. Chem.*, **240**, 1258–1266.

Moore, R. Y., Bhatnagar, R. K., and Heller, A. (1971). Anatomical and chemical studies of a nigro-striatal projection in the cat, *Brain Res.*, **30**, 119–135.

Morden, B., Conner, R., Mitchell, G., Dement, W., and Levine, S. (1968). Effects of rapid eye movement (REM) sleep deprivation on shock induced fighting, *Physiol. Behav.*, **3**, 425–432.

Morgan, C. T. (1965). *Physiological Psychology*, New York: McGraw-Hill.

Morgan, M., and Mandell, A. J. (1969). Indole(ethyl)amine *N*-Methyltransferase in the brain, *Science*, **165**, 492–493.

Morgane, P. (1964). Limbic-hypothalamic-midbrain interaction in thirst and thirst-motivated behavior. In M. J. Wayner (ed.), *Thirst*, Pergamon: London, 94–126.

Morrell, F. (1964). Modification of RNA as a result of neural activity. In M. A. B. Brazier (ed.), *Brain Function*, Vol. 2. *RNA and Brain Function*, Univ. of California: Berkeley, California, 183–202.

Moruzzi, G. (1964). Reticular influences on the EEG, *EEG. clin. Neurophysiol.*, **16**, 2–17.

Moruzzi, G., and Magoun, H. W. (1949). Brain stem reticular formation and activation of the EEG, *Electroencephalog clin. Neurophysiol.*, **1**, 455–473.

Mouret, J., Bobillier, P., and Jouvet, M. (1968). Insomnia following para-cholorophenyl-alanine in the rat, *Europ. J. Pharmacol.* **5**, 17–22.

Myers, R. D., and Cicero, T. J. (1968). Are the cerebral ventricles involved in thirst produced by a cholinergic substance, *Psychon. Sci.*, **10**, 93–94.

Nakamura, C. Y., and Anderson, N. H. (1962). Avoidance of behaviour differences within and between strains of rats, *J. comp. physiol. Psychol.*, **55**, 740–747.

Nichols, J. R. (1963). A procedure which produces sustained opiate-directed behavior (morphine addiction) in the rat, *Psychol. Rep.*, **13**, 895–904.

Nichols, J. R., and Hsiao, S. (1967). Addiction liability of albino rats: breeding for quantitative differences in morphine drinking, *Science*, **157**, 561–563.

Nicolaïdas, P. (1969). Early systemic responses to oro-gastric stimulation in the regulation of food and water balance. Functional and electrophysiological data, *Ann. N.Y. Acad. Sci.* , **159**, 1176–1203.

Nissen, Th., Røigaard-Petersen, H. H., and Fjerdingstad, F. J. (1965). Effect of ribonucleic acid (RNA) extracted from the brain of trained animals on learning in rats (11). Dependence of RNA effect on training conditions prior to RNA extraction, *Scand. J. Psychol.*, **6**, 265–272.

Norberg, K. A. (1965). Drug-induced changes in monoamine levels in the sympathetic adrenergic ganglion cells and terminals, *Acta physiol. scand.*, **65**, 221–234.

Nyswander, M., Winick, C., Bernstein, A., Brill, L., and Kaufer, G. (1958). Treatment of the narcotic addict, Workshop, 1957; the treatment of drug addicts as voluntary outpatients; a progress report, *Amer. J. Orthopsychiat.*, **28**, 714–727.

Ochs, S. (1972). Fast transport of materials in mammalian nerve fibres, *Science*, **176**, 252–259.

Olds, J. (1958). Effects of hunger and male hormones on self stimulation of the brain, *J. comp . physiol. Psychol.*, **51**, 320–324.

Olds, J. (1962). Hypothalamic substrates of reward, *Physiol. Rev.*, **42**, 554–604.

Olds, J., and Olds, M. E. (1964). The mechanisms of voluntary behavior. In R. G. Heath (Ed.), *The Role of Pleasure in Behavior*. New York: Harper and Row, 23–54.

Oliverio, S. (1968). Effects of scopolamine on avoidance conditioning and habituation of mice, *Psychopharmacologia (Berl.)*, **12**, 214–226.

242

Osmond, H., and Smythies, J. R. (1952). Schizophrenia: A new approach, *J. ment. Sci.*, **98**, 309–315.

Ostfeld, A. M., and Aruguete, A. (1962). Central nervous system effects of hyoscine in man, *J. Pharmacol.*, **137**, 133–139.

Ostfeld, A. M., Machne, X., and Unna, K. R. (1960). The effects of atropine on the electroencephalogram and behaviour in man, *J. Pharmacol.*, **128**, 265–272.

Oswald, I. (1962). *Sleeping and Waking*. Amsterdam: Elsevier.

Oswald, I. (1968). Drugs and Sleep, *Pharmacol. Rev.*, U.S.A., **20**, 273–297.

Oswald, I., and Thacore, V. R. (1963). Amphetamine and phenmetrazine addiction, *Brit. med. J.*, **11**, 427–431.

Othmer, E., Hayden, M. P., and Segelbaum, R. (1969). Encephalic cycles during sleep and wakefulness in humans: a 24 hour pattern, *Science*, **164**, 447–449.

Overall, J. E., Hollister, L. E., Johnson, M., and Pennington, V. (1965). Computer classification of depressions and differential effects of anti-depressants, *J. amer. med. Assoc.*, **192**, 561.

Overall, J. E., Hollister, L. E., Meyer, F., Kimbell, I., and Shelton, J. (1964). Impramine and thioridazine in depressed and schizophrenic patients. Are there specific anti-depressant drugs?, *J. amer. med. Assoc.*, **189**, 605–608.

Overton, D. A. (1964). State-dependent or 'dissociated' learning produced with pentobarbital, *J. comp. physiol. Psychol.*, **57**, 3–12.

Overton, D. A. (1966). State-dependent learning produced by depressant and atropine like drugs, *Psychopharmacologia (Berl.)*, **10**, 6–31.

Pagano, R. P., and Lovely, R. H. (1972). Diurnal cycle and ACTH facilitation of shuttlebox avoidance, *Physiol. Behav.*, **8**, 721–723.

Palaič, D., Page, I. H., and Khairallak, P. A. (1967). Uptake and metabolism of ^{14}C serotonin in rat brain, *J. Neurochem.*, **14**, 63–69.

Paré, C. M. B. (1968). Recent advances in the treatment of depression. In A. Coppen and A. Walk (Eds.), *Recent developments in affective disorders, Brit. J. Psychiat. Spec. Publ.*, **2**, 137–149.

Paré, C. M. B., Rees, L., and Sainsbury, M. J. (1962). Differentiation of two genetically specific types of depression by the response to antidepressants, *Lancet*, **2**, 1340–1343.

Paré, C. M. B., and Sandler, M. A. (1959). A clinical and biochemical study of a trial of iproniazid in the treatment of depression, *J. neurol. neurosurg. Psychiat.*, **22**, 247–251.

Parry, H. (1968). Use of psychotropic drugs by U.S. adults. *Public Health Reports*, **83**, 799–810.

Pazzagli, A., and Pepeu, G. (1964). Amnesic properties of scopolamine and brain acetylcholine in the rat, *Int. J. Neuropharmacol.*, **4**, 291–299.

Pearlman, C. A., Sharpless, S. K., and Jarvik, M. E. (1961). Retrograde amnesia produced by anaesthetic and convulsant agents, *J. comp. Physiol. Psychol.*, **54**, 109.

Persky, H., Smith, K. D., and Basu, G. K. (1971). Effect of corticosterone and hydrocortisone on some indicators of anxiety, *J. clin. Endocrinol. Metab.*, **33**, 467–474.

Pfaff, D. W., Silva, M. T. A., and Weiss, J. M. (1971). Telemetered recording of hormone effects on hippocampal neurons, *Science*, **172**, 394–395.

Pfeiffer, C. C., and Jenney, E. H. (1957). The inhibition of the conditioned response and the counteraction of schizophrenia by muscarinic stimulation of the brain, *Ann. N.Y. Acad. Sci.*, **66**, 753–764.

Phillis, J. W. (1970). *The Pharmacology of Synapses*. London: Pergamon.

Pletscher, A., Brossi, A., and Gey, K. F. (1962). Benzoquinolizine derivatives: a new class of monoamine decreasing drugs with psychotropic action, *Int. Rev. Neurobiol.*, **4**, 275–306.

Plotnikoff, N. (1966). Magnesium pemoline: enhancement of learning and memory of a conditioned avoidance response, *Science* , **151**, 703–704.

Poirier, L. J., and Sourkes, T. L. (1965). Influence of the substantia nigra on the catecholamine content of the striatum, *Brain*, **88**, 181–192.

Polidora, V. J., Cunningham, R. F., and Waisman, H. A. (1966). Dosage parameters of a behavioral deficit associated with phenylketonuria in rats, *J. comp. physiol.*, **61**, 436–441.

Polley, E. H., Vick, J. A., Cinchta, H. P., Fischetti, D. A., Macchitelli, F. J., and Montarelli, N. (1965). Botulium toxin, type A: effect on central nervous system, *Science*, **147**, 1036–1037.

Pollin, W., Cardon, P. V., and Kety, S. S. (1961). Effects of amino acid feedings in schizophrenic patients treated with iproniazid, *Science*, **133**, 104–105.

Popham, R. E. (1970). Indirect methods of alcoholism prevalence estimation: a critical evaluation. In R. E. Popham (Ed.), *Alcohol and Alcoholism*. Toronto: University of Toronto Press, 294–306.

Poser, C. M., and Bogaest, L. V. (1959). Neuropathologic observations in phenylketonuria, *Brain*, **82**, 1–9.

Potter, L. T., Glover, V. A. S., and Saelens, J. K. (1968). Choline acetyltransferase from rat brain, *J. Biol. Chem.*, **243**, 3864–3870.

Prescott, R. G. W. (1966). Estrous cycle in the rat: effects on self-stimulation behaviour, *Science*, **152**, 796–797.

Pribram, K. H. (1967). The new neurology and biology of emotion: a structural approach, *Amer. Psychol.*, **22**, 830–838.

Pribram, K. H. (1969). The neurobehavioral analysis of the limbic forebrain mechanisms: revision and progress report. In D. Lehrman (Ed.), *Advances in the Study of Behavior*, Vol. 2. New York: Academic, 297–332.

Proctor, R. C. (1962). Clinical use of chlordiazepoxide. In J. H. Nodine and J. H. Mayer (Eds.), *Psychosomatic Medicine*. Philadelphia: Lea and Febiger, 480–488.

Pujol, J. F., Mouret, J., Jouvet, M., and Glowinski, J. (1968). Increased turnover of cerebral norepinephrine during rebound of paradoxical sleep in the rat, *Science*, **159**, 112–114.

Purpura, D. P., Shofer, R. J., Housepian, E. M., and Nobach, C. R. (1964). Comparative ontogenesis of structure-function relations in cerebral and cerebellar cortex, *Progr. Res.*, **4**, 187–221.

Rafaelson, O. J., Bech, P., Christiansen, J., Christrup, H., Nyboe, J., and Rafaelson, L. (1973). Cannabis and alcohol: effects on simulated car driving, *Science*, **179**, 920–923.

Raisman, G., Cowan, W. M., and Powell, T. P. S. (1966). An experimental analysis of the efferent projection of the hippocampus, *Brain*, **89**, 83–108.

Randall, L. O., Schallek, W., Heise, G. A., Keith, E. F., and Bagdon, R. E. (1960). The psychosedative properties of methaminodiazepoxide, *J. pharm. exp. Therap.*, **129**, 163–171.

Ratner, A., and McCann, S. M. (1971). Effect of reserpine on plasma LH levels in ovariectomised and cycling proestrous rats, *Proc. Soc. Biol. Med.*, **138**, 763–767.

Ray, O. S. (1972). *Drugs, Society and Human Behavior*. St. Louis, Miss: C. V. Mosby.

Rechtschaffen, A., Cornwall, P., and Zimmerman, W. (1965). Brain temperature variations with paradoxical sleep in the cat, *Assoc. Physiol Study Sleep:* Washington, D.C.

Rechtschaffen, A., and Maron, L. (1964). The effect of amphetamine on the sleep cycle. *E.E.G. clin. Neurophysiol.*, **16**, 438–445.

Redlich, F. C., Ravitz, L. J., and Dession, G. H. (1951). Narcoanalysis and truth, *Amer. J. Psychiat.*, **107**, 586–593.

Rethy, C. R., Smith, C. B., and Villarreal, J. E. (1971). Effects of narcotic analgesics upon the locomotor activity and brain catecholamine content of the mouse, *J. Pharmacol. exp. Therap.*, **176**, 472–479.

Reynolds, R. W. (1965). An irritative hypothesis concerning the hypothalamic regulation of food intake, *Psychol. Rev.*, **72**, 105–116.

Richter, C. P. (1927). Animal behavior and internal drives, *Quart. Rev. Biol.*, **2**, 307–343.

Richter, C. P. (1942–43). Total self-regulating functions in animals and human beings, *Harvey Lect.*, **38**, 63–103.

Rickels, K., Clark, T. W., Ewing, J. H., Klingensmith, W. C., Morris, H. M., and Smock,

C. D. (1959). Evaluation of tranquillizing drugs in medical outpatients, *J. Amer. med. Assoc.*, **171**, 1659–1656.

Rinaldi, R., and Himwich, H. E. (1955). A cholinergic mechanism involved in the function of the mesodiencephalic activating system, *Arch. Neurol. Psychiat.*, **173**, 396–402.

Roberts, E. (1966). Models for correlative thinking about brain, behavior and biochemistry, *Brain*, **2**, 109–144.

Roberts, J. M. (1959). Prognostic factors in electroshock treatment of depressive states. Application of specific tests, *J. ment. Sci.*, **105**, 703–713.

Robertson, J. D. (1970). The ultrastructure of synapses. In F. O. Schmitt (Ed.), *The Neurosciences. Second Study Program.* New York: The Rockefeller University Press, 715–728.

Rockwell, D. (1968). Amphetamine use and abuse in psychiatric patients. *Arch. gen. Psychiat.*, **18**, 612–616.

Roderick, T. H. (1960). Selection for cholinesterase activity in the cerebral cortex of the rat, *Genetics*, **45**, 1123–1140.

Roffwarg, H., Muzzio, J., and Dement, W. (1966). The ontogenetic development of the sleep-dream cycle in the human, *Science*, **152**, 604–619.

Roos, B. E., and Steg, G. (1964). The effect of L-3, 4-dihydroxyphenylalanine and OL-5-hydroxytryptophan on rigidity and tremor induced by reserpine, chlorpromazine and phenoxybenzamine, *Life Sci.*, **3**, 351–360.

Rosenblatt, F. (1970). Induction of behavior by mammalian brain extracts. Georges Ungar (Ed.), *Molecular Mechanisms in Memory and Learning.* London: Plenum Press, 103–147.

Rosenblatt, F., Farrow, J. T., and Herblin, W. F. (1966). Transfer of conditioned responses from trained rats to untrained rats by means of a brain extract, *Nature*, **209**, 46–48.

Rosenblatt, F. R., Farrow, J. T., and Rhine, S. (1966). The transfer of learned behavior from trained to untrained rats by means of a brain extract. I and II., *Proc. Nat. Acad. Sci.*, **55**, 548–555, 787–792.

Rosenblatt, F., and Miller, R. G. (1966). Behavioral assay procedures for transfer of learned behavior by brain extracts, *Proc. nat. Acad. Sci.*, **56**, 1423–1430.

Rosenzweig, M. R. (1966). Changes in brain chemistry as consequences of differential experience. In Symposium 7, *Biochemical Basis of Behavior.* Proceedings of 18th International Congress of Psychology, Moscow.

Rosenzweig, M. R., Krech, D., and Bennett, E. L. (1960). A search for relations between brain chemistry and behavior, *Psychol. Bull.*, **57**, 476–492.

Rosenzweig, M. R., Krech, D., and Bennett, E. L. (1964). Cerebral effects of environmental complexity and training among adult rats, *J. comp. Physiol. Psychol.*, **57**, 438–439.

Rosenzweig, M. R., Krech, D., Bennett, E. L., and Diamond, M. C. (1962). Effects of environmental complexity and training on brain chemistry and anatomy. A replication and extension, *J. comp. physiol. Psychol.*, **55**, 429–437.

Rosenzweig, M. R., Krech, D., Bennett, E. L., and Diamond, M. C. (1968). Modifying brain chemistry and anatomy by enrichment or impoverishment of experience. In G. Newton and S. Levine (Eds.), *Early Experience and Behavior.* Springfield, Illinois: Charles Thomas, 258–298.

Rosvold, H. E. (1968). The prefrontal cortex and caudate nucleus: a system for effecting correction in response mechanisms. In C. Rupp (Ed.), *Mind as Tissue.* New York: Hocher Medical, Harper and Row, 21–38.

Roueché, B. (1965). Annals of medicine: Something a little unusual, *New Yorker*, May, 15th, 180–198.

Routtenberg, A. (1967). Drinking induced by carbachol: thirst circuit or ventricular modification?, *Science*, **157**, 838–839.

Routtenberg, A. (1968a). The two-arousal hypothesis: reticular formation function and limbic system, *Psychol. Rev.*, 51–80.

Routtenberg, A. (1968b). Hippocampal correlates of consummatory and observed behavior, *Physiol. Behav.*, **3**, 533–535.

Rowntree, D. W., Nevin, S., and Wilson, A. (1950). The effects of diisopropylfluoro-phosphonate in schizophrenia and manic depressive psychosis, *J. Neurol. Neurosurg. Psychiat.*, **13**, 47–62.

Rozin, P. (1968). Are carbohydrate and protein intakes separately regulated?, *J. comp. physiol. Psychol.*, **65**, 1, 23–29.

Rubin, R. T., Mandell, A. J., and Crandall, P. H. (1966). Corticosteroid responses to limbic stimulation in man: localization of stimulus, *Science*, **153**, 767–768.

Ruch, T. C., Patton, H. O., Woodbury, J. W., and Towe, A. L. (1965). *Neurophysiology*, New York: Saunders.

Ruf, K., and Steiner, F. A. (1967). Steroid-sensitive single neurons in rat hypothalamus and midbrain: identification by microelectrophoresis, *Science*, **156**, 667–668.

Russell, R. W. (1960). Drugs as tools in behavioural research. In L. Uhr and J. G. Miller (Eds.), *Drugs and Behavior*. New York: Wiley, 19–40.

Russell, R. W. (1964). Psychopharmacology, *Ann. Rev. Psychol.*, **15**, 87–114.

Russell, R. W. (1966). Biochemical substrates of behavior. In Russell, R. W. (Ed.), *Frontiers in Physiological Psychology*. New York: Academic, 185–246.

Russell, R. W., Singer, G., Flanagan, F., Stone, M., and Russell, J. W. (1968). Quantitative relations in amygdaloid modulation of drinking, *Physiol. Behav.*, **3**, 871–875.

Russell, R. W., Vasquez, B. J., Overstreet, D. H., and Dalglish, F. W. (1971a). Effect of cholinolytic agents on behavior following development of tolerance to low cholinesterase activity, *Psychopharmacol. (Berl.)*, **20**, 32–41.

Russell, R. W., Vasquez, B. J., Overstreet, D. H., and Dalglish, F. W. (1971b). Consummatory behavior during tolerance to and withdrawal from chronic depression of cholinesterase activity, *Physiol. Behav.*, **7**, 523–528.

Russell, R. W., and Warburton, D. M. (1973). Biochemical correlates of behavior. In B. Wolman (Ed.), *Handbook of General Psychology*. New York: Prentice-Hall, 165–187.

Russell, R. W., Warburton, D. M., and Segal, D. S. (1969). Behavioral tolerance during chronic changes in the cholinergic system, *Comm. behav. biol.*, **4**, 121–128.

Russell, R. W., Warburton, D. M., Vasquez, B. J., Overstreet, D. H., and Dalglish, F. W. (1971a). Acquisition of new responses by rats during chronic depression of acetyl-cholinesterase activity, *J. comp. physiol. Psychol.*, **77**, 228–233.

Russell, R. W., Watson, R. H. J., and Frankenhaeuser, M. (1961). Effects of chronic reductions in brain cholinesterase activity on acquisition and extinction of a conditioned avoidance response, *Scand. J. Psychol.*, **2**, 21–29.

Russell, W. R., and Newcombe, F. (1966). Contribution from clinical neurology. In D. Richter (Ed.), *Aspects of Learning and Memory*. New York: Basic, 15–24.

Saavedra, J. M., and Axelrod, J. (1972). Psychomimetic *N*-methylated tryptamines formation in brain in vivo and in vitro, *Science*, **175**, 1365–1366.

Sachar, E. J. (1970). Psychological factors relating to activation and inhibition of the adrenocortical stress response in man: a review, *Prog. Brain Res.*, **32**, 316–324.

Sandison, R. A. (1954). Psychological aspects of LSD, treatment of neuroses, *J. ment. Sci.*, **100**, 508–515.

Sargant, W., and Dally, P. (1962). Treatment of anxiety states by antidepressant drugs, *Brit. Med. J.*, **1**, 6–9.

Sassin, J. F., Parker, D. C., Mace, J. W., Gotlin, R. E. W., Johnson, L. C., and Rossman, L. G. (1969). Human hormone release: relation to slow-wave sleep and sleep-waking cycles, *Science*, **165**, 513–515.

Savage, C., Jackson, D., Terrill, J. (1962). LSD transcendence and the new beginning, *J. nerv. and ment. Dis.*, **135**, 5, 425–439.

Scapagnini, U., Moberg, G. P., Van Loon, G. R., De Groot, J., and Ganong, W. F. (1971). Relation of brain 5-hydroxytryptamine content to the diurnal variation in plasma corticosterone in the rat, *Neuroendocrinology*, **7**, 90–96.

Scapagnini, U., Van Loon, G. R., Moberg, G. P., and Ganong, W. F. (1970). Effect of a methyl *p*-tyrosine on the circadian variation of plasma corticosterone in rats, *Europ. J. Pharmacol.* **11**, 266–268.

246

Schachter, S. (1968). Obesity and eating. Internal and external cues differentially affect the eating behavior of obese and normal subjects, *Science*, **161**, 751–756.

Schallek, W., Kuehn, A., and Jew, N. (1962). Effects of chlordiazepoxide (Librium) and other psychotropic agents on the brain stem of the brain, *Ann. N.Y. Acad. Sci.*, **96**, 303–312.

Scheckel, C. L., and Boff, E. (1964). Behavioral effects of interacting imipramine and other drugs with *d*-amphetamine, cocaine and tetrabenazine, *Psychopharmacologia (Berl.)*, **5**, 198–208.

Scheckel, C. L., and Boff, E. (1966). Pharmacological control of behavior. In *Proceedings XVIII Int. Congress of Psychology Symposium 7, Moscow.*

Scheckel, C. L., Boff, E., Dahlen, P., and Smart, T. (1968). Behavioral effects in monkeys of racemates of two biologically active marijuana constituents, *Science*, **160**, 1467–1469.

Schiff, B. B. (1964). The effects of tegmental lesions on the reward properties of septal stimulation, *Psychonom. Sci.*, **1**, 397–398.

Schildkraut, J. J. (1967). *Neuropsychopharmacology and the Affective Disorders*. Boston: Little Brown.

Schildkraut, J. J. (1969). Rationale of some approaches used in biochemical studies of the affective disorders: the pharmacological bridge. In A. J. Mandell and M. P. Mandell (Eds.), *Psychochemical Research in Man*. New York: Academic, 113–126.

Schildkraut, J. J., and Kety, S. S. (1967). Biogenic amines and emotion, *Science*, **156**, 21–30.

Schmitt, F. O. (1962). In F. O. Schmitt (Ed.), *Macromolecular Specificity and Biological Memory*. Cambridge. Mass: M.I.T. Press, 1–6.

Schoemaker, W. J., and Wurtman, R. J. (1971). Perinatal undernutrition: accumulation of catecholamines in rat brain, *Science*, **171**, 1017–1019.

Schreiner, L. H., and Kling, A. (1953). Behavioral changes following rhinencephalic injury in the cat, *J. Neurophysiol.*, **16**, 643–659.

Schuberth, J., and Sundwall, A. (1967). Effects of some drugs on the uptake of acetylcholine in cortex slices of mouse brain, *J. Neurochem.*, **14**, 807–812.

Schultes, R. E. (1969). Hallucinogens of plant origin, *Science*, **163**, 245–254.

Scott, E. M., and Quint, E. (1946). Self-selection of diet. The effect of flavor, *J. Nutr.*, **32**, 113–119.

Scott, P. D., and Bucknell, M. (1971). Delinquency and amphetamines, *Brit. J. Psychiat.*, **119**, 179–182.

Scott, P., and Willcox, D. (1965). Delinquency and the amphetamines, *Brit. J. Psychiat.*, **111**, 865–875.

Scrimshaw, N. S., and Gordon, J. E. (1968). *Malnutrition, Learning and Behavior*. In Scrimshaw and Gordon (Eds.), Cambridge: Mass: M.I.T. Press.

Seevers, M., and Deneau, G. A. (1963). Physiological aspects of tolerance and physical dependence. In W. S. Root and F. G. Hofman (Eds.), *Physiological Pharmacology*, Vol. 1. New York. Academic, 565–640.

Segal, D. S., Cox, R. H., Stern, W. C., and Maickel, R. P. (1967). Stimulatory effects of pemoline and cyclopropylpemoline on continuous avoidance behaviour. Similarity to effects of D-amphetamine, *Life Sci.*, **6**, 2567–2572.

Segal, D. S., and Mandell, A. J. (1969). Behavioural activation of rats during intraventricular infusion of norepinephrine, *Proc. nat. Acad. Sci.*, **66**, 289–293.

Selling, L. S. (1955). Clinical study of a new tranquillizing drug, *J. Amer. med. Assoc.*, **159**, 1584–1596.

Sharma, K. N., and Nasset, E. S. (1962). Electrical activity in mesenteric nerves after perfusion of gut limen, *Amer. J. Physiol.*, **202**, 725–730.

Shepherd, M., Lader, M., and Rodnight, R. (1963). *Clinical Psychopharmacology*. London: English Universities.

Sherwood, S. L. (1952). Intraventricular medication in catatonic stupor, *Brain*, **75**, 68–75.

Sherwood, S. L. (1958). Consciousness, adaptive behaviour and schizophrenia. In D. Richter (Ed.), *Schizophrenia: somatic aspects*. London: Pergamon, 131–146.

Shute, C. C., and Lewis, P. R. (1967). The ascending cholinergic reticular system; neocortical, olfactory and subcortical projections, *Brain*, **90**, 497–520.

Sidman, M. (1953). Two temporal parameters of the maintenance of avoidance behavior by the white rat, *J. comp. physiol. Psychol.*, **46**, 253–261.

Siegal, P. S., and Sterling, T. D. (1959). The anorexigenic action of dextro-amphetamine sulfate upon feeding responses of differing strength, *J. comp. physiol. Psychol.*, **52**, 179–182.

Sim, J. (1963). *Guide to Psychiatry.* London: Livingstone.

Simpson, J. B., and Routtenberg, A. (1973). Subfornical organ: site of drinking elicitation by angiotensin, *Science*, **181**, 1172–1174.

Singer, G., and Montgomery, R. B. (1968). Neurohumoral interaction in the rat amygdala after central chemical stimulation, *Science*, **160**, 1017–1018.

Sjärne, L., Hedquist, P., and Bygdeman, S. (1969). Neurotransmitter quantum released from sympathetic nerves in cats' skeletal muscle, *Life Sci.*, **8**, 189–196.

Sjoerdsma, A. (1965) Conjoint clinic on serotonin, norepinephrine and tyramine. *J. Chron. Dis.*, **18**, 429–441.

Slangen, J. L., and Miller, N. E. (1969). Pharmacological tests for the function of hypothalamic norepinephrine in eating behavior, *Physiol. Behavior*, **4**, 543–552.

Smart, R. G., and Bateman, K. (1967). Unfavorable reactions to LSD: a review and analysis of the available case reports, *Canad. med. Assoc. J.*, **99**, 805–810.

Smart, R. G., and Fejer, D. (1972). Relationships between parental and adolescent drug use. In Wolfram Keup (Ed.), *Drug Abuse.* Springfield: Charles C. Thomas, 146–153.

Smirnov, E. D., and Il'yutchenok, R. Yu. (1962). Cholinergic mechanism of cortical activation, *Sechenov. Physiol. J. USSR*, **49**, 127–139.

Smith, C. B. (1965). Effects of *d*-amphetamine upon brain amine content and locomotor activity in mice, *J. pharmacol. exp. Ther.*, **147**, 96–102.

Smith, C. B., Villarreal, J. E., Bednarczyk, J. H., Sheldon, M. I. (1970). Tolerance to morphine induced increases in (^{14}C) catecholamine synthesis in mouse brain, *Science*, **170**, 1106–1107.

Smith, D. E., and Rose, A. J. (1968). The use and abuse of LSD in Haight-Ashbury, *Clin. Pediat.*, **7**, 317–322.

Smith, G. M., and Beecher, H. K. (1959). Amphetamine sulfate and athletic performance: I Objective effects, *J. am. med. Assoc.*, **170**, 542–557.

Smith, G. M., and Beecher, H. K. (1960a). Amphetamine, secorbarbital and athletic performance: II Subjective evaluations of performances, mood states and physical states, *J. amer. med. Assoc.*, **172**, 1502–1514.

Smith, G. M., and Beecher, H. K. (1960b). Amphetamine, secobarbital and athletic performance: III Quantitative effects on judgment, *J. amer. med. Assoc.* **172**, 1623–1629.

Smith, S., and Blackley, P. (1966). Amphetamine usage by medical students, *J. med. Educ.*, **41**, 167–170.

Smythies, J. R., Bradley, R. J., Johnston, V. S., Benington, F., Morin, R. D., and Clark, L. C. (1967). Structure-activity relationship studies on mescaline, *Psychopharmacologia (Berl.)*, **10**, 379–387.

Sommer, S., Novin, D., and LeVine, M. (1967). Food and water intake following intrahypothalamic injections of carbachol in the rabbit, *Science*, **156**, 983–984.

Spector, S., Sjoerdsma, A., and Udenfriend, S. (1965). Blockade of endogenous norepinephrine synthesis by *d*-methyl-tyrosine, an inhibitor of tyrosine hydroxylase, *J. pharmacol. expt. Therap*, **147**, 86–95.

Spehlmann, R. (1969). Effect of acetylcholine and atropine upon excitation of cortical neurons by reticular stimulation, *Fed. Proc.*, **28**, 795.

Sperry, R. W. (1963). Chemonaffinity in the orderly growth of nerve fibre patterns and connections, *Proc. nat. Acad. Sci.*, **50**, 703–710.

Starzl, T. E., Taylor, C. W., and Magoun, H. W. (1951). Ascending conduction in the reticular activating system with special reference to the diencephalon, *J. Neurophysiol.*, **14**, 461–467.

248

Steck, H. (1954). Le syndrome extrapyramidal et di-encéphalique au cours des traitments au largactil et au serpasil, *Ann. med-psychol.*, **112**, 737–743.

Stefano, F. J. E., and Donoso, A. O. (1967). Norepinephrine levels in the rat hypothalamus during the estrous cycle, *Endocrinology*, **81**, 1405–1406.

Steg, G. (1964). Efferent muscle innervation and rigidity, *Acta physiol. scand.*, **61**, Suppl., **225**, 1–53.

Stein, D. G., and Chorover, S. L. (1968). Effects of posttrial electrical stimulation of hippocampus and caudate nucleus on maze learning in the rat, *Physiol. Behav.*, **3**, 787–791.

Stein, H. H., and Lewis, G. J. (1969). Noncompetitive inhibition of acetylcholinesterase by eserine, *Biochem. Pharmacol.*, **18**, 1679–1684.

Stein, H. H., and Yellin, T. O. (1967). Pemoline and magnesium hydroxide: lack of effect on RNA and protein synthesis, *Science*, **157**, 96–97.

Stein, L. (1962). Effects and interactions of imipramine, chlorpromazine, reserpine and amphetamine on self-stimulation. Possible neurophysiological basis of depression. In. J. Wortis (Ed.), *Recent Advances in Biological Psychiatry*. New York: Plenum, 288–308.

Stein, L. (1964a). Reciprocal action of reward and punishment mechanisms. In R. G. Heath (Ed.), *The Role of Pleasure in Behavior*. New York: Harper and Row, 113–139.

Stein, L. (1964b). Self-stimulation of the brain and the central stimulant action of amphetamines, *Fed. Proc.*, **23**, 836–849.

Stein, L., and Seifter, J. (1962). Muscarinic synapses in the hypothalamus, *Am. J. Physiol.*, **202**, 751–756.

Stein, L., and Wise, C. D. (1969). Release of norepinephrine from hypothalamus and amygdala by rewarding medial forebrain hundle stimulation and amphetamine, *J. comp. physiol. Psychol.*, **67**, 189–198.

Stein, L., and Wise, C. D., and Berger, B. D. (1973). Antianxiety action of benzodiazepines: decrease in activity of serotonin neurons in the punishment system. In S. Garattini, E. Mussini and L. O. Randall (Eds.), *The Benzodiazepines*. New York: Raven, 299–326.

Stellar, E. (1954). The physiology of motivation, *Psychol. Rev.*, **61**, 5–22.

Stern, M. S. (1966). Cited in 'Management of psychotic episodes in users of LSD posing new problems', *Psychiat. Prog.*, **1**, 8.

Stewart, W. C. (1952). Accumulation of acetylcholine in brain and blood of animals poisoned with cholinesterase inhibitors, *Brit. J. Pharmacol.*, **7**, 270–276.

Stewart, W. W. (1972). Comments on the chemistry of scotophobin, *Nature*, **238**, 202–209.

Stolerman, J. P., and Kumar, R. (1970). Preferences for morphine in rats. Validation of an experimental model of dependence, *Psychopharmacologia (Berl.)*, **17**, 137–150.

Stone, G. (1964). Effects of drugs on non-discriminated avoidance behaviour. 1. Individual differences in dose-response relationships, *Psychopharmacologia (Berl.)*, **6**, 245–255.

Storm, T., and Smart, R. G. (1965). Dissociation: a possible explanation of some features of alcoholism and implication for its treatment, *Quart. J. Stud. Alc.*, **26**, 111–115.

Stumpf, C. (1965). Drug action on the electrical activity of the hippocampus, *Int. J. Neurobiol.*, **8**, 77–138.

Sutherland, N. S. (1964). The learning of discriminations by animals, *Endeavour*, **23**, 148–152.

Szara, S. (1957). The comparison of psychotic effect of tryptamine derivatives with the effects of mescaline and LSD-25 in self-experiments. In S. Garattini and V. Ghetti (Eds.), *Psychotropic Drugs*. Amsterdam: Elsevier, 460–466.

Szara, S., Axelrod, J., and Perlin, S. (1958). Is adrenochrome present in the blood?' *Amer. J. Psychiat.*, **115**, 162–163.

Szatmari, A., and Schneider, R. A. (1955). Induction of sleep by autonomic drugs, *J. nerv. ment. Dis.*, **121**, 311–320.

Takagi, H., and Ban, J. (1960). Effect of psychotropic drugs on the limbic system of the cat, *Jap. J. Pharmacol.*, **10**, 7–18.

Takesada, M., Kakimoto, Y., Sano, I., and Koneko, Z. (1963). 3,4-dimethoxyphenyl-ethylamine and other amines in the urine of schizophrenic patients, *Nature*, **199**, 203.

Talwar, G. P., Gupta, S. L., Pandian, M. R., Sharma, S. K., Rao, K. N., Jaikhani, B. L., Sen, K. K., Sopori, M. L., Basu, A. K., and Jha, P. (1971). Similarities and differences in the mode of action of growth hormone and estradiol. In M. Hamburgh and E. J. W. Barrington (Eds.), *Hormones in Development*. New York: Appleton-Century-Crofts, 95–100.

Tanimukai, H., Ginther, R., Spaide, J., Bueno, J. R., and Himwich, H. E. (1968). Psychotogenic N,N-dimethylated indole amines and behavior in schizophrenics. In J. Wortis (Ed.), *Recent Advances in Biological Psychiatry*, Vol. 10, New York: Plenum.

Tapp. J. T., and Markowitz, H. (1963). Infant handling: effects on avoidance learning, brain weight and cholinesterase activity, *Science*, **140**, 486–487.

Taylor, K. M., and Laverty, R. (1969). The effect of chlordiazepoxide, diazepam, and nitrazepam on catecholamine metabolism in regions of the rat brain, *Europ. J. Pharmacol.*, **8**, 296–301.

Taylor, K. M., and Laverty, R. (1973). The interaction of chlordiazepoxide, diazepam and nitrazepam with catecholamines and histamine in regions of the rat brain. In S. Garattini, E. Mussini, and L. O. Randall (Eds.), *The Benzodaizepines*. New York: Raven, 191–202.

Taylor, K. M., and Snyder, S. H. (1970). Amphetamine: differentiation by d- and l-isomers of behavior involving brain norepinephrine or dopamine, *Science*, **168**, 1487–1489.

Teitelbaum, P. (1955). Sensory control of hypothalamic hyperphagia, *J. comp. Physiol. Phychol.*, **48**, 156–163.

Teitelbaum, P., and Epstein, A. N. (1962). The lateral hypothalamic syndrome. Recovery of feeding and drinking after lateral hypothalamic lesions, *Psychol. Rev.*, **69**, 74–90.

Tenen, S. (1967). Recovery as a measure of C.E.R. strength: effect of benzodiazepines, amobarbital, chlorpromazine and amphetamine, *Psychopharmacologia (Berl.)*, **12**, 1–17.

Terrace, H. S. (1966). Stimulus control. In W. K. Honig (Ed.), *Operant behavior: areas of research and application*. New York: Appleton-Century-Crofts.

Thesleff, S. (1960). Supersensitivity of skeletal muscle produced by botulinum toxin, *J. physiol.*, **151**, 598–607.

Thompson, R. F. (1967). *Foundations of physiological psychology*. New York: Harper and Row.

Thomson, C. W., McGaugh, J. L., Smith, C. E., Hudspeth, W. J., and Westbrook, W. H. (1961). Strain differences in the retroactive effects of convulsive shock on maze learning, *Canad. J. Psychol.*, **15**, 69–74.

Tobin, J. H., and Lewis, N.D.C. (1960). New psychotherapeutic agent, chlordiazepoxide: use in treatment of anxiety states and related symptoms, *J. amer. med. assoc.*, **174**, 1242–1245.

Tonge, S. R., and Leonard, B. E. (1969). The effect of some hallucinogenic drugs upon the metabolism of 5-hydroxytryptamine in the brain, *Life Sci.*, **8**, 805–812.

Torda, C. (1967). Effect of brain serotonin depletion on sleep in rats, *Brain Res.*, **6**, 375–377.

Towne, J. C. (1964). Effect of ethanol and acetaldehyde on liver and brain monoamine oxidase, *Nature*, **201**, 709–710.

Treisman, A. (1964). Selective attention in man, *British Medical Bulletin*, **20**, 12–16.

Tryon, R. C. (1942). In F. A. Moss, *Comparative Psychology*. Englewood Cliff, N. J.: Prentice Hall.

Udenfriend, S. (1958). Metabolism of 5-hydroxytryptamine. In G. P. Lewis (Ed.), *5-Hydroxytryptamine*. London: Pergamon, 43–49.

Ungar, G. (1970). Role of proteins and peptides in learning and memory. In George Ungar (Ed.), *Molecular Mechanisms in Memory and Learning*. London: Plenum Press, 149–174.

Ungar, G., Galvan, L., and Clark, R. H. (1968). Chemical transfer of learned fear, *Nature*, **217**, 1259.

Ungerleider, J. T., Fisher, D. D., and Fuller, M. (1966). The dangers of LSD. Analysis of seven months' experience in a university hospital's psychiatric service, *J. amer. med. Assoc.*, **197**, 389–391.

Ungerstedt, U. (1971). Stereotaxic mapping of the monamine pathways in the rat brain, *Acta. physiol. Scand. suppl.*, **367**, 1–48.

Van Abeelan, J., Gilissen, L., Hanssen, Th., and Lenders, A. (1972). Effects of intrahippocampal injections with methylscopolamine and neostigmine upon exploratory behaviour in two inbred mouse strains, *Psychopharmacologia (Berl.)*, **24**, 470–475.

Van Abeelan, J. H. F., Smits, A. J. M., and Raaijmakers, W. (1971). Central location of a genotype-dependent cholinergic mechanism controlling exploratory behaviour in mice, *Psychopharmacologia (Berl.)*, **19**, 324–328.

Van Loon, G. R., Scapagnini, U., Cohen, R., and Ganong, W. F. (1971). Effect of intraventricular administration of adrenergic drugs on the adrenal venous M-hydroxycorticosteroid response to surgical stress in the dog, *Neuroendocrinol.*, **8**, 257–272.

Velasco, F., and Velasco, M. (1973). A quantitative evaluation of the effects of *l*-dopa on Parkinson's disease, *Neuropharmacology*, **12**, 89–99.

Velluti, R., and Hernandez-Peon, R. (1963). Atropine blockade within a cholinergic hypnogenic circuit, *Exp. Neurol.*, **80**, 20–29.

Verhave, T. (1958). The effect of metamphetamine on operant level and avoidance behavior, *J. exp. anal. Behav.*, **1**, 207–219.

Verney, E. B. (1947). The antidiuretic hormone and the factors which determine its release, *Proc. Roy. Soc. (Lond.), B*, **135**, 25–106.

Vogel, J. R., Hughes, R. A., and Carlton, P. L. (1967). Scopolamine, atropine and conditioned fear, *Psychopharmacologia*, **10**, 409–416.

Vogt, M. (1954). The concentration of sympathin in different parts of the central nervous system under normal conditions and after the administration of drugs, *J. Physiol.*, **123**, 451–481.

Wagman, W. D., and Maxey, G. C. (1969). The effects of scopolamine hydrobromide and methyl scopolamine hydrobromide upon the discrimination of interoceptive and exteroceptive stimuli, *Psychopharmacologia (Berl.)*, **15**, 280–288.

Wagner, J. W., and de Groot, J. (1963). Changes in feeding behaviour after intracerebral injections in the rat, *Amer. J. Physiol.*, **204**, 483–487.

Waisman, H. (1970). Disorders of amino acid metabolism and mental retardation. In K. E. Bowman, and S. P. Datta (Eds.), *Biochemistry of Brain and Behavior*. New York: Plenum Press, 223–242.

Wallgren, H., and Barry, H. (1971). *Actions of Alcohol.* Elsevier: London.

Walsh, J. (1964). Psychotoxic drugs, *Science*, **145**, 1418–1420.

Walsh, J. (1970). Methadone and heroin addiction: rehabilitation without a cure, *Science*, **168**, 684–686.

Walshe, F. M. R. (1952). *Diseases of the Nervous System.* Edinburgh: Livingston.

Wang, G. H. (1923). Relation between 'spontaneous' activity and oestrous cycle in the white rat, *Comp. Psychol. Monog.*, **2**, 1–27.

Warburton, D. M. (1967). *Some behavioral effects of central cholinergic stimulation with special reference to the hippocampus.* Unpublished Doctoral Thesis. Indiana University.

Warburton, D. M. (1968). Modified sensitivity to pentobarbital in a continuous avoidance situation, *Psychopharmacologia*, **13**, 387–393.

Warburton, D. M. (1969). Behavioral effects of central and peripheral changes in acetylcholine systems, *J. comp. Physiol. Psych.*, **68**, 56–64.

Warburton, D. M. (1972). The cholinergic control of internal inhibition In R. Boakes and M. S. Halliday (Eds.), *Inhibition and learning.* London: Academic.

Warburton, D. M. (1974). The effect of scopolamine on a two cue discrimination, *Quart. J. exp. Psychol.*, **26**, 395–404.

Warburton, D. M., and Brown, K. (1971). Scopolamine-induced attenuation of stimulus sensitivity, *Nature*, **230**, 126–127.

Warburton, D. M., and Brown, K. (1972). The facilitation of discrimination performance by physostigmine sulphate, *Psychopharmacologia (Berl.)*, **27**, 275–284.

Warburton, D. M., and Greeno, J. G. (1970). A general shape function model of learning with applications in psychobiology, *Psychol. Rev.*, **77**, 348–352.

Warburton, D. M., and Groves, P. (1969). The effects of scopolamine on habituation of acoustic startle in rats, *Comm. Behav. Biol.*, **3**, 289–293.

Warburton, D. M., and Heise, G. A. (1972). The effects of scopolamine on spatial double alternation in rats, *J. comp. physiol. Psychol.*, **81**, 3, 523–532.

Warburton, D. M., and Russell, R. W. (1968). Effects of 8-azaguanine on acquisition of a temporal discrimination, *Physiol. Behav.*, **3**, 61–63.

Warburton, D. M., and Russell, R. W. (1969). Some behavioural effects of cholinergic stimulation of the hippocampus, *Life Sci.*, **8**, 617–627.

Warburton, D. M., and Segal, D. S. (1971). Stimulus control during chronic reduction of cholinesterase activity, *Physiol. Behav.*, **7**, 539–543.

Ward, A. (1968). Function of the basal ganglia. In P. J. Vinken and G. W. Bruyn (Eds.), *Handbook of Clinical Neurology*, Vol. 6. 'Diseases of the Basal Ganglia. Amsterdam: North Holland, 367–408.

Weeks, J. R. (1962). Experimental morphine addiction: method for automatic intravenous injections in unrestrained rats, *Science*, **138**, 143–144.

Weeks, J. R. (1964). Experimental narcotic addiction, *Sci. Amer*, **211**(3), 46–52.

Weil, A. T., Zinberg, N. E., and Nelson, J. M. (1968). Clinical and psychological effects of marijuana in man, *Science*, **162**, 1234–1242.

Weiner, S., Dorman, D., Persky, H., Stach, T. W., Norton, J., and Levitt, E. E. (1963). *Psychosom. Med.* **25**, 158.

Weiss, B., and Laties, V. G. (1962). Enhancement of human performance by caffeine and the amphetamines, *Pharmacol. Rev.*, **14**, 1–36.

Weiss, J. M., McEwen, B. S., Silva, M. T. A., Kalkut, M. F. (1969). Pituitary-adrenal influences on fear responding, *Science*, **163**, 197–99.

Weiss, P. (1968). *Dynamics of Development: Experiments and Inferences*. New York: Academic Press.

Weitzman, E. D., Rapport, M. M., McGregor, P., and Jacoby, J. (1968). Sleep patterns of the monkey and brain serotonin concentration: effect of *p*-chlorophenylalanine, *Science*, **160**, 1361-1363.

Wertheim, G. A., Conner, R. L., and Levine, S. (1967). Adrenocortical influences on free operant behavior, *J. exp. anal. Behav.*, **10**, 555–563.

Wertheim, G. A., Conner, R. L., and Levine, S. (1969). Avoidance conditioning and adrenocortical function in the rat, *Physiol. Behav.*, **4**, 41–44.

West, E. D., and Dally, P. J. (1959). Effects of iproniazid in depressive syndromes, *Brit. Med. J.*, **1**, 1491–1494.

Westermann, E. O., Maickel, R. P., and Brodie, B. B. (1962). On the mechanism of pituitary adrenal stimulation by reserpine, *J. pharmacol. exp. Ther.*, **138**, 208–217.

White, A., Handler, P., Smith, E. L., and Stetten, D. (1959). *Principles of Biochemistry*. New York: McGraw-Hill.

White, R. P., Rinaldi, F., and Himwich, H. E. (1956). Central and peripheral nervous effects of atropine sulfate and mepiperphenidal bromide (Darstine) on human subjects, *J. applied Physiol.*, **8**, 635–642.

Whitehouse, J. M. (1966). The effect of physostigmine on discrimination learning, *Psychopharmacologia (Berl.)*, **9**, 183–188.

Whittaker, V. P. (1970). The investigation of synaptic function by means of subcellular fractionation techniques. In F. O. Schmidt (Ed.), *The Neurosciences Second Study Program*. New York: Rockefeller University Press, 761–768.

Wickler, A. (1952). Mechanisms of action of drugs that modify personality function, *Amer. J. Psychiat.*, **108**, 590.

Wickler, A. (1968). *The Addictive States*. Baltimore: Williams and Wilkins.

Wiener, N. I., and Deutsch, J. A. (1968). The temporal aspects of anticholinergic and anticholinesterase produced amnesia for an appetitive habit, *J. comp. physiol. Psychol.*, **66**, 613–617.

Wieser, S. (1970). Treatment of the chronic neuropsychiatric complications of alcholism.

In R. E. Popham (Ed.), *Alcohol and Alcocholism*. Toronto: University of Toronto Press, 278–280.

Wilson, C. W. M. (1968). *Adolescent Drug Dependence*. Oxford: Pergamon.

Winick, M., and Noble, A. (1966). Cellular response in rats during malnutrition at various ages, *J. Nutr.*, **89**, 300–306.

Winick, M., and Rosso, P. (1969). The effect of severe early malnutrition on cellular growth of human brain, *Pediat. Res.*, **3**, 181–184.

Wise, C. D., Berger, B. D., and Stein, L. (1972). Benzodiazepines: anxiety-reducing activity by reduction of serotonin turnover in the brain, *Science*, **177**, 180–183.

Wise, C. D., and Stein, L. (1969). Facilitation of brain self-stimulation by central administration of norepinephrine, *Science*, **163**, 229–301.

Wittenborn, J. R. (1966). *The Clinical Psychopharmacology of Anxiety*. Springfield, Ill.: Thomas.

Woods, M. N., and McCormick, D. B. (1964). Effects of dietary phenylalanine on activity of phenylalanine hydroxylase from rat liver, *Proc. Soc. exp. Biol. Med.*, **116**, 427–430.

Wolley, D. H., and Van der Hoeven, T. (1964). Serotonin deficiency in infancy as one cause of mental defect in phenylketonuria, *Science*, **144**, 883–884.

World Health Organization. (1952). *Alcohol Subcommittee Second Report*. World Health Organization Technical Report Series No. 48, Geneva.

World Health Organization (1964). *Evaluation of dependence producing drugs*. Report of a WHO Scientific group. World Health Organization Technical Report No. 287, Geneva.

Wurtman, R. J., and Axelrod, J. (1963). A sensitive and specific assay for the estimation of monoamine oxidase, *Biochem. Pharmacol.*, **12**, 1439–41.

Yablonski, L. (1965). *Synanon: The Tunnel Back*. Baltimore: Pelican.

Yahr, M. D., Duvoisin, R. C., Schear, M. J., Barrett, R. E., and Hoehn, M. M. (1969). Treatment of Parkinsonism with levodopa, *Arch. Neurol.*, **21**, 4, 343–354.

Yates, F. E., and Urquhart, J. (1962). Control of plasma concentrations of adrenocortical hormones, *Physiol. Rev.*, **42**, 359–443.

Young P. T. (1941). The experimental analysis of appetite, *Psychol. Bull.*, **38**, 129–164.

Young, W. C., and Fish, W. R. (1945). The ovarian hormones and spontaneous running activity in the female rat, *Endocrinology*, **36**, 181–189.

Yules, R. B., Freedman, D. X., and Chandler, K. A. (1966). The effect of ethyl alcohol on man's electroencephalographic sleep cycle, *E. E. G. clin. Neurophysiol.*, **20**, 109–111.

Zacks, S. I., Metzger, J. F., Smith, C. W., and Blumberg, J. M. (1962). Localization of ferritin-labelled botulinus toxin in the neuromuscular function of the mouse, *J. Neuropath. exp. Neurol.*, **21**, 610–633.

Zarcone, V., Gulevich, G., Pivik, T., and Dement, W. (1968). Partial R. E. M. phase deprivation and schizophrenia, *Arch. gen. Psychiat.*, **18**, 194–202.

Zbinden, M. D. (1960). Pharmacodynamics of tetrabenazine and its derivatives. In J. Nodine and J. Moyer, *Psychosomatic Medicine*. Baltimore: Lea and Fibiger, 443–454.

Zeman, D., and House, B. J. (1963). The role of attention in retardate discrimination learning. In N. R. Ellis (Ed.), *Handbook of Mental Deficiency*. New York: McGraw-Hill, 159–223, 529–531.

Zemp, J. W., Wilson, J. E., and Glassman, E. (1967). Brain function and macromolecules. Site of increased labelling of RNA in brains of mice during a short-term training experience, *Proc. nat. Acad. Sci.*, **58**, 1120–1126.

Zemp, J. W., Wilson, J. E., Schlesinger, K., Boggan, W. O., and Glassman, E. (1966). Brain function and macromolecules. 1. Incorporation of uridine into RNA of mouse brain during short-term training experience, *Proc. nat. Acad. Sci.*, **55**, 1423–1432.

Zolman, J. F., and Morimoto, H. (1962). Effects of age of training on cholinesterase activity in the brains of maze-bright rats, *J. comp. physiol. Psychol.*, **55**, 794–800.

Zolman, J. F., and Morimoto, H. (1965). Cerebral changes related to duration of environmental complexity and locomotor activity, *J. comp. physiol. Psychol.*, **60**, 382–387.

Zuckerman, M. (1972). Drug usage as one manifestation of a 'sensation-seeking' trait. In Wolfarm Kemp (Ed.), *Drug Abuse*. Springfield: Charles C. Thomas, 54–163.

Index

254

Subject Index

264